Lecture Notes in Computer Science

Lecture Notes in Computer Science

Edited by G. Goos and J. Hartmanis

453

J. Seberry   J. Pieprzyk   (Eds.)

Advances in Cryptology – AUSCRYPT '90

International Conference on Cryptology
Sydney, Australia, January 8–11, 1990
Proceedings

Springer-Verlag

Berlin Heidelberg New York London
Paris Tokyo Hong Kong Barcelona

Editors

Jennifer Seberry
Josef Pieprzyk
Department of Computer Science, University College
The University of New South Wales
Austirlian Defence Force Academy
Canberra, ACT 2600, Australia

CR Subject Classification (1987): E.3

ISBN 3-540-53000-2 Springer-Verlag Berlin Heidelberg New York
ISBN 0-387-53000-2 Springer-Verlag New York Berlin Heidelberg

© Springer-Verlag Berlin Heidelberg 1990
Printed in Germany

Printing and binding: Druckhaus Beltz, Hemsbach/Bergstr.
2145/3140-543210 – Printed on acid-free paper

AUSCRYPT'90

A Workshop on the Theory and Application of Cryptographic Techniques

January 8-11, 1990

The University of New South Wales, Sydney

Sponsored by

The International Association for Cryptologic Research

in cooperation with

The IPACE Institute at UNSW

and

Department of Computer Science, University College, ADFA

Centre for Communication Security Research

General Chair: Jennifer Seberry, The University of NSW, Canberra, Australia

Program Chairs: Josef Pieprzyk, The University of NSW, Canberra, Australia
Rainer Rueppel, Crypto AG, Zug, Switzerland
Scott Vanstone, University of Waterloo, Waterloo, Canada

Program Committee:

Gordon Agnew, University of Waterloo, Waterloo, Canada
Ingemar Ingemarsson, University of Linkoping, Linkoping. Sweden
Ron Mullin, University of Waterloo, Waterloo, Canada
Wyn Price, National Physical Laboratory, Teddington, United Kingdom
Rei Safavi-Naini, The University of NSW, Canberra, Australia

PREFACE

This book is the proceedings of AUSCRYPT '90, the first conference sponsored by the International Association for Cryptological Research to be held in the southern hemisphere and the first outside the EUROCRYPT series held in European countries each northern spring and the CRYPTO series held in Santa Barbara, USA each August.

The proceedings from these earlier conferences have been published in the Springer-Verlag Lecture Notes in Computer Science series since 1986.

Papers in this volume are organized into eleven sections. The first ten sections comprise all of the papers on the regular program, including a few papers on the program which, unfortunately, were not presented at the meeting. The last section contains some of the papers presented at the "Rump Session" organized by Josef Pieprzyk.

AUSCRYPT '90 was attended by 95 people representing sixteen countries.

For the first time a skills workshop was held the day before the conference, with such eminent cryptographers as D. Gollmann (West Germany), I. Ingemarsson (Sweden), K. Ohta (Japan), R. Rueppel (Switzerland) and S. Vanstone (Canada) teaching in the area of their expertise. J. Seberry represented Australia at this event. Forty-one people attended the workshop.

It gives us great pleasure to express our thanks here to the members of the program committee: Dr R. Rueppel, Dr W.Price, and Dr I. Ingemarsson from Europe; Dr S. Vanstone, Dr R. Mullin, and Dr G. Agnew from North America; and Dr R. Safavi-Naini, and the other members of the Centre for Computer Security Research, for the rest of the world. They were all efficient, pleasant and wonderful co-workers.

Local organization was arranged by Ms C. Burke of the IPACE Institute, on the Sydney campus of The University of NSW. The conference dinner was held on a ferry on Sydney Harbour and was a spectacular success.

Many thanks to all the members of the Centre for Computer Security Research: Dr R. Safavi-Naini; Mr L. Brown; Ms L. Condie; Mr T. Hardjono; Mrs C. Newberry; Mr M. Newberry; Mr D. Rubie; Mrs E. Trott and Mrs E. Tait, all of whom readily worked at every task they were given to make the conference function smoothly and ensure its scientific success.

This conference was made possible by the support of the International Association for Cryptologic Research, The Centre for Computer Security Research, Prentice Hall Pty Ltd, and Telecom Australia.

Jennifer Seberry
Josef Pieprzyk

CONTENTS

CONTENTS

SECTION 1: PUBLIC-KEY CRYPTOSYSTEMS

SECTION 2: PSEUDORANDOMNESS AND SEQUENCES I

SECTION 3: NETWORK SECURITY

SECTION 4: AUTHENTICATION

SECTION 8: THEORY

SECTION 9: APPLICATIONS

SECTION 10: IMPLEMENTATIONS

SECTION 11: RUMP SESSION

SECTION 1

PUBLIC-KEY CRYPTOSYSTEMS

The Implementation of Elliptic Curve Cryptosystems

Alfred Menezes   *Scott Vanstone*

Dept. of Combinatorics and Optimization, University of Waterloo

Waterloo, Ontario, N2L 3G1, Canada

Abstract

Elliptic curves have been studied for many years. Recent interest has revolved around their applicability to factoring integers and to
cryptography. In this paper we explore the feasibility of implementing
in hardware an arithmetic processor for doing elliptic curve computations over fields of characteristic two. The elliptic curve analogue of
the ElGamal cryptosystem is also analyzed.

1   Introduction

The points of an elliptic curve E over a field K form an abelian group. The
addition operation of this abelian group involves a few arithmetic operations in the underlying field K, and is easy to implement in both hardware
and software. Hence the group E can be used to implement the Diffie-
Hellman key passing scheme [Diff76], and the ElGamal public key cryptosystem [ElG85], as explained in [Kob87]. In this report, we consider various
issues that arise in the efficient hardware implementation of the ElGamal
public key cryptosystem.

We begin with a brief review of elliptic curves. For an elementary introduction to elliptic curves the reader is referred to Chapter 6 of the book by
Koblitz [Kob87b], while for a more thorough treatment of the subject we
refer the reader to [Sil86]. Section 3 mentions how arithmetic in $GF(2^m)$
can be efficiently implemented. This discussion helps in understanding why
we choose elliptic curves over fields of characteristic two, and this is done
in Section 4. The elliptic curve analogue of the ElGamal cryptosystem is
studied in Section 5. In Sections 6 and 7, we present two alternate schemes
for adding points on an elliptic curve also suitable for the implementation of

the ElGamal cryptosystem. Finally in Section 8, we predict the performance of the cryptosystem.

2 Review of Elliptic Curves

Assume first that K is a field of characteristic neither 2 nor 3. An elliptic curve over K (in affine coordinates), denoted by E or $E(K)$, is the set of all solutions $(x, y) \in K \times K$ to the equation

$$E_1 : y^2 = x^3 + ax + b, \tag{1}$$

where $a, b \in K$, and $4a^3 + 27b^2 \neq 0$, together with a special point O, called the point at infinity.

It is well known that $E(K)$ is an abelian group, and O is its identity. The rules for the addition are summarized below.

Addition Formula for (1)

If $P = (x_1, y_1) \in E_1$, then $-P = (x_1, -y_1)$. If $Q = (x_2, y_2) \in E_1$, $Q \neq -P$, then $P + Q = (x_3, y_3)$, where

$$
\begin{aligned}
x_3 &= \lambda^2 - x_1 - x_2 \\
y_3 &= \lambda(x_1 - x_3) - y_1,
\end{aligned}
$$

and

$$
\lambda = \begin{cases}
\dfrac{y_2 - y_1}{x_2 - x_1} & \text{if } P \neq Q \\[3mm]
\dfrac{3x_1^2 + a}{2y_1} & \text{if } P = Q.
\end{cases}
$$

If K is a field of characteristic 2, then there are two types of elliptic curves over K. An elliptic curve of zero j-invariant is the set of solutions to the equation

$$E_2 : y^2 + a_3 y = x^3 + a_4 x + a_6, \tag{2}$$

where $a_3, a_4, a_6 \in K$, $a_3 \neq 0$, together with the point at infinity O.

An elliptic curve of non-zero j-invariant is the set of solutions to the equation

$$E_3 : y^2 + xy = x^3 + a_2 x^2 + a_6, \tag{3}$$

where $a_2, a_6 \in K$, $a_6 \neq 0$, together with the point at infinity O.

The addition formulae for the two types of curves is given below.

Addition Formula for (2)

Let $P = (x_1, y_1) \in E_2$; then $-P = (x_1, y_1 + a_3)$. If $Q = (x_2, y_2) \in E_2$ and $Q \neq -P$, then $P + Q = (x_3, y_3)$, where

$$
x_3 = \begin{cases} \left(\dfrac{y_1 + y_2}{x_1 + x_2}\right)^2 + x_1 + x_2 & P \neq Q \\[3mm] \dfrac{x_1^4 + a_4^2}{a_3^2} & P = Q \end{cases}
$$

and

$$
y_3 = \begin{cases} \left(\dfrac{y_1 + y_2}{x_1 + x_2}\right)(x_1 + x_3) + y_1 + a_3 & P \neq Q \\[3mm] \left(\dfrac{x_1^2 + a_4}{a_3}\right)(x_1 + x_3) + y_1 + a_3 & P = Q. \end{cases}
$$

Addition Formula for (3)

Let $P = (x_1, y_1) \in E_3$; then $-P = (x_1, y_1 + x_1)$. If $Q = (x_2, y_2) \in E_3$ and $Q \neq -P$, then $P + Q = (x_3, y_3)$, where

$$
x_3 = \begin{cases} \left(\dfrac{y_1 + y_2}{x_1 + x_2}\right)^2 + \dfrac{y_1 + y_2}{x_1 + x_2} + x_1 + x_2 + a_2 & P \neq Q \\[3mm] \left(\dfrac{y_1}{x_1}\right)^2 + \dfrac{y_1}{x_1} + x_1^2 + x_1 + a_2 & P = Q \end{cases}
$$

and

$$
y_3 = \begin{cases} \left(\dfrac{y_1 + y_2}{x_1 + x_2}\right)(x_1 + x_3) + x_3 + y_1 & P \neq Q \\[3mm] x_1^2 + \left(x_1 + \dfrac{y_1}{x_1}\right)x_3 + x_3 & P = Q. \end{cases}
$$

3   Field Arithmetic in $GF(2^m)$

Since we will be choosing K to be a field of characteristic two, we briefly discuss how the arithmetic in $GF(2^m)$ can be efficiently accomplished.

Recall that the field $GF(2^m)$ can be viewed as a vector space of dimension m over $GF(2)$. Once a basis of $GF(2^m)$ over $GF(2)$ has been chosen, the elements of $GF(2^m)$ can be conveniently represented as 0–1 vectors of length m. In hardware, a field element is stored in a shift register of length m. Addition of elements is performed by bitwise XOR-ing the vector representations, and takes one clock cycle. A normal basis representation of $GF(2^m)$ is preferred because then squaring a field element is accomplished by a simple rotation of the vector representation, an operation that is easily implemented in hardware; squaring an element also takes one clock cycle.

To minimize the hardware complexity in multiplying field elements (i.e. to minimize the number of connections between the cells of the shift registers holding the multiplicands), the normal basis chosen has to belong to a special class called optimal normal bases. A description of these special normal bases can be found in [Mull88], where constructions are given, together with a list of fields for which these bases exist. An associated architecture for a hardware implementation is given in [Agnew89b]. Using this architecture, a multiplication can be performed in m clock cycles. For fields for which optimal normal bases do not exist, the so-called low complexity normal bases described in [Ash89] may be useful.

Finally, the most efficient techniques, in terms of minimizing the number of multiplications, to compute an inverse was proposed by Itoh, Teechai and Tsujii, and is described in [Agnew89]. The method requires exactly $\lfloor log_2(m-1) \rfloor + \omega(m-1) - 1$ field multiplications, where $\omega(m-1)$ denotes the Hamming weight of the binary representation of $m - 1$. However it is costly in terms of hardware implementation in that it requires the storage of several intermediate results. An alternate method for inversion which is slower but which does not require the storage of such intermediate results is also described in [Agnew89].

Recently Newbridge Microsystems Inc. in conjunction with Cryptech Systems Inc. has manufactured a single chip device that implements various public and conventional key cryptosystems based on arithmetic in the field $GF(2^{593})$. Since the field size is quite large, a slower two-pass multiplication technique was used in order to reduce the number of cell interconnections (see [Agnew89b] or [Ros89]). Also, to reduce the number of registers, the slower method mentioned in the previous paragraph to compute inverses was used. Multiplication of two elements takes 1,300 clock cycles, while an inverse computation takes 50,000 clock cycles. The chip has a clock rating of 20 MHz, and so the multiplication and inverse computation take .065ms and 2.5ms respectively.

4 Choice of Curve and Field K

From the addition formulae, we see that two distinct points on an elliptic curve can be added by means of three multiplications and one inversion of field elements in the underlying field K, while a point can be doubled in one inversion and four multiplications in K. Additions and subtractions are not considered in this count, since these operations are relatively inexpensive. Our intentions are to select a curve and field K so as to minimize the number of field operations involved in adding two points. We choose curves over $K = GF(2^m)$ with zero j-invariant for the following three reasons.

(i) The arithmetic in $GF(2^m)$ is easier to implement in computer hardware than the arithmetic in finite fields of characteristic greater that 2.

(ii) When using a normal basis representation for the elements of $GF(2^m)$, squaring a field element becomes a simple cyclic shift of the vector representation, and thus reduces the multiplication count in adding two points.

(iii) For curves of zero j-invariant over $GF(2^m)$, the inverse operation in doubling a point can be eliminated by choosing $a_3 = 1$, further reducing the operation count.

We thus prefer curves over $GF(2^m)$ of the form $y^2 + y = x^3 + a_4 x + a_6$. Furthermore, the case when m is odd is advantageous, because it is then easy to recover the y-coordinate of a point given its x-coordinate plus a single bit of the y-coordinate. This is useful in message embedding, and in reducing message expansion in the ElGamal scheme, as will be explained in the next section.

It is shown in [Men90] that there are precisely 3 isomorphism classes of elliptic curves with zero j-invariant over $GF(2^m)$, m odd. A representative curve from each class is $y^2 + y = x^3$, $y^2 + y = x^3 + x$, and $y^2 + y = x^3 + x + 1$. To simplify the addition formula even further, we choose the curve

$$E : y^2 + y = x^3.$$

The addition formula for E simplifies to

$$x_3 = \begin{cases} \left(\dfrac{y_1 + y_2}{x_1 + x_2}\right)^2 + x_1 + x_2 & P \neq Q \\ \\ x_1^4 & P = Q \end{cases}$$

and

$$y_3 = \begin{cases} \left(\dfrac{y_1 + y_2}{x_1 + x_2}\right)(x_1 + x_3) + y_1 + 1 & P \neq Q \\ y_1^4 + 1 & P = Q. \end{cases}$$

If a normal basis representation is chosen for the elements of $GF(2^m)$, we see that doubling a point in E is "free", while adding two distinct points in E can be accomplished in two multiplications and one inversion.

The addition formulae for the curves $y^2 + y = x^3 + x$ and $y^2 + y = x^3 + x + 1$ is similar to the addition formula for E, except that the formula for doubling a point becomes

$$x_3 = x_1^4 + 1$$
$$y_3 = y_1^4 + x_1^4 + 1.$$

In Table 1, we list the number of points on each of these curves over $GF(2^m)$, where m is odd. For the remainder of this report we focus our attention on

Curve	m	Order
$y^2 + y = x^3$	odd	$2^m + 1$
$y^2 + y = x^3 + x$	$m \equiv 1,7 \pmod 8$	$2^m + 1 + (\sqrt{2})^{m+1}$
$y^2 + y = x^3 + x$	$m \equiv 3,5 \pmod 8$	$2^m + 1 - (\sqrt{2})^{m+1}$
$y^2 + y = x^3 + x + 1$	$m \equiv 1,7 \pmod 8$	$2^m + 1 - (\sqrt{2})^{m+1}$
$y^2 + y = x^3 + x + 1$	$m \equiv 3,5 \pmod 8$	$2^m + 1 + (\sqrt{2})^{m+1}$

Table 1: Orders of curves of j-invariant zero over $GF(2^m)$, m odd.

the curve $y^2 + y = x^3$ over $GF(2^m)$, where m is odd. Note that this curve has $2^m + 1$ points.

5 ElGamal cryptosystem

Let E be the curve $y^2 + y = x^3$ over $GF(2^m)$, m odd, and let P be a publicly known point on E. The elements of $GF(2^m)$ are represented with respect to a normal basis. User A randomly chooses an integer a and makes public the point aP, while keeping a itself secret. The multiple aP of the point P can be computed by the usual method of repeated doubling and adding. Since doubling a point is "free", the procedure takes $2t$ field multiplications and t inversions, where $t+1$ is the number of ones in the binary representation of a. We assume that the messages are ordered pairs of elements in $GF(2^m)$. To transmit the message (M_1, M_2) to user A, sender B chooses a random integer k and computes the points kP and $akP = (\overline{x}, \overline{y})$. Assuming $\overline{x}, \overline{y} \neq 0$ (the event $\overline{x} = 0$ or $\overline{y} = 0$ occurs with very small probability for random k), B then sends A the point kP, and the field elements $M_1\overline{x}$ and $M_2\overline{y}$. To read the message, A multiplies the point kP by his secret key a to obtain $(\overline{x}, \overline{y})$, from which he can recover M_1 and M_2 in two divisions.

In the above scheme, four field elements are transmitted in order to convey a message consisting of two field elements. We say that there is message expansion by a factor of 2. The message expansion can be reduced to $3/2$ by only sending the x-coordinate x_1 of kP and a single bit of the y-coordinate y_1 of kP. y_1 can easily be recovered from this information as follows. First $\alpha = x_1^3$ is computed by a single multiplication of x_1 and x_1^2. Since the Trace of x_1^3 must be 0, we have that either

$$y_1 = \alpha + \alpha^{2^2} + \alpha^{2^4} + \cdots + \alpha^{2^{m-1}},$$

or

$$y_1 = \alpha + \alpha^{2^2} + \alpha^{2^4} + \cdots + \alpha^{2^{m-1}} + 1.$$

The identity 1 is represented by the vector of all 1's, and so the single bit of y_1 that was sent enables one to make the correct choice for y_1. Notice that the computation of y_1 is inexpensive, since the terms in the formula for y_1 are obtained by successively squaring α.

The drawback of the method described above is that if an intruder happens to know M_1 (or M_2), he can then easily obtain M_2 (M_1). This attack can be prevented by only sending $(kP, M_1\overline{x})$, or by embedding M_1 on the curve. If the user wishes to embed messages on the elliptic curve, the following deterministic scheme may be used. We assume that messages are $(m-1)$-bit strings $M = (M_0, M_1, ..., M_{m-2})$. We can consider M as an element of $GF(2^m)$ (where $M_{m-1} = 0$). To embed M on the curve, M^3 is first computed and then the Trace of M^3 is evaluated. If $Tr(M^3) = 0$, then we set $x_M = M$, otherwise we set $x_M = M + 1$. In either case, we have that $Tr(x_M^3) = 0$. As in the preceding paragraph, one can now easily find y_M such that $P_M = (x_M, y_M)$ is a point on E. Sender B can now send A the pair of points $(kP, akP + P_M)$. With this scheme the message expansion is by a factor of 4. The message expansion can be reduced to 2 by sending only the x-coordinate and a single bit of the y-coordinate of each point. Note that after user A recovers x_M, he can decide whether the message sent is x_M or $x_M + 1$, by checking whether the last bit of x_M is 0 or 1 respectively.

The security of the ElGamal scheme depends on the intractability of the discrete logarithm problem. For a general group G, the problem can be stated as follows: given $g \in G$ and $\alpha = g^x \in G$, find x. The best know methods for computing discrete logarithms in an elliptic curve group are those general methods which apply to any group, and run in time $O(\sqrt{n})$, where n is the order of the group. The more powerful index calculus algorithms that are used to compute logarithms in the multiplicative group of a finite field do not appear to extend to elliptic curve groups, as Miller argues in [Mill86]. In addition, to guard against a Pohlig-Hellman type attack, the order of the elliptic curve, or more particularly, the order of P, should have a large prime divisor. In Section 8, we will list some curves that satisfy this criterion.

6 Projective Coordinates

Even though there are special techniques for computing inverses in $GF(2^m)$, an inversion is still far more expensive than a field multiplication (see Section 3). The inverse operation needed when adding two points can be eliminated by resorting to projective coordinates.

Let E be the curve $y^2 + y = x^3$ over $K = GF(2^m)$. E can be equivalently viewed as the set of all points in $P^2(K)$ which satisfy the homogeneous cubic equation $y^2 z + y z^2 = x^3$. Here $P^2(K)$ denotes the projective plane over K. The points of $P^2(K)$ are all of the non-zero triples in K^3 under the equivalence relation $\sim$, where $(x, y, z) \sim (x', y', z')$ if and only if there exists $\alpha \in K^*$ such that $x' = \alpha x$, $y' = \alpha y$, and $z' = \alpha z$. The representative of an equivalence class containing (x, y, z) will be denoted by $(x : y : z)$. Note that the only projective point in E with z-coordinate equal to 0 is the point $(0 : 1 : 0)$; this point is the point at infinity O of E. If $O \neq (x : y : z) \in E$, then $(x : y : z) = (x/z : y/z : 1)$, and so the projective point $(x : y : z)$ corresponds uniquely to the affine point $(x/z, y/z)$.

Let $P = (x_1 : y_1 : 1) \in E$, $Q = (x_2 : y_2 : z_2) \in E$, and suppose that $P, Q \neq O$, $P \neq Q$ and $P \neq -Q$. Since $Q = (x_2/z_2 : y_2/z_2 : 1)$ we can use the addition formula for E in affine coordinates to find $P + Q = (x_3' : y_3' : 1)$. We obtain

$$x_3' = \frac{A^2}{B^2} + x_1 + \frac{x_2}{z_2},$$

$$y_3' = 1 + y_1 + \frac{A}{B}\left(\frac{A^2}{B^2} + \frac{x_2}{z_2}\right),$$

where $A = (y_1 z_2 + y_2)$ and $B = (x_1 z_2 + x_2)$.

To eliminate the denominators of the expressions for x_3' and y_3', we set $z_3 = B^3 z_2$, $x_3 = x_3' z_3$ and $y_3 = y_3' z_3$, to get $P + Q = (x_3 : y_3 : z_3)$, where

$$x_3 = A^2 B z_2 + B^4$$
$$y_3 = (1 + y_1)z_3 + A^3 z_2 + AB^2 x_2$$
$$z_3 = B^3 z_2.$$

This addition formula requires 9 multiplications of field elements, which is more that the 2 multiplications required when using affine coordinates. We save by not having to perform a costly inversion. This gain occurs at the expense of space however, as we need extra registers to store P and Q, and also to store intermediate results when doing the addition.

One can now compute the multiple kP, where P is the affine point $(x_1, y_1, 1)$, by repeatedly doubling P, and adding the result into an accumulator. The result $kP = (x_3, y_3, z_3)$ can be converted back into affine coordinates by dividing each coordinate by z_3.

7 Montgomery's Method

To reduce the number of registers needed to add points on an elliptic curve, we use a method for addition that is similar to that used by Montgomery in [Mont87,§10.3.1].

Let $P = (x_1, y_1)$ and $Q = (x_2, y_2)$ be two points on E, with $P \neq -Q$. Then $P + Q = (x_3, y_3)$ satisfies

$$x_3 = \frac{(y_1 + y_2)^2}{(x_1 + x_2)^2} + x_1 + x_2. \tag{4}$$

Similarly, since $-Q = (x_2, y_2 + 1)$, $P - Q = (x_4, y_4)$ satisfies

$$x_4 = \frac{(y_1 + y_2)^2 + 1}{(x_1 + x_2)^2} + x_1 + x_2. \tag{5}$$

Adding (4) and (5), we get

$$x_3 = x_4 + \frac{1}{(x_1 + x_2)^2}. \tag{6}$$

Notice that to compute the x-coordinate x_3 of $P + Q$, we only need the x-coordinates of P, Q, and $P - Q$, and this can be accomplished with a single inversion.

We can now compute kP from P using the double and add method. First $2P$ is computed, and then we repeatedly compute either $(2mP, (2m+1)P)$ or $((2m+1)P, (2m+2)P)$ from $(mP, (m+1)P)$, depending on whether the corresponding bit in the binary representation of k is 0 or 1. Notice however that we have to use the addition formula (6) each time a new pair of points is computed, and this is done exactly $\log_2 k$ times. In the methods of Section 4 and Section 6, the corresponding addition formulae were only used t times when computing kP, where t is the number of 1's occurring in the binary representation of k. Thus the improvement in storage requirements when using the Montgomery method is at a considerable expense of speed.

8 Implementation

We estimate the throughput rate of encryption using the elliptic curve analogue of the ElGamal public key cryptosystem. The elements of $GF(2^m)$ are represented with respect to an optimal normal basis. We assume that a multiplication in $GF(2^m)$ takes m clock cycles, while an inversion takes $I(m) = \lfloor \log_2(m-1) \rfloor + \omega(m-1) - 1$ multiplications. For simplicity, we ignore the cost of field additions and squarings.

It was noted in Section 5 that computing logarithms in an elliptic curve group is believed to be harder that computing logarithms in $GF(2^m)^*$. We can thus achieve a high level of security using the elliptic curve ElGamal cryptosystem, but by a using a significantly smaller field size. For this reason, we can assume that the number of registers used is not a crucial factor in an efficient implementation. We will thus represent points using projective coordinates.

In the ElGamal system, the computation of kP and kaP requires m additions of points on average, for a randomly chosen k. To increase the speed of the system, and to place an upper bound on the time for encryption, we limit the Hamming weight of k to some integer d, where $d \leq m$. A similar technique is used in RSA (see [Hast86]) and in [Agnew89b]. The integer d should be chosen so that the key space is large enough to prevent exhaustive attacks. For the present discussion, we choose $d = 30$.

The computation of kP and kaP takes 58 additions of points, 2 field inversions and 4 field multiplications. Computing $m_1 \overline{x}$ and $m_2 \overline{y}$, where

$kaP = (\overline{x}, \overline{y})$, takes another 2 multiplications. Thus two field elements can be encrypted in $528+2I(m)$ field multiplications. For concreteness we select the fields $GF(2^{191})$ and $GF(2^{251})$. These fields are appropriate because optimal normal bases exist. Also since the factorizations of $2^{191} + 1$ and $2^{251} + 1$ are

$$2^{191} + 1 = 3 \cdot P58$$
$$2^{251} + 1 = 3 \cdot 238451 \cdot P70$$

(where Pn denotes an n digit prime number), a point P of high order can be easily chosen from E by choosing a point on the curve at random, and checking that $3P \neq O$ (in the case $m = 191$), and $(3 \cdot 238451)P \neq O$ (in the case $m = 251$). Finally, noting that $I(191) = I(251) = 12$, and assuming a clock rate of 20 MHz, we get an encryption rate of 72K bits/s.

References

[Agnew89] G. Agnew, T. Beth, R. Mullin, and S. Vanstone, "Arithmetic operations in $GF(2^m)$", submitted to *Journal of Cryptology*.

[Agnew89b] G. Agnew, R. Mullin, I. Onyszchuk and S. Vanstone, "An implementation for a fast public key cryptosystem", to appear.

[Ash89] D. Ash, I. Blake and S. Vanstone, "Low complexity normal bases", *Discrete Applied Mathematics*, **25** (1989), 191-210.

[Diff76] W. Diffie and M. Hellman, "New directions in cryptography", *IEEE Transactions on Information Theory*, **22** (1976), 644-654.

[ElG85] T. ElGamal, "A public key cryptosystem and a signature scheme based on discrete logarithms", *IEEE Transactions on Information Theory*, **31** (1985), 469-472.

[Hast86] J. Hastad, "On using RSA with low exponent in a public key network", *Advances in Cryptology: Proceedings of Crypto '85*, Lecture Notes in Computer Science, **218** (1986), Springer Verlag, 403-408.

[Kob87] N. Koblitz, "Elliptic curve cryptosystems", *Mathematics of Computation*, **48** (1987), 203-209.

[Kob87b] N. Koblitz, *Course in Number Theory and Cryptography*, Springer-Verlag, New York, 1987.

[Men90] A. Menezes and S. Vanstone, "Isomorphism classes of elliptic curves over finite fields", Research Report CORR 90-1, Department of Combinatorics and Optimization, University of Waterloo, January 1990.

[Mill86] V. Miller, "Uses of elliptic curves in cryptography", *Advances in Cryptology: Proceedings of Crypto '85*, Lecture Notes in Computer Science, **218** (1986), Springer Verlag, 417-426.

[Mont87] P. Montgomery, "Speeding the Pollard and elliptic curve methods of factorization", *Mathematics of Computation*, **48** (1987), 243-264.

[Mull88] R. Mullin, I. Onyszchuk, S. Vanstone and R. Wilson, "Optimal normal bases in $GF(p^n)$", *Discrete Applied Mathematics*, **22** (1988/89), 149-161.

[Ros89] T. Rosati, "A high speed data encryption processor for public key cryptography", *Proceedings of IEEE Custom Integrated Circuits Conference*, San Diego (1989), 12.3.1 – 12.3.5.

[Sil86] J. Silverman, *The Arithmetic of Elliptic Curves*, Springer-Verlag, New York, 1986.

Direct Demonstration of the Power to Break Public-Key Cryptosystems

Kenji Koyama

NTT Basic Research Laboratories
Nippon Telegraph and Telephone Corporation
3-9-11, Midori-cho, Musashino-shi, Tokyo, 180 Japan

Abstract : This paper describes a method of proving that a prover (or a cryptanalyst) really knows a secret plaintext or a new code-breaking algorithm for a particular public-key cryptosystem, without revealing any information about the plaintext or algorithm itself. We propose a secure direct protocol which is more efficient than the conventional protocols. This protocol requires only two transmissions between a prover and a verifier. A general form of the secure direct protocol is shown. The explicit forms for the RSA cryptosystem and the discrete logarithm problem are also proposed.

1. Introduction

Let an encryption function for the target public-key cryptosystem be denoted by $C = E(M)$, where C is a ciphertext and M is the plaintext. Breaking the cryptosystem is generally defined as obtaining M from any C by computing $E^{-1}(C)$. A cryptanalyst may succeed in breaking a cryptosystem by inventing a new efficient algorithm. He may want to convince potential buyers that he knows an efficient code-breaking method without revealing the knowledge and the value of M. What protocol is needed for a trade between a doubtful cryptanalyst and a potential buyer? The protocol must be non-cheatable and revealing no-knowledge.

It was shown by [GMW, BC] that a zero-knowledge proof protocol can be constructed for all problems in NP under certain conditions. The problem of breaking public-key cryptosystems, such as the RSA system, belongs to the NP class. Especially, if the encryption function E has the *homomorphism* property, the practical zero-knowledge protocols for demonstrating the breaking of public-key cryptosystems can be constructed. For example, several practical protocols have been shown for the Rabin system [FS], the RSA system [KV], the discrete logarithm problem [CEG, K]. These protocols are interactive, that is, indirect. These conventional indirect protocols (with sequential or parallel versions) require large transmission information to prevent prover's cheating. Subsequently, some researchers have presented "non-interactive" protocols to improve transmission efficiency [BFM, DMP]. However, so far there has been no general method of constructing a direct protocol for demonstrating the breaking of public-key cryptosystems. The main purpose of this paper is to show a *secure direct* protocol for demonstrating the breaking of public-key cryptosystems. The proposed direct protocol requires only one interaction or two transmissions between a prover and a verifier. It is more efficient than the conventional indirect protocols.

A general form of the secure direct protocol is shown in Section 2. The explicit forms for the RSA system and the discrete logarithm problem are proposed in Sections 3 and 4, respectively.

2. General Form of Direct Protocol

2.1. Protocol

Before specializing in a certain cryptographic scheme, we describe a very general form of direct protocol for demonstrating the power of breaking the public key cryptography. Hereafter, the prover (cryptanalyst) is referred to A, and the verifier (buyer) is referred to B. Protocol 1 shows prover A convinces verifier B that he knows M such that $C = E(M)$.

[Protocol 1]

step 1 : A and B share random input C, and agree on functions E, f and g such that

$$E(M) = g(f(M)).$$

step 2 : A computes M from C using an efficient attacking method as

$$M = E^{-1}(C),$$

computes P easily from M using

$$P = f(M),$$

and sends P to B.

step 3 : B checks easily whether or not

$$C = g(P).$$

If the check is passed, B accepts the claim that A can break the target public-key cryptosystem.

Otherwise, B detects cheating.

Remark:

- The ciphertext C must be chosen uncontrollably by A and B. Oblivious transfer protocol [B] can be used to generate C.

2.2. Existence Condition of Secure Direct Protocol

We can define the security of the direct protocol 1 from three viewpoints: completeness, zero-knowledgeness and soundness.

Definition. A direct protocol (such as Protocol 1) is *secure* if the following conditions hold:

Completeness:

> If the proof is correct, B should accept A's proof with probability 1. That is, if A can find M from any C, then B always accepts A's claim.

Zero-knowledgeness:

> The proof does not yield knowledge such as the value of M and the code-breaking method. That is, the difficulty that B has in finding M with the protocol is equal to the difficulty that B will have in finding M from C without the protocol.

Soundness:

> If the proof is incorrect, B should reject A's proof with overwhelming probability. That is, the difficulty that A has in cheating B with the protocol is equal to the difficulty that A will have in finding M from C.

The above security conditions are explicitly expressed in Theorem 1.

Theorem 1. *A secure direct protocol exists if the following three conditions hold:*
(1) A one-way encryption function E is divided as follows.

$$E(M) = g(f(M)).$$

where f and g are also one-way functions.
(2) Computing M from C and P such that $C = E(M)$ and $P = f(M)$ is as hard as computing M from only C such that $C = E(M)$.
(3) Computing P from C using $C = g(P)$ is as hard as computing P from C using $C = E(M)$ and $P = f(M)$. That is, computing $g^{-1}(C)$ is as hard as computing $f(E^{-1}(C))$.

Proof. It is clear from the definition of the security of the direct protocol. The relationship $E(M) = g(f(M))$ in condition (1) guarantees the completeness condition. Condition (2) guarantees the zero-knowledgeness condition. Condition (3) guarantees the soundness condition.

Remarks:
- Conditions (2) and (3) requires that f and g be one-way functions, respectively.
- Condition (2) implies that f^{-1} is not computationally easier than E^{-1}.
- Since f is easy to compute, condition (3) implies that g^{-1} is not computationally easier than E^{-1}. If A has the power to compute g^{-1}, then he has the power to compute E^{-1}.

3. Protocol for the RSA Cryptosystem

Let an encryption function for the target RSA cryptosystem be defined by $C = E(M) = M^e \bmod n$. The security of a protocol for the RSA system is based on the following Conjectures 1 and 2:

Conjecture 1. *The difficulty of breaking the RSA system is not affected by its exponent value e.*

Note: Conjecture 1 seems to be true in circumstantial evidence, because it was proved that the most efficient attack found to date are all computationally equivalent to factoring n.

Conjecture 2. *If $e = km$, $k \geq 3$, $m \geq 3$, then computing M from (C, e, n, P, k) satisfying*

$$C = E(M) = M^e \bmod n$$

$$P = f(M) = M^k \bmod n$$

is as hard as computing M from (C, e, n) satisfying

$$C = E(M) = M^e \bmod n.$$

Note: Conjecture 2 seems to be true.

If Conjectures 1 and 2 hold, we can construct a secure direct protocol for demonstrating the power of breaking the RSA cryptosystem as follows.

[Protocol for the RSA]

step 1: A and B share random input integer C, where $0 \leq C \leq n - 1$ and agree on functions E, f and g.

step 2: A computes M from (C, e, n) such that

$$C = E(M) = M^e \bmod n,$$

computes P easily from (M, k, n) using

$$P = f(M) = M^k \bmod n,$$

and sends P to B.

step 3: B checks easily whether or not (C, P, m, n) satisfies

$$C = g(P) = P^m \bmod n.$$

If the check is passed, B accepts A's claim.

Otherwise, B detects cheating.

4. Protocol for the Discrete Logarithm Problem

Let an encryption function for the target discrete logarithm problem be defined by $C = E(M) = a^M \mod p$, where a is a primitive element modulo p, and p is a prime or prime power. The security of a protocol for the discrete logarithm problem is based on the following Lemmas 1 and 2:

Lemma 1. *The difficulty of computing the discrete logarithm is not affected by its base value a. That is, computing X_1 from (Y_1, a_1, p) satisfying*

$$Y_1 = a_1^{X_1} \mod p$$

is as hard as computing X_2 from $(Y_2, a_2 (\neq a_1), p)$ satisfying

$$Y_2 = a_2^{X_2} \mod p.$$

Proof. Without loss of generality, we prove that solvability of the former discrete logarithm problem implies that of the latter discrete logarithm problem. Assume it is easy to obtain X_1 from (Y_1, a_1, p) satisfying

$$Y_1 = a_1^{X_1} \mod p.$$

Thus it is easy to obtain Z from (a_2, a_1, p) satisfying

$$a_2 = a_1^{Z} \mod p,$$

and it is also easy to obtain ZX_2 from (Y_2, a_1, p) satisfying

$$Y_2 = a_1^{ZX_2} \mod p.$$

Therefore it is easy to obtain X_2 from Z and ZX_2 by

$$X_2 = ZX_2/Z \mod p - 1.$$

Lemma 2. *Computing the simultaneous discrete logarithms is not easier than the basic discrete logarithm. That is, computing X from (Y_1, a_1, Y_2, a_2, p) satisfying*

$$Y_1 = a_1^{X} \mod p,$$

$$Y_2 = a_2^X \bmod p$$

is as hard as computing X from (Y_1, a_1, p) satisfying

$$Y_1 = a_1^X \bmod p.$$

Proof. See [CEG].

Since Lemmas 1 and 2 hold, we can construct a secure direct protocol for demonstrating the power of solving the discrete logarithm problem as follows.

[Protocol for the DLP]

step 1: A and B share random input C, and agree on the values of base a and modulus p. B generates random R relatively prime to $p-1$, and computes S and k such that

$$RS \equiv 1 \pmod{p-1},$$

$$k = E(R) = a^R \bmod p.$$

B keeps S secret, and sends k to A.

step 2: A computes M from (C, a, p) such that

$$C = E(M) = a^M \bmod p,$$

computes P easily from (M, k, p) using

$$P = f(M) = k^M \bmod p,$$

and sends P to B.

step 3: B checks easily whether or not (C, P, S, p) satisfy

$$C = g(P) = P^S \bmod p.$$

If the check is passed, B accepts A's claim.

Otherwise, B detects cheating.

5. Conclusion

We have proposed a secure direct protocol. Its general form and the explicit forms for the RSA and DLP have been shown.

Acknowledgements: We wish to thank Professor S. A. Vanstone at the University of Waterloo for valuable discussions on this work.

References

[B] M. Blum : "Three applications of the oblivious transfer: 1. coin flipping by telephone, 2. how to exchange secrets, 3. how to send certified electronic mail,", Dept. EECS, Univ. of California, Berkeley, Calif. (1981).

[BC] G. Brassard, and C. Crepeau, : "Non-transitive transfer of confidence: A perfect zero-knowledge interactive protocol for SAT and beyond", *Proc. 27th Symp. on Foundations of Computer Science (FOCS)*, pp.188-195, (1986).

[BFM] M. Blum, P. Feldman, and S. Micali: "Non-interactive zero-knowledge proof systems and applications", *Proc. of 18th Symp. on Theory of Computing (STOC)*, pp.103-112, (1988).

[CEG] D. Chaum, J. Evertse and J. Graaf, : "An improved protocol for demonstrating possession of a discrete logarithm and some generalizations", *Proc. of EUROCRYPT'87, Lecture Notes in Computer Science Vol. 304*, pp.127-142, Springer-Verlag, (1987).

[DMP] A. DeSantis, S. Micali, and G. Persiano, : "Non-interactive zero-knowledge proof systems", *Proc. of Crypto'87, Lecture Notes in Computer Science Vol. 293* pp.52-72, Springer-Verlag, (1987).

[FS] A. Fiat and A. Shamir, : "How to prove yourself: Practical solutions to identification and signature problems", *Proc. of CRYPTO'86, Lecture Notes in Computer Science Vol. 263*, pp.186-194, Springer-Verlag, (1986).

[GMW] O. Goldreich, S. Micali, and A. Wigderson, : "Proofs that yield nothing but their validity and a methodology of cryptographic protocol design", *Proc. of 27th Symp. on Foundations of Computer Science (FOCS)*, pp.174-187 (1986).

[K] K. Kizaki : "A note on zero-knowledge proof for the discrete logarithm problem", *Research Reports on Information Sciences at Tokyo Institute of Technology*, March (1987).

[KV] K. Koyama and S. A. Vanstone : "How to demonstrate the breaking of public-key cryptosystems", *Proc. of the 1987 Workshop on Cryptography and Information Security,* Noda, Japan, pp.161-170, July, (1987).

SECTION 2

PSEUDORANDOMNESS AND SEQUENCES I

Continued Fractions and the Berlekamp-Massey Algorithm

Zongduo Dai
Department of Electronic Engineering
University of Linköping
ISY 581-83 Linköping, Sweden

Kencheng Zeng *
Center for Advanced Computer Studies
University of Southwestern Louisiana
P.O. Box 4330, Lafayette, LA 70504

In this paper, we show how the famous linear synthesis algorithm of Berlekamp-Massey for binary sequences follows naturally from solving the classical problem of rational approximation to arbitrary elements in the field

$$F_2((x)) = \{\alpha = \sum_{i=m}^{\infty} a_i x^{-i} \mid m \in Z, a_i \in F_2\}$$

of formal Laurent series over the binary field F_2, a fact which makes clear the algebraic essence of the algorithm.

(I) Rational Approximation and Linear Synthesis

1) We write every $\alpha \in F_2((x))$ as a sum of two parts:

$$\alpha = \sum_{i=m}^{0} a_i x^{-i} + \sum_{i=1}^{\infty} a_i x^{-i}$$

$$= [\alpha] \; (\text{ the integral part }) + \{\alpha\} \; (\text{ the decimal part }) ,$$

and define the function $Ord\,(\alpha)$ over $F_2((x))$ by putting

---

* Both authors on leave from the Graduate School of Academia Sinica, Beijing, People's Republic of China.

$$Ord\,(\alpha) = \begin{cases} -\infty\,, & \text{if } \alpha = 0\,, \\ w\,, & \text{if } \alpha \neq 0 \text{ and } [x^{-w}\alpha] = 1\,. \end{cases}$$

It is easy to verify that
   i) $Ord\,(\alpha\beta) = Ord\,(\alpha) + Ord\,(\beta)$,
   ii) $Ord\,(\alpha + \beta) \leqslant \max\{Ord\,(\alpha), Ord\,(\beta)\}$, *with equality iff $Ord\,(\alpha)$ and Ord (β) are distinct.*
This means $Ord\,(\alpha)$ is a valuation of the field $F_2((x))$.

The field $F_2((x))$ contains the field $F_2(x)$ of rational fractions as a subfield, in such a way that the latter is dense in the former in the sense of the topology induced by the valuation $Ord\,(\alpha)$. So we can raise the rational approximation problem (RAP) in $F_2((x))$: Given $\alpha \in F_2((x))$ and $n \in N$, find all rational fractions $\dfrac{v(x)}{u(x)}$, with denominators $u(x)$ having the least possible degree, which approximate α to the order of precision n, i.e.,

$$Ord\,(\frac{v(x)}{u(x)} - \alpha) < -n\,. \tag{1}$$

Evidently, in considering the RAP we need only pay attention to find the non-zero polynomial $u(x)$, for we must have $v(x) = [u(x)\alpha]$, and (1) can be reformulated in terms of $u(x)$ alone as

$$Ord\,(\{u(x)\alpha\}) - deg\ u(x) < -n\,. \tag{1'}$$

We simply call $u(x)$ a solution to our problem.

2) In considering the linear synthesis problem for the binary sequence

$$\mathbf{a} = \{a_i \mid i \geqslant 0\}\,,$$

we identify it with the formal Laurent series

$$\alpha = \sum_{i=0}^{\infty} a_i x^{-i-1}\,,$$

an element belonging to the maximal ideal of the corresponding local ring.

The RAP in $F_2((x))$ is related to the linear synthesis problem through the following simple observation: The polynomial $u(x)$ is a minimal polynomial for the initial segment

$$\mathbf{s}_n = (a_0, a_{1,...}, a_{n-1})$$

of the binary sequence $\mathbf{a}$ if and only if it is a solution to the RAP to the order of precision n, raised with regard to the corresponding field element α.

(II) Continued Fractions and the RAP

The tool for solving the RAP is a ready one which can be borrowed directly from Diophantine analysis in the real field. Convert α into a simple continued fraction

$$\alpha \sim [q_0(x), q_1(x),..., q_k(x), \cdots]$$

by putting

$$\alpha_0 = \alpha, \quad q_0(x) = [\alpha_0], \quad \alpha'_1 = \{\alpha_0\}, \tag{2}$$

and

$$\alpha_k = \alpha'^{-1}_k, \quad q_k(x) = [\alpha_k], \quad \alpha'_{k+1} = \{\alpha_k\}, \tag{3}$$

whenever $\alpha'_k \neq 0$, otherwise the procedure terminates. Compute the successive convergents of the continued fraction according to the recursive relations

$$u_k(x) = q_k(x)u_{k-1}(x) + u_{k-2}(x) \tag{4}$$

and

$$v_k(x) = q_k(x)v_{k-1}(x) + v_{k-2}(x)$$

with the initial setting

$$u_0(x) = 1, u_{-1}(x) = 0; \quad v_0(x) = q_0(x), v_{-1}(x) = 1, \tag{5}$$

and define

$$d_0 = 0; d_k = \deg u_k(x) = \sum_{i=1}^{k} \deg q_i(x), \quad k \geqslant 1$$

with the convention to put $d_{N+1} = \infty$ in case $\alpha'_{N+1} = 0$. Observe that we always have $d_k < d_{k+1}$ since $\deg q_k(x) \geqslant 1$ for $k \geqslant 1$.

Lemma 1. We have for every $k \geqslant 1$

$$Ord(\{u_k(x)\alpha\}) = -d_{k+1}.$$

Proof. Making use of (2), (3), (4), and (5), it is easy to show by induction on k that

$$u_k(x)\alpha - v_k(x) = \alpha'_{k+1}(u_{k-1}(x)\alpha - v_{k-1}(x)),$$

so we have

$$Ord\,(u_k(x)\alpha - v_k(x)) = Ord\,(\alpha'_{k+1}\alpha'_k \cdots \alpha'_1)$$

$$= -\sum_{i=1}^{p+1} deg\ q_i(x) = -d_{k+1} < 0,$$

which means $u_k(x)\alpha - v_k(x)$ is a "pure decimal" coincident with $\{u_k(x)\alpha\}$.

Theorem 1. If $d_{k-1} + d_k \leqslant n < d_k + d_{k+1}$, then $u(x)$ is a solution to the RAP iff

$$u(x) = u_k(x) + h(x)u_{k-1}(x),\quad deg\ h(x) < 2d_k - n.$$

Proof. First, suppose $u(x) \neq u_k(x)$ is a solution to the RAP. Since

$$Ord\,(\{u_k(x)\alpha\}) - deg\ u_k(x) = -d_k - d_{k+1} < -n,$$

we contend that

$$d = deg\ u(x) \leqslant d_k.$$

After dividing $u(x)$ by $u_k(x)$ and then dividing the remainder by $u_{k-1}(x)$, etc., we can write

$$u(x) = \sum_{i=\tau}^{k} h_i(x)u_i(x)$$

where $h_k(x)$ is a constant, $h_\tau(x) \neq 0$ and

$$deg\ h_i(x) < d_{i+1} - d_i,\quad \tau \leqslant i \leqslant k-1. \tag{6}$$

We have

$$\{u(x)\alpha\} = \sum_{i=\tau}^{k}\{h_i(x)\{u_i(x)\alpha\}\},$$

But we see from lemma 1 and (6) that

$$-d_{i+1} \leqslant Ord\,(\{h_i(x)\{u_i(x)\alpha\}\}) < d_{i+1} - d_i - d_{i+1} = -d_i,$$

so it follows from the second property of the valuation function that

$$Ord\left(\{u(x)\alpha\}\right) = Ord\left(\{h_\tau(x)\{u_\tau(x)\alpha\}\}\right) = deg\ h_\tau(x) - d_{\tau+1}$$

and hence

$$deg\ h_\tau(x) - d_{\tau+1} - d_k \leqslant Ord\left(\{u(x)\alpha\}\right) - d < -n \leqslant -d_{k-1} - d_k$$

or

$$deg\ h_\tau(x) < d_{\tau+1} - d_{k-1}\ .$$

The latter inequality means we must have $\tau = k - 1$ and, after coming back to (1') again,

$$deg\ h_{k-1}(x) < d + d_k - n \leqslant 2d_k - n\ .$$

On the other hand, were $h_k(x) = 0$, then $u(x) = h(x)u_{k-1}(x)$ and we would come to the contradiction

$$-d_{k-1} - d_k = (d - d_{k-1}) - d_k - d$$

$$= Ord\left(\{u(x)\alpha\}\right) - d < -n \leqslant -d_{k-1} - d_k\ .$$

This proves the only if part. To prove the if part is a matter of direct verification.

(III) The Berlekamp-Massey Algorithm

Theorem 1 gives in terms of the polynomials $u_k(x)$ a complete description of the solution set of the RAP to any order of precision, but provides no recursive algorithm to find the solutions. However, a closer examination shows it can be refined to reach at one which coincides with that of Berlekamp-Massey.

First, we refine each

$$q_k(x) = \sum_{r=1}^{w_k} x^{t_k - j_{k,r}}\ ,\quad k \geqslant 1$$

$$0 = j_{k,1} < j_{k,2} < \cdots < j_{k,w_k} \leqslant t_k$$

into a series of w_k polynomials

$$q_{k,r}(x) = \sum_{i=1}^{r} x^{t_k - j_{k,i}}, \quad 1 \leqslant r \leqslant w_k,$$

and then correspondingly refine the polynomials $u_k(x)$, $v_k(x)$ into two series of polynomials $u_{k,r}(x)$, $v_{k,r}(x)$ by putting

$$u_{k,r}(x) = q_{k,r}(x)u_{k-1}(x) + u_{k-2}(x)$$

and

$$v_{k,r}(x) = q_{k,r}(x)v_{k-1}(x) + v_{k-2}(x).$$

It is easy to see that all the polynomials $u_{k,r}(x)$ with the same k are of the same degree d_k. We have

Lemma 2. For every $k \geqslant 1$ and $1 \leqslant r \leqslant w_k$,

$$Ord\left(\{u_{k,r}(x)\alpha\}\right) = -d_{k-1} - j_{k,r+1}, \tag{7}$$

with the convention that $j_{k,w_k+1} = t_k + t_{k+1}$.

Proof. As in the proof of Lemma 1, we have

$$u_{k,r}(x)\alpha - v_{k,r}(x) = (\alpha_k + q_{k,r}(x))(u_{k-1}(x)\alpha - v_{k-1}(x)),$$

and hence

$$Ord(u_{k,r}(x)\alpha - v_{k,r}(x)) = Ord(\alpha_k + q_{k,r}(x)) + Ord(u_{k-1}(x)\alpha - v_{k-1}(x))$$

$$= (t_k - j_{k,r+1}) - d_k$$

$$= -d_{k-1} - j_{k,r+1} < 0,$$

which establishes (7).

Now, let $W_n(\alpha)$ denote the set of all polynomials which are solutions of the RAP of order of precision n to α.

Theorem 2. Define a sequence of polynomials $f_n(x)$ by putting

$$f_n(x) = \begin{cases} 1, & 0 \leqslant n < d_1 \\ u_{k,r}(x), & k \geqslant 1, \ d_{k-1}+d_k+j_{k,r} \leqslant n < d_{k-1}+d_k+j_{k,r+1} \end{cases}$$

and write $l_n = deg\, f_n(x)$. Then we have
1) $f_n(x) \in W_n(\alpha)$,
2) For any $n \geqslant d_1$, if $f_n(x) \in W_{n+1}(\alpha)$, then $f_{n+1}(x) = f_n(x)$;

otherwise

$$f_{n+1}(x) = \begin{cases} f_n(x) + x^{(m-l_m)-(n-l_n)} f_m(x), & \text{if } (m-l_m)-(n-l_n) \geq 0, \\ x^{(n-l_n)-(m-l_m)} f_n(x) + f_m(x), & \text{if } (m-l_m)-(n-l_n) < 0, \end{cases}$$

where m is the largest integer such that $l_m < l_n$.

Proof. To see 1), we need only observe that $f_n(x) = u_{k,r}(x)$ is of degree d_k which is common to all polynomials of $W_n(\alpha)$ and, by (7), we have

$$Ord(\{f_n(x)\alpha\}) - deg\, f_n(x) = -d_{k-1} - d_k - j_{k,r+1} < -n.$$

To see 2), observe that $f_n(x) \in W_{n+1}(\alpha)$ implies

$$n + 1 < d_{k-1} + d_k + j_{k,r+1},$$

and so, according to our definition,

$$f_{n+1}(x) = u_{k,r}(x) = f_n(x).$$

Now suppose $f_n \notin W_{n+1}(\alpha)$. In this case we have

$$n + 1 = d_{k-1} + d_k + j_{k,r+1},$$

and accordingly,

$$f_{n+1}(x) = \begin{cases} u_{k,r+1}(x), & \text{if } r < w_k, \\ u_{k+1,1}(x), & \text{if } r = w_k. \end{cases}$$

Moreover, we see in this case

$$m = d_{k-1} + d_k - 1, \quad f_m(x) = u_{k-1}(x), \quad l_m = d_{k-1}$$

and

$$(m-l_m)-(n-l_n) = \begin{cases} t_k - j_{k,r+1} \geq 0, & \text{if } r < w_k \\ -t_{k+1} < 0, & \text{if } r = w_k. \end{cases}$$

So, if $(m-l_m)-(n-l_n) \geq 0$, then

$$f_{n+1}(x) = u_{k,r+1}(x) = q_{k,r+1}(x)u_{k-1}(x) + u_{k-2}(x)$$

$$= (q_{k,r}(x) + x^{t_k - j_{k,r+1}})u_{k-1}(x) + u_{k-2}(x)$$

$$= u_{k,r}(x) + x^{t_k - j_{k,r+1}}u_{k-1}(x)$$

$$= f_n(x) + x^{(m-l_m)-(n-l_n)}f_m(x) \; ;$$

if $(m-l_m)-(n-l_n) < 0$, then

$$f_{n+1}(x) = u_{k+1,1}(x) = q_{k+1,1}(x)u_k(x) + u_{k-1}(x)$$

$$= x^{t_{k+1}}u_k(x) + u_{k-1}(x)$$

$$= x^{(n-l_n)-(m-l_m)}f_n(x) + f_m(x) \; ,$$

just as stated in the theorem.

Thus we see, if somebody tries to synthesize a binary sequence by help of the Berlekamp-Massey algorithm, then he will be doing nothing else than computing recursively the polynomials $f_n(x)$ we defined in the above in an apriori way.

Acknowledgements

The authors would like to express their thankfulness to Professors J. Massey and T.R.N. Rao for their interest in the present work and the useful suggestions they contributed.

References

[1] J. W. Cassels, *Diophantine Approximation*, Cambridge University Press, London, 1965.

[2] J. L. Massey, "Shift-Register Synthesis and BCH Decoding," *IEEE Trans. Info. Th.*, Vol. IT-15, pp. 122-127, Jan. 1969.

[3] W.H. Mills, "Continued Fractions and Linear Recurrence," *Math. Comp.*, 29(1975), pp. 173-180.

Nonlinear Generators of Binary Sequences with Controllable Complexity and Double Key

Gong Guang
Applied Mathematics Department
Electronic Science and Technology University
Cheng Du, Sichuan, PRC

Abstract

This paper discusses analysis and synthesis of non- linear generators which consist of a single linear feedback shift register (LFSR) with a primitive connection polynomial, a random select generator and a non-linear feedforward logic. It is shown that the linear complexity of the generated keystream can be determined, realization is easy and its security is much better than the original feedforward system.

1 Introduction

In this paper, we present a new kind of nonlinear generator which consists of a single LFSR with a primitive connection polynomial, a random select generator and a nonlinear feedforward logic (called double key generator, denoted as DKG, see Fig. 1), and discuss its analysis, synthesis and realization by means of results of [1] and [2] or [3].

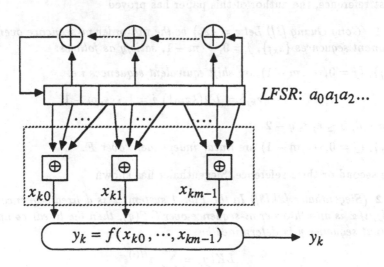

Fig. 1

2 Preliminaries

Definition 1 *Let* $x = \{x_k\}$, $x_k \in GF(q)$, *then* x *is called a sequence in* $GF(q)$; *meanwhile,* $x_k = (x_{k0}, \cdots, x_{km-1})$, $x_{kj} \in GF(2) = F$, $\{x_{kj}\}$ *called* x *'s component sequence,* $j = 0, \cdots, m-1$ $(q = 2^m)$.

Definition 2 *Let* $a = \{a_k\}$ *be a binary lth-order m-sequence (that is a maximal length sequence),* $x = \{x_k\}$ *be a sequence in* $GF(q)$, x *'s component sequences satisfy* $\{x_{kj}\} = L^{v_j}(a)$, $0 \le v_j \le 2^l - 2$, *then* x *is called a parallel sequence generated by* $\{a_k\}$, $(v_0, \cdots v_{m-1})$ *called* x *'s character [4].*

Definition 3 *Let* $x = \{x_k\}$ *be a parallel sequence generated by* $\{a_k\}$, *if* x *'s character* $(v_0, \cdots v_{m-1})$ *satisfies*

1. $v_j = k_j d$, $0 \le k_j \le q - 2$, $d = (q^n - 1)/(q - 1)$

2. $\{x_{kj}\}$, $j = 0, \cdots, m-1$ *are linear dependent over* F *then* x *is called a base sequence generated by* $\{a_k\}$ *in* $GF(q)$ *(or base sequence).*

See Fig. 1. Let LFSR be a linear feedback shift register with primitive connection polynomial:

$$g(t) = \sum_{i=0}^{e} c_i t^i$$

$(l = mn, c_i \in F)$ $\oplus$ be exclusive or (EXOR), the part in dotted lines be a random select generator which generates the base sequence $x = \{x_k\}$, $x_k = (x_{k0}, \cdots, x_{km-1})$, $x_{kj} \in F$, $f(t_1, \cdots, t_m)$ be a nonlinear feedforward logic and $y = \{y_k\}$, $y_k = f(x_{k0}, \cdots, x_{km-1})$, the output sequence.

Remark: All symbols in this paper are defined only if they are first used.

3 DKG's linear complexity

In the first reference, the author of this paper has proved

Lemma 1 *(Gong Guang [1]) Let* $z = \{z_k\}$ *be the nth-order m-sequence over* $GF(q)$, *then* z *'s component sequences* $\{z_{kj}\}$, $j = 0, \cdots m - 1$, *satisfy as follows:*

1. $\{z_{kj}\}$, $(j = 0, \cdots, m-1)$ *are shift equivalent sequences i.e.*

$$\{z_{kj}\} = L^{v_j}(\{z_{k0}\}) \; j = 1, \cdots, m - 1.$$

2. $v_j = e_j d$, $0 \le e_j \le q - 2$

3. $\{z_{kj}\}$, $(j = 0, \cdots, m-1)$ *are linear independent over* F.

In the second or third reference, Siegenthaler has shown

Lemma 2 *(Siegenthaler [2],[3]) In the Fig. 1 system,* x *is a fixed sequence (not random generated), if* x *is an nth-order m-sequence over* $GF(q)$, *then the linear complexity* $LK(y)$ *of the output sequence* y *is determined by*

$$LK(y) = \sum_{f_i \ne 0} n^{H(i)}$$

where f_i is the coefficient of $f(t_1, \cdots, t_m)$'s polynomial representation $f(t) = \sum_{i=0}^{q-1} f_i t^i$, $f_i \in GF(q)$, $H(i)$ is the Hamming weight of integral i.

Theorem 1 Let T be an arbitrary invertible linear transformation from $GF(q)$ to $GF(q)$, $fT(t_1, \cdots t_m)$'s polynomial representation be

$$fT(t) = \sum_{i=0}^{q-1} b_i t^i, \quad b_i \in GF(q)$$

then DKG's linear complexity $LK(y)$ is determined by

$$LK(y) = \sum_{b_i \neq 0} n^{H(i)}$$

Proof: In DKG, $x = \{x_k\}$ is a base sequence generated by $a = \{a_k\}$. Let $z = \{z_k\}$ be the nth-order m-sequence over $GF(q)$ and z's component sequences satisfy $\{z_{kj}\} = L^{u_j}(a)$, then z is a base sequence generated by a from Lemma 1. Hence there exists an invertible linear transformation T from $GF(q)$ to $GF(q)$, such that $x = T(z)$. Therefore $y_k = f(x_k) = f(T(z_k)) = fT(z_k)$. From Lemma 2 there is

$$LK(y) = LK(\{fT(z_k)\}) = \sum_{b_i \neq 0} n^{H(i)}$$

4   DKG's synthesis and realization

The main point of DKG's realization is how to random generate a base sequence, that is how to generate l-tuples $(c_{0j}, \cdots, c_{(l-1)j})$, $c_{ij} \in F$, such that

$$\{x_{kj}\} = \sum_{i=0}^{l-1} c_{ij} L^i(a), \quad j = 0, \cdots, m-1$$

Let e_j, $j = 0, \cdots, m-1$ be random numbers, $0 \leq e_j \leq q-2$, $\{x_{kj}\} = L^{e_j}(a)$. Since the number of different base sequences is

$$\prod_{i=0}^{m-1} (2^m - 2^i)$$

it is more than 0.29 that the probability p which $\{x_{kj}\}$, $j = 0, \cdots, m-1$ are linear independent over F, i.e.

$$p \geq 0.29$$

Let w be a primitive element in $GF(q^n)$ and w's minimal polynomial over F is the minimal polynomial $g(t)$ of $\{a_k\}$, then $u = w^d$ is a primitive element in $GF(q)$. Hence we have the following algorithm to random generate a base sequence:

Step 1 Generating $m \times m$ order 0-1 matrix $A = (w_{ij})_{m \times m}$ if $det(A) \neq 0$ going on to Step 2; if $det(A) = 0$, repeating Step 1.

Step 2 Let $Q_j = \sum_{i=0}^{m-1} w_{ij} u^i$, computing Q_j's vector representation in $GF(q^n)$, obtain

$$Q_j = \sum_{i=0}^{l-1} c_{ij} w^i, \quad j = 0, \cdots, m-1$$

hence

$$\{x_{kj}\} = \sum_{i=0}^{l-1} c_{ij} L^i(a), \quad j = 0, \cdots, m-1$$

and $x = \{x_k\}$ is a base sequence, the nonzero c_{ij}, $j = 0, \cdots l-1$, represent the phases of input sequences in the jth-EXOR.

The above algorithm can be realized in two ways, one is using hardware - we can use the encoder of BCH codes [5] with additive determining function, another way is using software - we can use the division algorithm in $FG(q^n)$ and an algorithm of computing $m \times m$ order determinant, where the computational complexity is $O(m^3 + nm)$.

5 Discussion

(1). DKG's security is much better than that of the original feedforward system. In fact, if the LFSR and $f(t_1, \cdots, t_m)$ have been broken, DKG still maintains higher security since the number of the second layer keys is $\prod_{i=0}^{m-1}(2^m - 2^i)$. This is a very big number, for example, when $m=32$, we have

$$\prod_{i=0}^{m-1}(2^m - 2^i) \geq 2^{32^2} 0.29 \geq 2^{1022}$$

(2). If we fix the base sequence x, then the security of this system is much better than that of Siegenthaler's [2],[3] system (see Lemma 2). Because if x is the nth-order m-sequence over $GF(q)$ with minimal polynomial $G(t) = \sum_{i=0}^{n} g_i t^i$, $g_i \in GF(q)$, then

$$g(t) = \sum_{i=0}^{m-1} \sigma^i(G(t))$$

where $\sigma^i(G(t)) = \sum_{j=0}^{n} g_j^{2^i} t^j$, there are only m shift equivalent m-sequences, but there are $\prod_{i=0}^{m-1}(2^m - 2^i)$ shift unequivalent base sequences.

At last we conclude that the DKG system is simple and easy to realize, if we select fitting nonlinear feedforward logic then we can obtain the system which is a balance between linear complexity and correlation immune.

References

1. Gong Guang, *m-Sequence Runs, component sequences and vector sequences in Galois field*, ACTA ELECTRONICA SINICA, Vol. 14, No. 4, July 1986, pp94-100.

2. Th. Siegenthaler, *Methoden fur den entwurf von stream cipher system*, Diss. ETH, No. 8185, Dec. 1986.

3. Th. Siegenthaler and Rejane Forre, *Generation of binary sequences with controllable complexity and ideal r-tupel distribution*, Advances in Cryptology - EUROCRYPT '87.

4. Gong Guang, *Analysis of the parallel sequences*. Was selected for the Proceedings of the 5th Meeting of a Workshop on Applied Mathematics, Oct. 1988, China.

5. F.J. MacWilliams and N.J.A. Sloane, *The Theory of Error- Correcting Codes*, North Holland, 1977.

K-M Sequence is Forwardly Predictable

Yang Yi Xian
Dept. of Information Engineering
Beijing University of Posts and Telecommunication
Beijing, 100088, PRC

1 Introduction

In a recent paper [1] K. Kurosawa and K. Matsu showed a new kind of stream key sequence $(b_0, b_1, \cdots)$ (called K-M sequence here) with an exact proof that the *backwardly* prediction of the K-M sequence is very very difficult. But we will point out in the following that the K-M sequence can be easily *forwardly* predicted. Hence the K-M sequence is cryptographically insecure.

2 Restatement of the K-M sequence and its difficulty for backwardly prediction

Let $R = pq$ where p and q are prime numbers.

$$x_{i+1} = x_i - c_i/x_i \quad (\bmod\ R) \tag{1}$$

$$c_{i+1} = 4c_i \tag{2}$$

$$b_i = \begin{cases} 0 & \text{if } x_i < -c_i/x_i \\ 1 & \text{if } x_i > -c_i/x_i \end{cases} \tag{3}$$

where

$$(c_0/p) = (c_0/q) = -1 \quad ((a/p) \text{ denotes the Legendre symbol})$$
$$x_0^2 + c_0 \neq 0 \quad (\bmod\ p) \text{ and } x_0^2 + c_0 \neq 0 \quad (\bmod\ q).$$

The sequence $\{b_i\}$ is the K-M sequence.

About the cryptographic security of the K-M sequence, the following theorems were shown in reference [1].

Theorem 1 *If b_i is predicted from $b_{i+1}, b_{i+2}, \cdots$ in polynomial time, then there exists a polynomial time algorithm for the c-quadratic residue problem. (Note: a c-quadratic residue problem is assumed as a very difficult problem.)*

Theorem 2 *There exists a probabilistic polynomial time algorithm solving the c-quadratic residue problem if there exists a polynomial time algorithm which predicts b_i from $b_{i+1}, b_{i+2}, \cdots$ with probability $\frac{1}{2} + e$ for $e > 0$.*

The above theorems told us that it is very difficult to backwardly predict the K-M sequence, but this does not mean that the K-M sequence is cryptographically secure, so the theorem 3 in reference [1] is not in order. In fact in the next section we will show that the K-M sequence can be easily forwardly predicted.

3 How to forwardly predict the K-M sequence ?

From formula 1 and the lemma 2 of reference [1], it is known that the formula 3 can be rewritten as:

$$b_i = \begin{cases} 0 & \text{if } x_i^2 < -c_i \\ 1 & \text{if } x_i^2 > -c_i \end{cases} \tag{4}$$

From the formula 2, we have $c_i = 4^i c_0$ ($i = 0, 1, 2, \cdots$) hence the formula 3 is equivalent to:

$$b_i = \begin{cases} 0 & \text{if } x_i^2 < -4^i c_0 \\ 1 & \text{if } x_i^2 > -4^i c_0 \end{cases} \tag{5}$$

Now we begin to forwardly predict the K-M sequence in two cases:

Case 1: If $c_0 > 0$, then $x_i^2 > -4^i c_0$ for any i. This means that $(b_0, b_1, \cdots) = (1, 1, \cdots)$. It is clear that the sequence 1,1,1... is forwardly predictable.

Case 2: If $c_0 < 0$ then $-c_0 = d_0 > 0$. Because $1 \leq x_i \leq R - 1$ (i.e. $1 \leq x_i^2 \leq (R-1)^2$), when i is large enough to hold $x_i^2 < 4^i d_0 = -4^i c_0$. This is to say that when i is large enough to hold $b_i = 0$. In practice the integer $R = pq$ is a number with 200 bits (i.e. R is around 2^{100}). In this case $x_i^2 < 4^{i d_0}$ when i is larger than 200.

Combine the above two cases ($c_0 > 0$ and $c_0 < 0$), it is clear that the K-M sequence can be easily forwardly predicted. In fact let R be an m-bit integer (generally m=200). If the enemy knows $b_{m+1} = b$ ($b \in \{0, 1\}$) then they know that $b_i = b$ for all successive i (i.e. for $i \geq m + 1$).

Note: Maybe the formula 3 should have been replaced by the following formula:

$$b_i = \begin{cases} 0 & \text{if } x_i \pmod{R} < -c_i/x_i \pmod{R} \\ 1 & \text{if } x_i \pmod{R} > -c_i/x_i \pmod{R} \end{cases} \tag{6}$$

But in the latter case the next question remains open. Does the original proof of theorem 1 in reference [1] remain suitable?

Reference

[1] K. Kurosawa and K. Matsu, *Cryptographically secure pseudorandom sequence generator based on reciprocal number cryptosystem*, Electron. Lett. Vol. 24, No. 1, pp16-17, 1988.

Lower Bounds on the Weight Complexities of Cascaded Binary Sequences

Cunsheng Ding[*]
Department of Applied Mathematics
Northwest Telecommunication Engineering Institute
Xian, China

Abstract

The stability of linear complexity of sequences is a basic index for measuring the quality of the sequence when employed as a key stream of a stream cipher. Weight complexity is such a quantity which can be used to measure the stability of a sequence. Lower bounds on the weight complexities of a kind of cascaded binary sequences are presented in this correspondence.

1 Introduction

Clock controlled sequences employed as running-key sequences were investigated by T. Beth and F.C. Piper[1], Kjeldsen Andresen[2], Rainer Vogel[3], Dieter Gollman[4], C.G. Gunther[5], and Rainer A. Rueppel[6]. The general results about clock controlled sequences are not easy to obtain, but the results are clear for some special cases. Rainer Vogel proposed the following sequence generator and gave its properties[3](See Figure 1).

Where the output sequence of SRG1 is (a_i), and that of SRG2 is (b_i), and

$$G(0) = 0, \ G(k) = \sum_{i < k}^{\infty} a_i,$$

Weight complexity or sphere surface complexity and u-sphere complexity were introduced to measure the stability of linear complexity of sequences [7]. Weight complexity and lower bounds for the weight complexities of binary sequences with period 2^n and that of binary ML-sequences were given by Ding [8] [9]. In order to measure the stability of linear complexity of the output sequence in Fig.1, we shall derive lower bounds for the cascaded sequence in this paper.

---

[*]This work is one part of the research of the Stability of Stream Ciphers supported by the Chinese Natural Science Foundation under No.6882007.

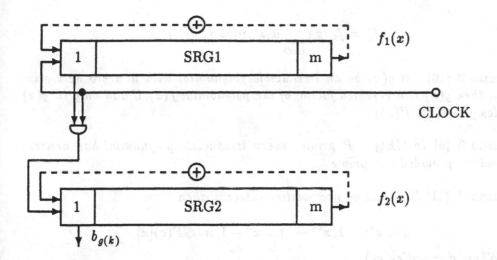

Figure 1: a kind of cascaded sequence generator proposed by Vogel

2 Lower Bounds for the Cascaded Sequence

In this section, we shall derive lower bounds for the output sequences of the sequence generator depicted in Fig.1.

Let S^∞ and T^∞ be two binary sequences of period $N, S = (s_0 s_1 \ldots s_{N-1})$ and $T = (t_0 t_1 \ldots t_{N-1})$ be the first period segments of S^∞ and T^∞ respectively. The weight complexity and u-sphere complexity of S^∞ are defined as [7],[8].

$$WC_u(S) = \min_{W_H(T)=u} L(S^\infty + T^\infty)$$

$$SC_u(S) = \min_{0 < W_H(T) \le u} L(S^\infty + T^\infty)$$

Where $W_H(.)$ denote the Hamming weight, $L(.)$ the linear complexity.

Lemma 1 [8] *Let f_s and f_w be the minimal polynomial of S^∞ and W^∞, then*

$$WC_u(S^\infty) = \min_{W_H(W)=u} deg(f_s f_w / gcd(f_s f_w, f_s f_w^* + f_s^* f_w))$$

Especially, we have

$$WC_1(S^\infty) = \min_{0 \le i \le N-1} deg((x^N + 1)/gcd(x^N + 1, x^i + f_s^* g))$$

where $g(x) = (x^N + 1)/f_s(x)$, and

$$f_s^* = f_s(x) \sum_{i=o}^{\infty} s_i x^i (mod \ x^{deg f_s})$$

$$f_w^* = f_w(x) \sum_{i=O}^{\infty} w_i x^i (mod \ x^{deg f_w})$$

Lemma 2 [10] *Let $g(x)$ be an irreducible polynomial with nonzero derivative $g'(x)$, then $g(x)$ is a repeated factor of the polynomial $f(x)$ if and only if $g(x)$ divides $gcd(f(x), f'(x))$.*

Lemma 3 [9] *In $GF(p)$, P prime, every irreducible polynomial has nonzero derivative, provided p is prime.*

Lemma 4 [11] *If e_1 and e_2 are positive integers, then*

$$gcd(x^{e_1} - 1, x^{e_2} - 1) = x^d - 1, in \ GF(q)[x]$$

Where $d = gcd(e_1, e_2)$.

Lemma 5 [3] *Let the feedback polynomials $f_1(x)$ and $f_2(x)$ be primitive polynomials of degree m. Then*

1. *$f_2(x^{2^m-1})$ is the minimal polynomial of the sequence $(b_{G(k)})$, thus the linear complexity of $(b_{G(k)})$ is $m(2^m - 1)$.*

2. *The sequence $(b_{G(k)})$ has minimum period $(2^m - 1)^2$, and $f_2(x^{2^m-1})$ has exponent $(2^m - 1)^2$.*

Theorem 1 *Denote the output sequence $(b_{G(k)})$ as S^{∞}. Then*

$$WC_1(S^{\infty}) \geq (2^m - 1)(2^m - m - 1)$$

Proof Let $N = (2^m - 1)^2$. By Lemma 2 we see that $X^N + 1$ has no repeated factor. By Lemma 5 we know $f_s(x) = f_2(x^{2^m-1})$. If $h(x)$ divides $x^N + 1$, and $f_s(x)$ does not divides $h(x)$, then $h(x)$ must divides $(x^N + 1)/f_s(x)$. Therefore

$$gcd(x^N + 1, x^i + f_s^* \ g(x)) = f_2(x^{2^m-1})^e, O \leq e \leq 1.$$

Hence, it follows from Lemma 1 that

$$WC_1(S^{\infty}) \geq (2^m - 1)^2 - m(2^m - 1)$$

$$= (2^m - 1)(2^m - m - 1).$$

Theorem 2 *Let M be the maximum proper factor of $2^m - 1$. Then*

$$WC_2(S) \geq (2^m - 1)(2^m - M - m - 1).$$

Proof Let W^∞ be a binary sequence of period $N = (2^m - 1)^2$, and $W_H(W) = 2, w_i = w_j = 1; w_k = 0, k \neq i, j$. Let $G = gcd(|j - i|, N)$, then by Lemma 4 we get

$$f_w = (x^N + 1)/(x^G + 1)$$

$$f_w^* = (x^i + x^j)/(x^G + 1).$$

Notice that the exponent of $f_2(x^{2^m-1})$ is N and $G < N$, so $f_2(x^{2m-1})$ does not divides $x^G + 1$, therefore it must divides f_w. On the other hand, f_w has no repeated factor (because $x^N + 1$ has no repeated factor) and $gcd(f_w, f_w^*) = 1$. whence

$$gcd(f_w, f_w^* + f_s^* f_w/f_s) = f_2(x^{2^m-1})^e, 0 \leq e \leq 1.$$

hence, it follows from Lemma 1 that

$$WC_2(S^\infty) = N - \max_{0 \leq i < j \leq N-1} deg((x^G + 1)gcd(f_w, f_w^* + f_s^* f_w/f_s))$$

$$\geq N - (M(2^m - 1) + deg(f_2(x^{2m-1}))$$

$$= (2^m - 1)(2^m - M - m - 1).$$

Corollary 1 *If $2^m - 1$ is prime, then*

$$WC_2(S^\infty) \geq (2^m - 1)(2^m - m - 2).$$

Lemma 6 [9] *Let S^∞ be a binary sequence of period N, k be a positive integer. If the minimal polynomial f_s is irreducible, then*

$$WC_k(S^\infty) \geq [N/k] - deg\, f_s.$$

Proof *See the proof of Theorem 3 in [9].*

Theorem 3 *Let k be a positive integer, then*

$$WC_k(S^\infty) \geq [(2^m - 1)^2/k] - m(2^m - 1).$$

Proof By Lemma 6 the above assertion is apparently true.

It is too early to draw a conclusion about the stability of linear complexity of the foregoing cascaded sequences, but it seems by the above bounds that the stability of the sequences is relatively good.

References

[1] T. Beth and F.C. Piper, 'The Stop-and-Go Generator', Advances in Cryptology, Lecture Notes in Computer Science, Vol.209, pp.88-92.

[2] Kjeldsen Andresen, 'Some Randomness Properties of Cascaded Sequences', IEEE Infor. Th. Vol IT-26 No.2, 1980.

[3] Rainer Vogel, 'On the Linear Complexity of Cascaded Sequences', Advances in Cryptology — Proceedings of Eurocrypt'84, Lecture Notes in Computer Science, Vol.209, pp.99-109.

[4] Dieter Gollmann, 'Pseudo Random Properties of Cascaded Conncetions of Clock Controlled Shift Registers', Advances in Cryptology, Lecture Notes in Computer Science, Vol.209, pp.93-98.

[5] C.G. Gunther, 'A generator of pseudorandom sequences with clock controlled linear feedback shift registers', Presented at Eurocrypt 87, Amsterdam, Netherlands.

[6] Rainer A. Rueppel, 'When Shift Registers Clock Themselves', Presented at Eurocrypt 87, Amsterdam, Netherlands.

[7] Cunsheng Ding, Guozhen Xiao and Weijuan Shan, 'New Measure Index on the Security of Stream Ciphers', Northwest Telecommunication Engineering Institute, Xian, China.

[8] Cunsheng Ding, 'Weight Complexity and Lower Bounds for the Weight Complexities of Binary Sequences with Period 2' Unpublished.

[9] E.R. Berlekamp, 'Algebraic Coding Theory', McGraw-Hill, New York, 1968.

[10] Rudolf Lidl, 'Finite Fields', Encycolpedia of Mathematics and Applications, Vol.20.

SECTION 3

NETWORK SECURITY

Secure User Access Control for Public Networks

Pil Joong Lee

Bell Communications Research

Morristown, N.J. 07960-1910

ABSTRACT

Secure authentication of users' identity to deny illegitimate accesses is one of the most important steps for preventing computer crimes and losses of valuable network assets. The need for secure user access control for public networks increases rapidly as the open network architecture concept emerges and more network control is allowed to users. In this study a secure and convenient user identity authentication method is presented that provides the users with evidence of mutual authentication, based on public-key cryptographic techniques. This two-round protocol assumes that each user has a personalized card issued by a trusted certification center and uses it at a user access terminal. This protocol provides a signed session key for users who want protection for their communications with a conventional one-key cryptosystem at no extra cost. By the use of an identity certificate, the need for a trusted public-key directory is eliminated. For the card issuing process, we considered two initialization protocols, with dumb cards and with smart cards. For the smart card case, it is shown that users' secrets need not be exposed even to the trusted certification center.

I. INTRODUCTION

The need for secure user access control for public networks increases rapidly as the open network architecture concept emerges and more network control is allowed to users. Also, as more and more valuable information is stored in computer memories, computer crime and mischief become bigger concerns, for instance, as illustrated by the internet worm incident in November, 1988. One should know that current regular login procedures to remote computers over public network are not secure, since every key stroke is transmitted in cleartext. So, the login procedure based simply on a password is vulnerable with respect to passive wiretapping (eavesdropping). Even login procedures with calling id and/or device id can be attacked by active wiretapping (tampering). Call-back procedure is also vulnerable against active wiretapping. Therefore, for every remote access to a valuable resource, the owner of the resource must authenticate the identity of the accessing person in a secure way.

The authentication of identity should be based on person's unique possession, knowledge, and/or characteristics [1]. What one only has can be an ID badge, card, key, token, etc. What one only knows can be a personal identification number (PIN) or password. And, what one only is can be physiological or behavioral characteristics such as a hand signature, finger print, voice print, retina scan, key stroke pattern, DNA code, face picture, hand geometry, etc.

Notice that relying on only one method is often not secure enough, since each one has some weaknesses. Cards can be lost or stolen. And if the information contained in the card is transmitted over a public network, an eavesdropper can make a copy of it and impersonate the owner. PINs or passwords can be forgotten. If they are too short or too easy to remember, then they can be guessed [2] or anyone looking over the shoulder can get them very easily. Biometrics of a person's characteristics should not be transmitted over an unprotected communication medium. Also, one can use some spy movie escape techniques, such as secretly recording a legitimate voice for voice print checking systems, etc. In addition, some biometrics (e.g., signature, voice, photo) change with time. Hence, it is desirable to combine two or more methods. Since what we want is remote access control over a public network, and since there always is a threat of eavesdropping on lines or at switches, we cannot have secure access control without some form of encryption.

The purpose of this study is to find a secure and convenient remote user access control method over a public network for which a mutual audit trail is possible. Of course, there are some other aspects of access control, such as restricting access to resources according to what privilege a user has. But these will not be discussed here. The main focus is user access control by authenticating the user's identity.

The method is based on public-key cryptography and we assume the existence of a trusted certification center (TCC) and many easily accessible access control terminals (ACT), and we also assume that each user remembers a PIN and has a personal card with a certificate issued by the TCC. The need for a trusted public-key directory, which is very inconvenient in practice, is eliminated with the certificate.

We assume that the reader is familiar with concept of cryptography. Readers who are not familiar with the subject may refer a special section on cryptography in [3]. Many of the ideas in this work were inspired by [4], but are quite different in content and the differences will not be specified.

In the next section, assumptions will be described and notations will be given. The initialization protocols will be discussed with the assumption of a dumb card. Then, the authentication protocol of the identity for both ends will be described. In the following section, some improvements in the initialization protocols will be discussed, including the use of smart cards. Finally, discussion and conclusions will follow.

II. ASSUMPTIONS AND INITIALIZATION PROTOCOLS WITH DUMB CARDS

Let's assume that a computationally secure commutative public-key algorithm is publicly available such that $S_A(P_A(M)) = P_A(S_A(M)) = M$ where M is a message, S_A is the transform algorithm which can be either encryption or decryption with secret-key of user A (say, Alice), and P_A is the corresponding transform algorithm with her public-key. An example of such a commutative public-key algorithm is the RSA algorithm [5]. Recall that publishing a public-key of a secure public-key algorithm does not reveal the corresponding secret-key. For notational convenience, we will use S_A (P_A) to represent either the transform with Alice's secret- (public-) key or her secret- (public-) key itself.

In this section we will assume that the card is a dumb memory card, i.e., the card does not have the capability of performing any computation by itself. We will later consider the case of the smart card with direct user interface.

A. System Initialization

Trusted Certification Center (TCC) generates a pair consisting of its own secret key, S_T, and the corresponding public key, P_T, which is public only to access control terminals (ACT) but not to anyone else. TCC distributes "tamper-free" ACTs containing P_T to all users. ACT communicates with other ACTs, with a card by electrical contacts, as well as with a user by a key pad and a display.

B. New User Initialization

When Alice, who might be either a service provider or a service requester in a public network, wants to join the system, the following new user initialization protocol is performed.

{1} Alice provides TCC with her identifier I_A, her personal identification number, PIN_A, and her secret number SN_A.

{2} After verifying Alice's identifier, TCC generates a pair consisting of her public- and secret-keys, P_A and S_A. Then, TCC issues a card, Card A, which contains the authenticity certificate of Alice, $C_A = S_T(P_A; I_A; T_A)$, and signed S_A, SN_A, and PIN_A, $S_T(S_A; SN_A; PIN_A)$.

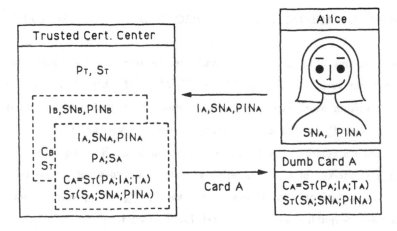

Figure 1. New User Initialization Protocol with Dumb Card

Figure 1 summarizes this protocol. In the notation, the semicolon (;) represents the concatenation with predetermined interleaving.

Alice's identifier, I_A, includes her identification information such as name, social security number, telephone number, address, date of birth, etc. TCC may verify I_A by checking passport, birth certificate, and/or reliable biometrics such as fingerprint, hand signature, DNA code, etc. Alice must remember her PIN, PIN_A, for ACT's verification of her, and secret number, SN_A, for her verification of ACT, as will be explained shortly.

The main purpose of the authenticity certificate is to provide the owner with a method to show her authentic public key to others. Recall that, for a public-key cryptosystem, it is necessary for users to find a public key of an intended person correctly. If the system keeps a public-key directory, it should be tamper proof and hence its maintenance when adding new users would be troublesome. With authenticity certificates it becomes unnecessary to maintain a trusted public-key directory. Notice that the authenticity certificate contains not only Alice's identification information and public-key but also the valid duration of her card, T_A, which is necessary when the card is lost or stolen to limit the number of currently valid cards to only one. The protocol for renewal of the card will be very similar.

C. Session Initialization

Whenever Alice wants to initiate a session with her card, Card A, she needs to carry out the following session initialization protocol.

{1} Alice inserts Card A into an ACT.

{2} ACT reads C_A and $S_T(S_A; SN_A; PIN_A)$ from Card A, and decrypts them with TCC's public-key P_T to recover P_A, I_A, T_A, S_A, SN_A, and PIN_A. If current time is within the valid duration T_A, ACT shows SN_A and T_A on its display.

{3} Only if she can verify that the displayed SN_A matches her remembered SN_A, she enters PIN_A at the keypad of the ACT.

{4} The session is ready to begin if the entered PIN_A is the same as PIN_A recovered from decrypting C_A.

In step {2}, if ACT finds the card is not valid, it shows that the card is expired.

If the test in step {4} fails due to incorrect PIN entry, the user is allowed to try a few more times. However, we protect against the usage of lost or stolen cards by limiting the number of incorrect PIN entries per certain unit time. An ACT may even erase the memory of the card if it thinks the card is not in the hands of the legitimate owner. Furthermore, the ACT notifies all other ACTs to erase the illegal copies of the stolen card. By this protection, illegitimate owners of lost or stolen cards can only have a infinitesimal chance of using them successfully. Therefore, it is very important to make ACT with a covered key-pad for hiding hand movements from curious observers.

Alice's verifying the ACT's display of her secret number in step {3} is also necessary to warn her against entering PIN_A at the keypad of a look-alike terminal which may be installed by a malicious person who wants to record PIN_A and plans to steal her card later. Without this step, after Alice inserts her card and enters PIN_A at its keypad, the phony terminal may record PIN_A and tell her that this ACT is out of order. But, with this step, since none but legitimate ACTs know P_T, the correct SN_A cannot be displayed. (T_A is displayed as a reminder of the expiration date.)

III. AUTHENTICATION PROTOCOL OF IDENTITIES

In this section we will describe a two-round protocol that authenticates the identification information of both ends over a public network. Here we assume Alice wants to initiate a communication session with another user (say, Bob).

{1} After initializing a session Alice first sends her certificate C_A to Bob by using an ACT.

{2} Bob's ACT decrypts C_A with TCC's public-key P_T to recover Alice's public-key, P_A, her identifier, I_A, and the valid duration of her card, T_A, and displays her name and T_A. If Bob wants to communicate with Alice, then, after his session initialization protocol, he sends his certificate C_B together with a randomly generated number R_b protected with Alice's public-key, $P_A(R_b)$, to her.

{3} Alice then recovers P_B, I_B, and T_B with P_T and checks Bob's name and T_B. Alice recovers Bob's random number R_b with her secret key S_A. Then Alice also generates her random number R_a and makes $K_{ab} = g(R_a; R_b)$ with a simple function g. Alice sends $\alpha = P_B(S_A(R_a; K_{ab}; I_B; T_a))$ to Bob, where T_a is Alice's current time to serve as a time stamp.

{4} Upon receiving α, Bob decrypts it with his own secret key S_B to recover a proof showing that at a particular time, T_a, Alice sent K_{ab} to Bob (and he saves the proof). By decrypting the proof with Alice's public-key, Bob recovers Alice's random number R_a, K_{ab}, his own identification information, and Alice's time stamp T_a. By checking the equality of $g(R_a; R_b) = K_{ab}$, Bob authenticates Alice's identification information. Similarly Bob sends $\beta = P_A(S_B(K_{ab}; I_A; T_b))$ to Alice.

{5} From β, first with S_A, Alice recovers the corresponding proof (and saves it), and then with P_B, she recovers K_{ab}, her own identification information I_A, and Bob's time stamp T_b. Alice finally checks the authentication of Bob's identification information by comparing whether the received K_{ab} is the same as what she generated by herself before.

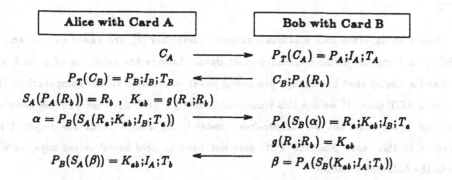

Figure 2. Authentication Protocol of Identities

This protocol is summarized in Figure 2. Notice that, except for the certificates, every piece of information that moves around via the public switched network is protected by the receiver's public-key, and, furthermore, the content varies each time due to the time stamps and random numbers. (We used lowercase subscripts for those items that differ every session.)

And for this and session initialization protocols, TCC needs not be involved. Hence, in theory, if no new user is expected and the valid duration of cards will outlast the useful life of the system, TCC can be closed. Also notice that if users A and B want to use a conventional (symmetric-key) cryptographic technique for the remaining session they can use K_{ab} as their one-time session key. And a user can keep the proof for a possible later dispute, since this proof can be verified by anyone who also has a valid card.

IV. INITIALIZATION PROTOCOLS WITH SMART CARDS

In section II we described initialization protocols assuming dumb cards and tamper-free ACTs. In that scenario, a user must trust TCC for everything since TCC even knows users' secret-keys. In other words, the system protects user's privacy with respect to other users, but allows TCC to be a 'big brother' that has a capability of monitoring all communications between users, even though it is not supposed to involve itself in the authentication protocol of identities. In this section we present other initialization protocols with smart cards which enable users just to trust TCC for its honesty in certification but not to give up their privacy to TCC.

Smart cards, often called *ultimate personal computers* [6], are easily carried and have the ability to compute, memorize, and protect data. Assume the existence of a card with a keypad and a display that has enough processing power to perform all the computations that a tamper-free ACT does. If we use this smart card, then an ACT does not have to know users' secrets and hence need not be tamper-free. Instead, the smart cards are required to be tamper-free in this case. Also the TCC does not have to hold users' secret keys, as will be shown in the following protocol.

A. System Initialization

Trusted Certification Center (TCC) generates a pair consisting of its own secret key, S_T, and the corresponding public key, P_T. TCC distributes ACTs containing P_T to all service providers and service requesters in public networks. Here, P_T is public to anyone. ACT communicates with other ACTs, with a card by electrical contacts, as well as with a user by a display.

B. New User Initialisation

When Alice wants to join the system, the following new user initialization protocol is performed.

{1} Alice generates her own pair of public- and secret-keys, P_A and S_A. Then she provides TCC with her identifier, I_A, and public-key, P_A.

{2} After verifying the identifier of Alice, TCC issues a smart card which contains her authenticity certificate, $C_A = S_T(P_A; I_A; T_A)$ to her.

{3} Upon receiving the smart card, Alice decrypts C_A with P_T to check whether it contains I_A and P_A as sent in step {1}. Alice puts S_A and PIN_A into the protected area of memory in her card.

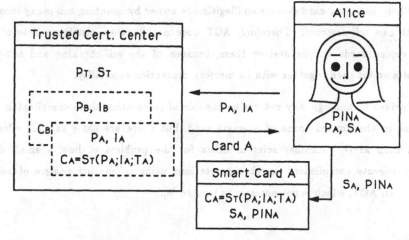

Figure 3. New User Initialization Protocol with Smart Card

This protocol is summarized in Figure 3. Alice must remember her PIN, PIN_A. For the generation of her own pair of public- and secret-keys in step {1}, she may use a personal computer and provide TCC with a diskette containing I_A and P_A, and keep S_A in a separate diskette safely. S_A should not be left in any other place than the smart card after finishing this protocol. Some contents of card memory will never leave the card such as S_A and PIN_A, while C_A will be released upon the user's approval.

C. Session Initialization

Whenever Alice wants to initiate a session with her smart card, she needs to perform the following session initialization protocol.

{1} Alice enters PIN_A on the keypad of Card A .

{2} Card A compares entered PIN_A to that in its memory. If no match, it asks for retry a few times. If it matches, it displays a sign of activation of Card A.

{3} Alice inserts Card A into an ACT.

{4} ACT reads C_A from Card A and decrypts it with TCC's public-key P_T to recover P_A, I_A, T_A. ACT displays T_A as a sign of protocol completion.

When the smart card detects an illegitimate owner by counting too many incorrect PIN entries, it can self-destruct. Therefore, ACT does not have to distribute lists of stolen or illegally copied cards and/or destroy them, because of the self-checking and self-destroying feature of a smart card together with its memory protection capability.

Current technology may not yet be capable of integrating these complicated public-key algorithms in the limited space of a smart card, but there are some existing efforts in this direction such as [7]. Another solution exists for the problem of how a small device can efficiently execute complicated secret computations using computing powers of an auxiliary device like an ACT, which is not necessarily trusted [8].

V. DISCUSSION AND CONCLUSION

We have discussed the use of biometrics for TCC's verifying new and/or renewing users' identities. For this purpose biometrics should be reliable even at high cost. But with less reliable and inexpensive biometrics, the security of the user-to-user authentication system can be enhanced when it is combined with the card (possession) and PIN (knowledge) as the third element of authentication. Furthermore, if the card memorizes the biometric template signed with TCC's secret key, S_T, then the biometrics can be verified by the tamper-free ACT or by the user at the other side of public network. One choice for the use with a card, which does not require an additional complicated device, is to use each person's unique key stroke dynamics [9] since a PIN must be entered into a keypad anyway. Another reasonably inexpensive biometric method is the use of a telephone for speaker identification by analyzing speech [10], since telephones are readily available in most cases. The biometric method for TCC's identification of a user should incorporate more accurate tests.

When we use the RSA algorithm (with, say, 512 bits per block and with common short public exponent) as the public-key algorithm, we may choose the size of the certificate C_A to be 2 blocks while R_b and $(R_s; K_{sb}; I_B; T_s)$ are 1 block each. Then, the overall number of secret transformations, public transformations, and communications are 5, 9, and 7 blocks, respectively.

In this study, we presented a secure user identity authentication protocol based on public-key cryptographic techniques in which the need for a trusted public-key directory is avoided through the use of an identification certificate issued by a Trusted Certification Center (TCC). This two-round protocol provides not only a proof of authentication but also a session key for users who want the rest of their communication session to be private.

This authentication protocol assumes that each user has a personalized card issued by the TCC and uses it at an access control terminal (ACT). For the initial card issuing process, we considered first the dumb card case and then the smart card case. For the dumb card case, we have to make an assumption that every ACT must be tamper-free and that the TCC is able to monitor all communications between users. But for the smart card case, it is shown that users' secrets need not be exposed, even to the trusted certification center.

ACKNOWLEDGEMENT

The author thanks R. F. Graveman for his valuable comments.

REFERENCES

[1] H. Wood, "The use of passwords for controlling the access to remote computer systems and services," *Computer and Security,* Vol. III. C. Dinardo, Ed., p. 137, Montvale, N.J., AFIPS Press, 1978.

[2] D. Feldmeier & P. Karn, "UNIX password security - ten years later," presented at Crypto '89, Santa Barbara, CA, 8/89.

[3] *Proceeding of the IEEE,* 5/88.

[4] J. Omura, "A computer dial access system based on public-key techniques," *IEEE Commun. Mag.,* pp. 73-79, 7/87.

[5] R. Rivest, A. Shamir & L. Adleman, "A method of obtaining digital signatures and public key cryptosystems," *Commun. of ACM,* pp. 120-126, 2/78.

[6] J. Svigals, *Smart Cards: The Ultimate Personal Computer,* Macmillian, 1985.

[7] J. Omura, "A new smart card chip for creating signature," *SCAT/ASIT '89 Conference Proceedings,,* pp. II.37-48, 5/89. 1989.

[8] T. Matsumoto, K. Kato & H. Imai, "Speeding up secret computations with insecure auxiliary devices," in *Proceedings of Crypto'88,* Santa Barbara, CA, 8/88.

[9] J. Legget & G. Williams, "Verifying identity via keystroke characteristics," *Int. J. Man-Machine Studies,* pp. 67-76, 1988.

[10] G. Velius, & T. Feustel, "Signal Detection Analysis of Speaker Identity Verification Systems," *Journal of the American Voice Input/Output Society,,* pp. 46-59. 1987.

Formal Specification and Verification of Secure Communication Protocols

Svein J. Knapskog

University of Trondheim
Division of Computer Systems and Telematics
NORWAY

1 Introduction

A secure communication protocol is a set of rules for establishing and maintaining secure services [1] for the users of communication systems. Secure services are supported by security mechanisms [1] e.g. encipherment, mechanisms for digital signatures, access control, data integrity and authentication exchange. Engineering of secure communcation protocols will demand the same kind of expertise as engineering of ordinary communication protocols, and in addition thorough knowledge of the different mechanisms and algorithms that are specific for the secure services. All protocol work need some formality in the description and construction phases to make more or less formal verification possible, but for the engineering of communication protocols in which security is the main objective,a good formal method for specification and verification becomes crucial. This paper will give an overview over existing methods that can be used for this purpose, discuss their capabilities in a given context, and give examples of protocols construction using these methods.

2 Security Modelling

The ISO seven layered model for Open Systems Interconnection (OSI) is an architectural framework that promotes modular construction of communication protocols [2]. The protocols which are already defined and developed within each layer of this framework are basicly communication oriented, but not concerned with security. Secure protocols must, however, be implemented in the same architectural context, since the basic communication needs are still to be met by using the same network and communication equipment as before, only with confidentiality, authenticity and signature capability of the transmissions added as an enhanced quality of service. The modeling of the secure protocols will therefore be along the lines of adding new mechanisms to existing protocols in each appropriate layer. This is a delicate matter, and careful considerations within the different standardization organizations are necessary to guarantee that changes that are made do not adversely affect existing protocols, and at the same time that the security mechanisms are implemented in a proper way so that the intended enhanced quality of service is indeed achieved, without compromise of message or system security.

3 Formal Description and Design Methods

A large number of description methods or high level languages have been proposed for different system engineering purposes. The common goal that these methods pursue, is to have a formal description of a complex system made in such a manner that the following engineering tasks of construction and implementation can be computerized, or at least in some way automated, and that the implementation realize the theoretical model to the extent that a theoretical verification of the intended functionality is possible. The set of rules and mechanisms which constitutes a secure communication protocol can readily be fit into such a formally described model, and the benefit of using a formal description method for the construction of secure protocols are thereby evident. The following methods are candidates to be considered in the further work:

- *Abstract Syntax Notation no. 1 - ASN.1.* This method is specified in a joint CCITT/ ISO - standard [3]. It is used as a semi-formal tool to define protocols in form of a high level language that has defined software constructs that make it possible to describe and link together communication primitives which are used to build those protocols. Some basic security primitives have already been proposed, but more research and development are needed (and expected) in this area. The use of the notation does not necessarily preclude ambiguous specification. It is the responsibility of the users of the notation to ensure that their specification is complete and not ambiguous.

- *Functional Specification and Description Language - SDL.* This method has developed from the well known Extended Finite State Machine (EFSM) modeling technique. It has been standardized for use within CCITT for modeling telecommunication systems in general. It uses either a graphic or textual phrase representation in the modeling phases, and a more or less automated generation of code for a software/firmware/hardware implementation of the graphic or textual abstract representations [4].

- *Numerical Petri Nets - NPN.* This method has been used extensively for general protocol description and construction [5]. Tools for verification purposes have been developed [6].

These formal description methods should be evaluated with respect to description and construction flexibility, completeness and provability. The SDL- and ASN.1- methods will be discussed in connection with design examples given in the later parts of this paper. There are other, newer description and design tools, e.g. [7], [8] and [9], but none of these have been generally accepted or adopted as recommendations or standards yet.

4 Design of Secure Protocols

The term protocol is general and must be further qualified to be uniquely understood. In the first part of this chapter, secure protocols implemented at (or above) the OSI application layer will be described. The security will be the responsibility of the communicating parties only, and the network between them will merely transport encrypted or authenticated traffic as a bit-stream. In the next part, the necessary services and mechanisms to allow secure protocols to be implemented by adding security mechanisms to the existing communication protocols in a given architectural layer of the network will be discussed. This will be seen as an enhanced quality of network service from the users, an optional facility that can be used or not depending on the nature of the information users are exchanging. In the final part of this chapter, a sketch of how security can be incorporated in the future Integrated Services Digital Network (ISDN), will be given. This network, apart from being fully digitized, will have properties that can be effectively utilized for security purposes.

5 Application Layer Secure Protocols

In principle, all security measures that users of a communication network deem necessary could be done outside the network by the users themselves. Simple, secure protocols for exchanging information over insecure channels do exist, and this fact, combined with the fact that there exist many most useful application layer services implemented by application layer protocols without any special precautions with regard to security having been taken, add up to the logical conclusion that there is lot of effort necessary to fill this void. When working in the application layer, it would be natural to consider ASN.1 as a natural choice for a description and design tool, and during the process try to evaluate the fitness-for-purpose regarding provability of the implemented protocol.

ASN.1 is used to enforce the necessary transformation of user data as they exist at the application layer in form of strings of characters, possibly from different character sets, into user data below the presentation layer in form of binary value of sequence octets, ref. Figure 1. Data-types will be identified by Tags and a Context. The Tags can be:

- Universal (specified in the Recommendation)
- Application (assigned by other standards)
- Private (enterprise specific)
- Context-specific (freely assigned by users)

The encoding rules for ASN.1 constitutes a Transfer Syntax for the Presentation Layer.

As an example, the ASN.1 specification of Remote Operations Service Element (ROSE), will be given in the following. ROSE supports interactive applications in a distributed

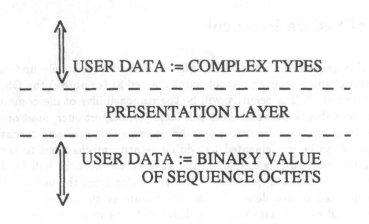

USER DATA := COMPLEX TYPES

- - - - - - - - - - - - - - - -

PRESENTATION LAYER

- - - - - - - - - - - - - - - -

USER DATA := BINARY VALUE
OF SEQUENCE OCTETS

ENCODING RULES := TRANSFER SYNTAX

Figure 1: ASN.1 Encoding of Data

Open Systems environment, and provides its services in conjunction with the Association
Control Service Element (ACSE) services and the ACSE protocol. The ROSE procedures
are defined in terms of the interactions between peer ROSE protocol machines, and the
interactions between the ROSE protocol machine and its service-user. The actual ASN.1
specification of the data types for the ROSE protocol is given in Figure 2, Figure 3, and
Figure 4.

As stated earlier, the ASN.1 specification is semi-formal, and gives in itself no guarantee
of completeness and correctness. The reason for considering this method at all, is that
the strict formalism that exist in the definition of datatypes and constructs, that makes it
possible to check the PDU's for compliance with their ASN.1 definition, both for structure
and value. Further more, particular PDU's can be constructed for testing purposes of
the actual user protocol in question [10]. The ASN.1-method is a likely candidate for
adaptation to a fully formalized description method that could be a helpful tool also when
it comes to prove correctness, completeness, and thereby also security. This is, however,
not a trivial task.

In contrast to ASN.1, the SDL method is not in any way connected to the layered architectural model. It can be equally useful for designing an application protocol and a
protocol defined for an N-layer entity. The second example of an implementation is an
experimental secure mail system implemented on segments of a local networks connected
via a 64 kbit/s link in the public domain, using SDL as the design tool, see Figure 5.

5.1 The Mail System

An electronic mail system is implemented along the guidelines of the Recommendations
X.400 and X.500 given by CCITT in 1984 [11]. Some new mechanisms regarding security

```
Remote-Operation-Notation {joint-iso-ccitt remote-operations(4) notation(0)}
DEFINITIONS ::=
BEGIN

EXPORTS BIND, UNBIND, OPERATION, ERROR;

-- macro definition for bind-operations

BIND MACRO ::=
BEGIN

TYPE NOTATION              ::= Argument Result Error
VALUE NOTATION             ::= Argument-value | Result-value | Error-value

Argument                   ::= empty |"ARGUMENT" Name type (Argument-type)
                               -- Expects any ASN.1 type and assigns it to the variable Argument-type

Result                     ::= empty |"RESULT" Name type (Result-type)
                               -- Expects any ASN.1 type and assigns it to the variable Result-type

Error                      ::= empty |"BIND-ERROR" Name type (Error-type)
                               -- Expects any ASN.1 type and assigns it to the variable Error-type

Name                       ::= empty | identifier

Argument-value             ::= empty |"ARGUMENT" value (Arg-value Argument-type)
                               -- Expects a value for the type in Argument-type, and assigns it to the
                               -- variale Arg-value
                               <VALUE [16] EXPLICIT Argument-type ::= Arg-value>
                               -- Returns the final value as explicitly tagged type

Result-value               ::= empty |"RESULT" value (Res-value Result-type)
                               -- Expects a value for the type in Result-type, and assigns it to the
                               -- variable Res-value
                               <VALUE [17] EXPLICIT Result-type ::= Res-value>
                               -- Returns the final value as explicitly tagged type

Error-value                ::= empty |"ERROR" value (Err-value Error-type)
                               -- Expects a value for the type in Error-type, and assigns it to the
                               -- variable Err-value
                               <VALUE [18] EXPLICIT Error-type ::= Err-value>
                               -- Returns the final value as explicitly tagged type

END
-- Remote Operations Notation continued
```

Figure 2: Formal Definition of Remote Operations Data Types

```
- - Remote Operations Notation continued

- - macro definition for unbind-operations


UNBIND MACRO :: -


BEGIN


TYPE NOTATION              :: - Argument Result Errors

VALUE NOTATION             :: - Argument-value | Result-value | Error-value


Argument                   :: - empty |"ARGUMENT" Name type (Argument-type)

                                - - Expects any ASN.1 type and assigns it to the variable Argument-type


Result                     :: - empty |"RESULT" Name type (Result-type)

                                - - Expects any ASN.1 type and assigns it to the Result-type


Error                      :: - empty |"UNBIND-ERROR" Name type (Error-type)

                                - - Expects any ASN.1 type and assigns it to the Error-type


Name                       :: - empty | identifier


Argument-value             :: - empty |"ARGUMENT" value (Arg-value Argument-type)

                                    - - Expects a value for the type in Argument-type, and assigns it to the
                                    - - variable Arg-value
                                    <VALUE [19] EXPLICIT Argument-type :: - Arg-value >
                                    - - Returns the final value as explicitly tagged type


Result-value               :: - empty |"RESULT" value (Res-value Result-type)

                                    - - Expects a value for the type in Result-type and assigns it to the
                                    - - variable Res-value
                                    <VALUE [20] EXPLICIT Result-type :: - Res-value >
                                    - - Returns the final value as explicitly tagged type


Error-value                :: - empty |"ERROR" value (Err-value Error-type)

                                    - - Expects a value for the type in Error-type and assigns it to the
                                    - - variable Err-value
                                    <VALUE [21] EXPLICIT Error-type :: - Err-value >
                                    - - Returns the final value as explicitly tagged type


END

- - Remote Operations Notation continued
```

Figure 3: Formal Definition of Remote Operations Data Types

```
– – Remote Operations Notation continued
– – macro definition for operations

OPERATION MACRO :: –

BEGIN

TYPE NOTATION          :: –   ArgumentResultErrorsLinkedOperations
VALUE NOTATION         :: –   value (VALUE CHOICE |
                                       localValue   INTEGER,
                                       globalValue OBJECT IDENTIFIER|)

Argument               :: –   "ARGUMENT" NamedType | empty

Result                 :: –   "RESULT" ResultType | empty

ResultType             :: –   NamedType | empty

Errors                 :: –   "ERRORS" "|"ErrorNames"|" | empty

LinkedOperations       :: –   "LINKED" "|"LinkedOperationsNames"|" | empty

ErrorNames             :: –   ErrorList | empty

ErrorList              :: –   Error | ErrorList "," Error

Error                  :: –   value (ERROR)        – – shall reference an error value
                            | type    – – shall reference an error type if no error value is specified

LinkedOperation-       :: –   OperationList | empty
Names

OperationList          :: –   Operation | OperationList "," Operation

Operation              :: –   value (OPERATION)        – – shall reference an operation value
                            | type    – – shall reference an operation type if no operation value is specified

NamedType              :: –   identifier type | type

END
– – macro definition for operations errors

ERROR MACRO :: –
BEGIN
TYPE NOTATION          :: –   Parameter
VALUE NOTATION         :: –   value (VALUE CHOICE |
                                       localValue   INTEGER,
                                       globalValue OBJECT IDENTIFIER|)

Parameter              :: –   "PARAMETER" NamedType | empty

NamedType              :: –   identifier type | type

END

END    – – end of Remote Operations Notation
```

Figure 4: Formal Definition of Remote Operations Data Types

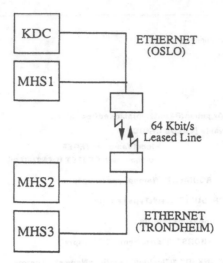

Figure 5: Experimental Secure Mail System Outline

have been standardized in the 1988 version of these recommendations, and these mechanisms will influence some of the choices that have been made for the implementation, but not on the principal level.

5.2 Encryption Algorithm

For information that is of value for a group or a large company, but is not in any way "classified" or "of national interest", a reasonable level of security is achievable by using publicly known standard encryption and decryption algorithms, like for instance DES. This has the advantage that cheap, readily available and reasonable fast hardware can be purchased and used in proprietary designs, or off-the-shelf hardware units containing DES can be installed in the communication path. The experimental mail system described here uses DES in cipher block-chaining mode for text encryption and decryption. Because speed in this experimental phase is not of great importance, the encipherment and decipherment will be done in software on the work-stations participating in the experiment.

5.3 Key Management

As long as the number of communication parties are reasonably small, generation and distribution of data encryption keys is a manageable task. As the number of communicators grow, however, this task becomes one of the major issues of the overall system management and security. A hierarchical model is the usual choice, and in addition a trusted third party can be introduced into the system to increase both security and ease of operation. For this particular experimental system, a key generation and distribution center are implemented. This center will serve all the users of the local networks. Data

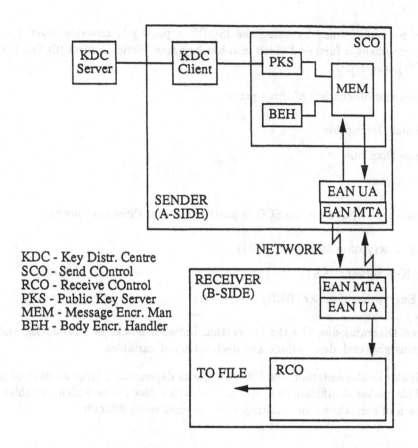

Figure 6: Secure Mail System Implementation

encryption keys will be distributed on a per-connection basis, and an RSA public key key-encryption scheme will protect the data encryption keys during transmission.

5.4 Authentication Protocol

During the key exchange phase between the communicating parties and the key distribution center, an authentication session will be performed by using the RSA public key algorithm.

6 SDL Description and Implementation

The structure of the implemented experimental system is shown in Figure 6. The software development of the Message Handling System (MHS) and the Key Distribution Center

(KDC) have been performed by using the ISODE - package[1], covering layer 5 - 7 in OSI. The development is further divided into blocks named Send Control (SCO), Receive Control (RCO), and KDC.

The SDL - description consists of three parts:

- Functional Description
- Sequence Diagrams
- State Diagrams

The Functional Description for the SCO is partitioned into three sub-blocks:

- Message Encryption Manager (MEM)
- Public Key Server (PKS)
- Body Encryption Handler (BEH)

The Sequence Diagrams describe the interaction between blocks or sub-blocks, and the State Diagrams give task descriptions and declaration of variables.

In this particular implementation, the different blocks depend on a large number of global variables. This makes it difficult to keep track of which block uses which variables, and the verification of correctness and security also becomes more difficult.

7 Characterization of the Formality of SDL

SDL uses a top - down approach in the sense that there is a systematic partition of systems into communicating blocks and further partitioning of blocks into sub-blocks. The blocks and sub-blocks will then be given either a graphic or textual formal description. The descriptions and declarations are in a formal language, and the formality is such that it shall be possible to analyse and interpret the decriptions unambiguously. This gives a strong reason to believe that given the right (computer based, automated) tool, SDL in it's present form lend itself amicably to formal methods for verification of security, but these methods do not exist implicit in the specification method itself. The conclusion is hence that there exists no standardized way of actually proving, in any mathematical sense, the security built into the implemented system.

However, the systematic approach and partitioning of blocks into sub-blocks, and partition of functionality (processes into sub-processes), makes it possible to thoroughly analyze a system, at least in a semi-formal way, and thus be convinced (or convince others) of the soundness of the SDL implementation. It should also be possible to employ more or less standardized software verification tools to verify the correctness of the software part of the implementation from the lowest level of specification and description.

---

[1]ISODE - ISO Development Environment

8   Protocols for Secure Networks

Several amendments to the existing communication protocols with the aim of adding security mechanisms to these protocols in the layered OSI-architecture are under preparations in ISO standardization committees. The following layers have been identified as appropriate for security mechanisms:

- Physical layer

- Network layer

- Transport layer

- Presentation layer

The proposal for the physical layer security mechanisms is the most advanced, as this was published as a standard in 1988 [12]. The work for the other layers have reached a level at which, even if changes must be anticipated, implementations can be made along the guidelines that the proposals contain. This paper will give a brief survey of the services and mechanisms proposed for use in the network layer protocols for connection mode and connectionless mode of operation [13].

9   The Network Layer

In the network layer, the peer entities are processes which see the network as a collection of communication processes. Each process will be contained in a "node" in the network, or in a host connected to the node. Some nodes will have host processors directly connected to them, other will only be connected to other nodes. A node will normally only keep track of what hosts or other nodes that are directly connected to itself, or in other words, its neighbours. The network protocol exchange data packets between nodes, containing source and destination addresses. Error detection mechanisms will normally be included in the data packets.

Node-to-node encryption and decryption functions are appropriate in the network layer, both for connection oriented and connectionless communication protocols.

10   Network Layer Interoperability Requirements

Conditions for practical operation of encipherment/decipherment techniques in the network layer exist in the form of a working document ISO/IEC/JTC1/SC20/WG3 Working Document [13], and will be further advanced to a Draft International Standard in the near future. The following security services have been identified for this layer:

- peer entity authentication

- data origin authentication

- connection confidentiality

- connectionless confidentiality

- traffic flow confidentiality

- connection integrity without recovery

- connectionless integrity

These services can be used by the network entities in end-to-end systems or relay systems. The standard will specify the use of cryptographic mechanisms and mode of operation, and give the requirements for establishing and terminating cryptographic protection within a network layer connection. Parameters and procedures for cryptographic processing of transferred data are specified, and methods for proper reactions to cryptographic exceptions are given. However, the standard will address key management issues only to the extent necessary to establish a secure network connection.

The peer entity authentication is based on verifying knowledge of a shared secret by the authenticating network entities, or by the entities and a trusted third party. A shared secret key may be established as part of the peer entity authentication, using the key distribution procedure. The equivalent of one-way peer entity authentication is the data origin authentication in connectionless mode. It does not offer any protection against replay of data units.

Confidentiality, both in connection and connectionless mode is achieved by invoking the encipherment/decipherment mechanisms. A limited amount of traffic confidentiality is achieved by traffic padding in conjunction with encipherment. The integrity of the the data on the connection is achieved by using a Message Integrity Code (MAC) and a Message Identification Number (MID), in conjunction with encipherment. In the connectionless mode, the MID offers a limited amount of replay protection. All the security mechanisms are assumed to operate before the normal operation of the network layer functions, that means that the Network Service Data Units (NSDU's) are enciphered, but not the network adress information.

The key selection procedures identifies Data Encipherment Keys (DEK) for use during a connection. For connectionless mode, a new key will be needed for each NSDU. Keys can be both symmetric and asymmetric. The procedure for encipherment either enciphers or deciphers the data it is given (plaintext or ciphertext) as directed, using the appropriate DEK selected for the connection, or NSDU. The encipherment algorithm to be used is to be negotiated between the two parties, to ensure that it is available to both, and to choose one if several are available. In the same manner, mode of operation and initializing values must be negotiated, if appropriate.

Data sequence integrity and message integrity are achieved by calculating and transmitting a MID and a MAC. The MAC is a vector of n-octets ($n \geq 4$), its means of calculation is given in [14].

Keys for encipherment/decipherment will be established by including in the network connection request message a key distribution method parameter. The different values of this parameter distinguish between:

- a mutual key encrypting key (KEK) is to be used

- asymmetric key techniques are to be used

- a key distribution service is to be used

The receiving end has the option of accepting the key distribution mechanism proposed in the parameter or of rejecting it with a reason. The connection request message will also contain a authentication parameter, when needed. This parameter contains two random or pseudo-random numbers. The first is used by the receiving end as the MID for the first user data that it returns to the initiating network entity, excluding expedited data. The second random number is used by the receiving end as the MID of the first expedited data it sends on the connection.

Generally, the cryptographic protocols to support connectionless operation are the same as those that support connection-mode operation. The major difference is that in the connectionless operation, each NSDU has to contain sufficient information to allow it to be deciphered. No negotiation of security services is possible, so the receiving end must have the option of discarding data units which indicate values of security parameters which are not acceptable.

Further details concerning practical security operations in the network layer are given in [13]. This is, as the title indicates, a working document, and changes must be anticipated until the proposed text has become "stable", and has been accepted by the ISO - member countries through the balloting procedure. For the actual protocol design, construction and implementation, SDL will be an appropriate tool. If the ambition, as this paper anticipates, is to realize a *provable* secure network layer protocol, a considerable research effort is still necessary to achieve that goal.

11 Protocol Testing in General

The CCITT Recommendation X.290 - Conformance Testing [15] contains general guidelines for protocol testing. Conformance to a specification is neither an adequate measure of security, nor a guarantee for robustness or reliability of an implementation, but standardized test methods will be play a very important role also for security verification. It is a fact that the complexity of most OSI protocols is such that exhaustive testing will be impractical both for technical and economic reasons. A successful conformance test will increase the confidence that an implementation has the required capabilities and that its behaviour conforms consistently in representative instances of communication. To evaluate the conformance of a particular implementation, it is necessary to have a statement of

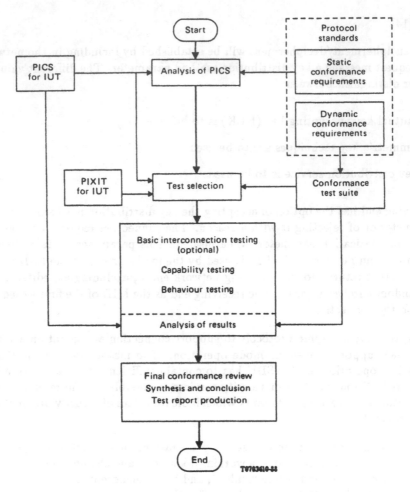

Figure 7: Conformance Assessment Process Outline

the capabilities and options that have been implemented for the Implementation Under Test (IUT). Such a statement is called a Protocol Implementation Conformance Statement (PICS). For specific test purposes, an additional statement, called Protocol Implementation extra Information for Testing (PIXIT). The outline for the assessment process using these documents is shown in Figure 7. The relevance of this general model for security testing rests on the fact that for protocols specified to implement a certain level of security, it should be possible to verify that security level.

12 Conclusion

The worlds growing dependency on telecommunications is a fact. The different networks give efficient and user friendly services to a large number of users. With adequate security measures in terms of confidentiality, authentication functions and signature capabilities, the functionality of the networks will be greatly enhanced, and the users would experience a better quality of service. The full benefit of the future ISDN will only be reachable with security features built into both the application layer protocols and protocols at one or more of the other architectural layers. The formal methods to be used to do the actual implementations will be topics for further research and development, because none of the existing formal description, construction and verification methods are fully adequate for this very complex task.

References

[1] ISO International Standard - Information Processing Systems - OSI Reference Model - Part 2: Security Architecture.

[2] Information Processing Systems - Open Systems Interconnection - Basic Reference Model.

[3] Information Processing Systems - Open Systems Interconnection - Specification of Abstract Syntax Notation One (ASN.1). CCITT Recommendation X.208. Blue Book 1988.

[4] Functional Specification and Description Language (SDL). CCITT Recommendations Z.101. Blue Book 1988.

[5] Wheeler, "Numerical petri nets - a definition," Tech. Rep. Technical Report 7780, Telecom Australia Research Laboratories, 1985.

[6] PROTEAN: A Specification and Verification Aid for Communication Protocols. Telecom Research Laboratories Branch Paper SS0113, 1987.

[7] M. Burrows, M. Abadi, and R. Needham, "Authentication: A practical study in belief and action," Tech. Rep. Technical Report 138, University of Cambridge Computer Laboratory, 1988.

[8] D. M. Berry, "Towards a formal basis for the formal development method and the ina jo specification language," *IEEE Transactions on Software Engineering*, vol. SE-13, Feb 1987.

[9] J. K. Millen, S. C. Clark, and S. B. Freedman, "The interrogator: Protocol security analysis," *IEEE Transactions on Software Engineering*, vol. SE-13, pp. 274–296, Feb 1987.

[10] U. Cleghorn, "ASNST: An abstract syntax notation - one support tool," *Computer Communications*, vol. 12, Oct 1989.

[11] S. Andresen, S. Haug, and T. Bechman, "Sikring i MHS," Tech. Rep. Report STF44 F88040, Elab, 1988.

[12] ISO International Standard - Information Processing - Data Encipherment - Physical layer Interoperability Requirements.

[13] ISO/IEC/JTC1/SC20/WG3 Working Document - Data Cryptographic Techniques - Conditions for Practical Operation in the Network Layer.

[14] Data Integrity Mechanism Using a Cryptographic Check Function Employing an N-Bit Algortihm with Truncation. ISO Draft International Standard 1988.

[15] Conformance Testing. CCITT Recommendation X.290. Blue Book 1988.

Network Security Policy Models

Vijay Varadharajan

Hewlett-Packard Labs., Bristol, U.K.

Abstract

This paper considers the development of security policy models for networks which can operate in a multilevel security environment. We outline an abstract security policy model which addresses the access control and information flow requirements in a multilevel network. The network security policy is formulated by drawing as much parallel as possible with the computer security policy. The model is defined and the associated security requirements are given and the model is used to prove that the security conditions are not violated by the defined network operations. In developing network security policies, it is necessary to relate them to specific layers of the network architecture. We conclude this paper by outlining another network security policy model with different network subjects and objects supported by different layers in the architecture.

1 Introduction

A security model essentially captures the notion of security in a given system. Recently, the question of the design of a security model has come under intense scrutiny, largely because of the publication of the Department of Defense Trusted Computer System Evaluation Criteria ([1]). A formal model of the applicable security policy is required for systems to be rated in the higher protection classes. To precisely describe the security requirements of the model and to provide a framework for the specification, implementation and verification of the security properties of the network, it is necessary to have a mathematical formulation of the model.

A secure network requires at least three types of control : Access Control, Information Flow Control and Cryptographic Security Control.

Network access control is basically concerned with any user or any process executing on behalf of a user wishing to gain access to and use of a resource on the network. As a basic premise, the access control mechanisms require reliable identification and authentication of the users in the network system. Information flow control suitably restricts the flow of information from one system to another so as not to violate a prescribed set of flow policies.

When access to information contained in an object is authorized and the flow of information between two entities is validated, the information can be transmitted over the network. If the communication itself is not protected, the purpose of enforcing both access control policy and the information flow control policy may become useless. Hence in addition, we need to use cryptographic techniques to provide confidentiality, integrity and authentication of communications. However, in this paper, we will not consider such cryptographic issues and they are not incorporated in the formal model. The security models described in this paper only address the notions of access control and information flow control. This agrees with "the current practice that a formal security policy model is required only for the access control policies to be enforced" ([2], page 226). However, the functionality providing the required communication security features can be incorporated as part of the trusted network base which ensures access control and information flow security control.

The aim of this paper is to consider the development of multilevel network security policy models by drawing as many parallels as possible from the computer security models. We first consider how access control and information flow security issues arise in the design of multilevel secure network systems. We analyse these issues using typical scenarios of the type considered in ([3]). Then we develop an outline of a simple abstract network security model which considers the access control and information flow security aspects in a multilevel network environment. The model addresses some of the important issues mentioned in the *Trusted Network Interpretation ([2])*. We give a formal definition of such a model and the associated security requirements and then derive suitable conditions for the system to meet the security requirements. By definition, the system model is said to be "secure" if these conditions are satisfied.

The appropriateness of the security policy depends on the layer to which the policy has been applied to in the network system. We illustrate this aspect by briefly outlining another network security policy model in the *Discussion* section. The fundamental concepts underlying the models described in this paper are based on a combination of some well-known security models [5], [6], [7] proposed in the context of computer system security.

2 Multilevel Network Security

Let us assume that we have a collection of trusted and untrusted computers and we would like to connect them to form a secure network system. A network is said to be multilevel secure if it is able to protect multilevel information and users. That is, the information handled by the network can have different classifications and the network users may have varying clearance levels.

Several approaches to network security have been proposed over the years that can be used to handle the problem of providing *multilevel security* in a computer network. Three such approaches include physical separation, temporal separation and cryptographic separation. Refer to [9] for details.

Before we look at the design of a suitable network security model, it is useful to consider the recent developments that have been taking place in the formulation of evaluation criteria for trusted network systems.

2.1 Trusted Network Interpretation

The US Department of Defense published in December 1985 the *Trusted Computer System Evaluation Criteria (TCSEC)*, commonly known as the Orange Book. The TCSEC "provides a means of evaluating specific security features and assurance requirements in trusted automatic data processing (ADP) systems". The basic philosophy of the protection described in the TCSEC requires that the access of subjects (i.e. human users or processes acting on their behalf) to objects (i.e. containers of information) be mediated in accordance with an explicit and well-defined security policy. The *Trusted Network Interpretation (TNI)*, commonly known as the Red Book, provides an interpretation of the TCSEC for networks. The TNI presents two views of the network : the interconnected view and the single trusted system view. Our approach is more closely related to the latter than to the former. In this case, the security policy for the network is enforced and implemented at a conceptual level by a Network Reference Monitor (NRM) which is analogous to the Reference Monitor in the TCB. The TNI considers the network security policy as an extension of the security policy of each component in the network. It assumes that the totality of security-relevant portions of the network reside in a Network Trusted Computer Base (NTCB). The NTCB itself is partitioned and is distributed over a number of network components. The network security architecture determines how the NTCB is to be partitioned.

The implementation of the Network Reference Monitor is dependent upon :

(a) whether there are trusted computer systems in the network

(b) their modes of operation

(c) the level of trust required for the network elements.

In the development of the security models, we will assume that the protection mechanisms reside in an NTCB analogous to a TCB. The TCB is concerned with restricting the access to information to legitimate users and processes as well as regulating the information flow between the processses and the users *within* a system. The NTCB controls the transfer of information *between* different network components. The NTCB is not concerned with determining whether the information stored in an object within a system can be accessed by a particular user or a process. Hence the network security policy model must take into account the distributed nature of the network system. That is, in addition to specifying "which users are authorised to have access to which objects", the network access control policy should identify the connection requirements for setting up a session between two network components. Further, the flow of information from one object to another residing in different systems must be addressed.

A major goal of a network security policy is to determine which pairs of entities can communicate. In trying to formulate a *network* security policy by drawing as much parallel as possible with the *computer* system policy, we are immediately faced with the problem of clearly identifying the subjects and objects in a network environment. In practice, subjects and objects are identified in the context of a particular system in conjunction with the access policy it is designed to support. In the case of networks, there are several options for the interpretation of subjects and objects. For instance, users, hosts, nodes, subnets can all be classified as subjects. At the application layer in the OSI model, the network subjects could be *users* whereas at the transport/network layer, the subjects may be *hosts* present in the network. Similarly, messages, files, packets, connections can all represent network objects. The type of subjects and objects supported by a secure network depends *on the layer at which security in the network* is to be considered. That is, whether a security policy makes sense depends on the service provided by the network, as specified by the user interface to a particular protocol layer ([8]).

3 Typical Network Configurations

Let us now consider some network configurations involving both trusted and untrusted systems and examine some typical connection scenarios. This will be helpful in developing specific network security requirements and the formulation of network security models. This section summarizes some of the discussions presented in [3], [4]. The four possible network configurations are as follows :

(1) Untrusted Systems on an Untrusted Network

(2) Trusted Systems on an Untrusted Network

(3) Untrusted Systems on a Trusted Network

(4) Trusted Systems on a Trusted Network

Case 1 : Untrusted Systems on an Untrusted Network

In this case, both the untrusted systems and the untrusted network operate in a Dedicated or in a System High mode. Neither the computer system nor the network have any access control policy and there are no labels associated with the information processed and transferred in the network. That is, there is no need for a Network Trusted Computing Base. In the case of a Dedicated mode network, the Dedicated and System High mode systems operate at the same security level as the network. In the case of a System High mode network, the Dedicated and System High mode systems can operate at some level lower than the maximum level of the network. The users are cleared to the maximum level but the information can range from some low level to the maximum level. Hence it is necessary to ensure that the information classified at a higher level should not combine with the lower level information.

Case 2: Trusted Systems on an Untrusted Network

In this scenario, the systems are trusted to operate in a Multilevel mode including the network security level. The access control mechanisms in the trusted systems extend to process to process communications across the network. The TCB is required to ensure that the information is properly labelled with the network high security level and that no information is put on the network at a security level higher than the level of the network. Further to establish connections with different systems, the TCB of each computer system must know the levels of other systems in the network.

Both the System High mode systems and the trusted systems have and manage their own discretionary access controls. Hence there is no need for the network to implement the discretionary access control. The granularity of the discretionary access controls influences the complexity of the implementation. For instance, the access control could be carried out at the "system level", to determine whether a remote system has an authorized access to a given process. Alternatively, the access control could be performed at the "process level", to check whether a remote process can have an authorized access to a given process.

Case 3 : Untrusted Systems on a Trusted Network

The systems connected to the network operate in either the Dedicated or System High mode. As the trusted network can carry information at different classifications, it is necessary to attach labels either to information units or to virtual circuits when sessions are established. Further the integrity of labels must be ensured. The level of each of the untrusted systems must be within the range of levels for which the network is trusted and a network mandatory security policy must be enforced.

Case 4 : Trusted Systems on a Trusted Network

This case consists of both trusted computer systems and trusted networks operating in a Multilevel mode. The systems connected to the network must have TCBs and the network itself must have a NTCB. Here not all users of the systems and the network are cleared for all information in the system or flowing on the network. Therefore, the range of security levels of the systems must overlap with the corresponding levels of the network. The systems must ensure that the network can only receive information in its range and similarly the network should ensure that each computer system can only receive information in its range. The network and the systems should also ensure that there is no illicit information flow and that the information is correctly labelled.

3.1 A Typical Connection

We will consider a trusted network consisting of the following components : hosts, input-output components (e.g. terminals) and output components (e.g. printers). Let us consider a typical interaction between two hosts in such a network.

Each network component possesses a security classification and every user and a process acting on a user's behalf possesses a clearance. It is important to say a few words about terminology. In the context of computer security, there is no difference between a classification and a clearance : classification is used for an object and clearance is used for a subject. In this paper, we will use the term *security class* for both and we write Scls(X) for the security class of a component X. In fact, Denning uses this term in her book ([11]). Other terms that have appeared in the literature include access class and security level.

We assume that a partial ordering $\geq$ is defined on the set of security classes. A partial ordering relation is reflexive, antisymmetric and transitive. Given two security classes $sc1$ and $sc2$, if $sc1 \geq sc2$, then we say that $sc1$ *dominates* $sc2$. It is not always possible to compare two security classes using the "dominate" relationship. In this case, they are said to be incomparable. In this paper, we will further assume that the set of security classes is a lattice with respect to the partial ordering $\geq$. That is, given any pair of elements $sc1$ and $sc2$:

- the set of all security classes dominated by both $sc1$ and $sc2$ is non-empty and contains a unique greatest lower bound (glb) that dominates all the others

- the set of all security classes that dominate both $sc1$ and $sc2$ is non-empty and contains a unique least upper bound (lub) that is dominated by all the others.

We require the following additional definitions before considering a typical interaction. Assume that entities X and Y (subjects or objects) are trusted to operate at more than one security class, e.g. Scls(X) = $\{sc_{x1}, sc_{x2}, . . ., sc_{xn}\}$ and Scls(Y) = $\{sc_{y1}, sc_{y2} , . . ., sc_{ym}\}$. Let sc be some security class. Then, we have the following :

- Scls(X) $\geq sc$, if and only if $sc_{xi} \geq sc, \forall i, 1 \leq i \leq n$.

- $sc \geq Scls(X)$, if and only if $sc \geq sc_{xi}, \forall i, 1 \leq i \leq n$.

- $Scls(X) \geq Scls(Y)$, if and only if for every sc_{xi} $(1 \leq i \leq n)$, $sc_{xi} \geq lub(sc_{y1}, ..., sc_{ym})$.

3.2 Host - Host

Let the range of security classes of the hosts H1 and H2 be Scls(H1) and Scls(H2) respectively. Let $NTCB_{H1}$ and $NTCB_{H2}$ be associated with the hosts H1 and H2 respectively. We have three cases to consider depending upon whether the hosts are trusted or not.

(a) H1 and H2 are Trusted

For the connection between H1 and H2 to be granted the security classes of the two hosts must at least have some overlap. That is, Scls(H1) $\cap$ Scls(H2) $\neq \emptyset$.

Now consider a process p in the host H1 which attempts to establish a session with the host H2. $NTCB_{H1}$ first checks to determine whether the security class of p is within the set of security classes of the host H2. If this is the case, then the request for the connection is granted. A similar procedure is followed when a process p in the host H2 wishes to establish a session with H1. After the connection is set up, $NTCB_{H1}$ enforces the security policy to prevent the flow of higher classified information in H1 into a lower level object in H2. Similarly, $NTCB_{H2}$ enforces an information flow control to prevent illicit flow from H2 to H1.

(b) H1 and H2 are Untrusted

In this situation, the connection requirement is that both H1 and H2 and both processes in H1 and H2 must have the same security class.

(c) A Trusted Host and an Untrusted Host

We will assume that H1 is a trusted host and H2 is an untrusted host. When a process p1 in H1 wishes to communicate with a process p2 in H2, $NTCB_{H1}$ checks to see if the security class of p1 dominates the security class of H2. If the condition is satisfied then $NTCB_{H1}$ passes the identity of the process and the security class of p1 to $NTCB_{H2}$. The process p1 can now write/read to/from any object in the host H2 as H2 is untrusted.

Further, $NTCB_{H1}$ is to enforce the flow policy that any object in H1 that could be read by p1 should not be allowed to flow into an object in H2 having lower security class than that in H1. $NTCB_{H2}$ should enforce similar flow control so that objects in H2 do not flow into objects in H1 with security classes lower than in H2.

Now if p2 in H2 tries to establish a connection with a process in H1, the request can only be granted if the security class of p2 is within the set of security classes of H1. Similar controls as described above are applied by $NTCB_{H1}$ and $NTCB_{H2}$ to regulate information transfer between H2 and H1.

4 A Network Security Policy Model

In this section, we describe a simple abstract network security policy model. From the above discussions, we see that the NTCB performs the following two primary functions :

(1) controlling the establishment of a connection between network entities
(2) regulating the flow of information between network entities.

Note that we will be addressing *only* these NTCB functions in our security policy model.

Informally, the network discretionary and mandatory access control policy can be described as follows :

Discretionary Access Control

We assume that the information required to provide discretionary access control resides within the NTCB associated with each network component, rather than in a centralized access control centre.

The network discretionary access control policy is based on the identity of the network components, implemented in the form of an authorization connection list. The authorization list determines whether a connection is allowed to be established between two network entities. The individual components may in addition impose their own controls over their users - e.g. the controls imposed when there is no network connection.

Mandatory Access Control

The network mandatory security policy requires appropriate labelling mechanisms to be present in the NTCB. One can either explicitly label the information transferred over the network or associate an implicit label with a virtual circuit connection. In our model we have the following scheme :

(a) Each network component is appropriately labelled. A mandatory policy based on the labels of the network components is imposed and it determines whether a requested connection between two entities is granted or not.

(b) Information transferred over the network is appropriately labelled. A mandatory security policy is used to control the flow of information between different subjects and objects, when performing different operations involving information transfer over the network.

4.1 Modelling Approach

The advantages of the use of formal techniques for modelling systems are well known. Not only do they provide a means for a precise and unambiguous representation of the security policy, but they also help us to reason about the security of the system.

The network security policy model we describe here is a state-machine based model. Essentially a state machine model describes a system as a collection of entities and values. At any time, these entities and values stand in a particular set of relationships. This set of relationships constitutes the state of the system. Whenever any of these relationships change, the state of the system changes. The common type of analysis that can be carried out using such a model is the reachability graph analysis. The reachability graph analysis is used to determine whether the system will reach a given state or not. For instance, we may identify a subset of states W which represent "insecure" states and if the system reaches a state within this subset W, then the system is said to be insecure.

In describing such a state machine based security model, we need to perform the following steps :

- Define *security related state variables* in the network system.
- Define the requirements of a *secure network state*.
- Define the *network operations* which describe the system state transitions.

In proving that a system represented using this model is secure, we need to show that

- each operation preserves the security requirements.
- the initial network state is secure.

We can also propose the *Basic Security Theorem* and formally show that the Theorem holds for the described model. This shows that security constraints on operations ensure that any state reachable from an initial secure state will always be secure. It is important to realize that the proof of such a theorem does not depend upon the particular definition of security provided by the model.

We make the following basic assumptions :

1. Reliable user authentication schemes exist within a host network component, implemented as part of the TCB.

2. Only a user with the role of a Network Security Officer can assign security classes to network subjects and network components, and roles to users. A host component is assumed to have a range of security classes in which it can operate.

3. Reliable transfer of information across the network.

4. Appropriate cryptographic techniques are incorporated within the NTCB which protect the information before transferring over the network. These cryptographic measures are not included in the model.

4.2   Model Representation

We define a network security model, MODEL, as follows :

$$MODEL = < S, O, A, s_0 >$$

where S is the set of States, O is the set of system Operations, A is the system Action function and s_0 is the initial system state.

The set S models the security related state variables in the network system. The set O describes the network operations and the action function A describes the transition from one state to another by applying one or a sequence of operations from the set O. The state s_0 refers to the initial state of the system.

Let us first define the basic sets used to describe the model.

- Sub : Set of all network subjects. This includes the set of all Users (Users) and all Processes (Procs) (executing on behalf of the users) in the network. That is,

  $Sub = Procs \bigcup Users$

- Obj : Set of all network objects. This includes both the set of Network Components (NC) and Information Units (IU). That is,

  $Obj = NC \bigcup IU$

  Typically, the set of Network Components includes Hosts (H), Input-Output Components (IOC) and Output Components (OC) whereas Information Units include files and messages. That is,

  $NC = H \bigcup IOC \bigcup OC$

- $SCls$: Set of Security Classes. We assume that a partial ordering relation $\geq$ is defined on the set of security classes as described in Section 3.1. Further the set $SCls$ is a lattice with respect to the partial ordering $\geq$.

- $Rset$: Set of user roles. This includes for instance the role of a Network Security Officer.

- $Strings$: Set of character strings.

Before describing the model, a brief note regarding notation is necessary. Sometimes we will use the notation x_s to denote the element x at state s and $f(x_s)$ to denote the value of f of element x at state s. At other times, where it is appropriate, we use x and x' to denote the element x at states s and s' respectively.

4.2.1 System State

We only consider the security relevant state variables.

Each state $s \in S$ can be regarded as a 13-tuple as follows :

$s = < Sub_s, Obj_s, authlist, connlist, accset, subcls, objcls, curcls, subrefobj, role, currole,$
$\quad term, contents >$

Let us now briefly describe the terms involved in the state definition.

Sub_s defines the set of subjects at the state s.

Obj_s defines the set of objects at the state s.

authlist is a set which consists of elements of the form (sub, nc) where $sub \in Sub_s$ and $nc \in Obj_s$. The existence of an element $(sub1, nc1)$ in the set indicates that the subject $sub1$ has an access right to connect to the network component $nc1$.

connlist is again a set of elements of the form (sub, nc). This set gives the current set of authorized connections at that state.

accset is a set which consists of elements of the form $(sub, iuobj)$, where $sub \in Sub_s$ and $iuobj \in Obj_s$. The existence of an element $(sub1, iuobj1)$ in the set indicates that the subject $sub1$ has an access right to bind to the object $iuobj1$. More will be said about binding in Section 4.2.4.

subcls : $Sub -> SCls$.

subcls is a function which maps each subject to a security class.

objcls : $Obj -> PS(SCls)$, where PS denotes the power set.

objcls is a function which maps each object to one or more security classes. We assume that an output device object and an information unit object have a single security class associated with them whereas a host object may have several security classes.

curcls : $Sub -> SCls$

curcls is a function which determines the current security class of a subject.

subrefobj : $Sub -> PS(Obj)$

subrefobj is a mapping which indicates the set of objects referenced by a subject at that state.

role : $Users -> PS(Rset)$

role gives the authorized set of roles for a user.

currole : $Users -> Rset$.

currole gives the current role of a user.

term : $Users -> IOC$

term is a function which gives the terminal in which a user is logged on.

contents : $IU -> Strings$

contents is a function which maps the set of Information Units (IU) Objects into the set of Strings. It gives the contents of information objects.

Finally, let us introduce another definition associated with the input - output components. This is similar to the one given in [7].

For a $nc \in IOC \cup OC$, $view(nc)$ is a set of ordered pairs $\{(x_1, y_1), (x_2, y_2), ..., (x_n, y_n)\}$, where each y_i is displayed on the component nc. Each x_i is an information unit object and y_i is the result of applying the *contents* function to x_i.

4.2.2   Secure State

To define the necessary conditions for a secure state, we need to consider the different phases gone through by the system during its operation.

Login Phase

From our assumptions, we have reliable and appropriate authentication and identification schemes within the TCB of the network component. These aspects are not modelled as part of the NTCB. Further, we require that if the user is logging through a terminal, he must have appropriate clearance with respect to the terminal. That is, the security class of the user must dominate the security class of the terminal in which the user is attempting to log on. In addition, the current security class of the user must not dominate the maximum security class of that user and the role of the user must belong to the authorized role set allocated to that user. So we have the following :

Login Constraint

Proposition 1 : A state s satisfies the Login Constraint if
$\forall x \in Users,$

- $subcls(x) \geq objcls(term(x))$
- $subcls(x) \geq curcls(x)$
- $currole(x) \in role(x)$

Connect Phase

Having logged-on to the network, a user may wish to establish a connection with another network component. In determining whether such a connection request is to be granted, both network discretionary and mandatory security policies on connections need to be satisfied.

The discretionary access control requirement is specified using the authorization list which should contain an entry involving the requesting subject and the network component.

If the network component in question is not an output component, then the current security class of the subject must at least be equal to the lowest security class of the network component. On the other hand, if the network component is an output component, then the security class of the subject must not be greater than the security class of that component.

Hence we have the following constraint.

Connect Constraint

Proposition 2 : A state s satisfies the Connect Constraint if $\forall (sub, nc) \in connlist$

- $(sub, nc) \in authlist$
- if $nc \notin OC$, then $curcls(sub) \geq glb(oc_1, oc_2, ..., oc_j)$, where $objcls(nc) = \{oc_1, oc_2, ...oc_j\}$
- if $nc \in OC$ then $objcls(nc) \geq subcls(sub)$

Other Conditions

We require two additional conditions :

(1) The classification of the information that can be "viewed" through an input-output device must not be greater than the classification of that device.

(2) The role of the users at a state belong to the set of authorized roles.

Now we can give the definition of a *secure state* as follows :

Definition : A state s is Secure if

1. s satisfies the *Login Constraint*
2. s satisfies the *Connect Constraint*
3. $\forall z \in (IOC_s \bigcup OC_s), \forall x \in IU_s, (x, contents(x)) \in view(z) => objcls(z) \geq objcls(x)$
4. $\forall u \in Users_s, currole(u) \in role(u)$

4.2.3 Initial State

We assume that the initial system state s_0 is defined in such a way that it satisfies all the conditions of the secure state described above.

4.2.4 Operations

Connect Operation

The operation *connect(sub,nc)* allows a subject sub to connect to a remote network entity nc. From the Connect Constraint (Proposition 2) given earlier, for this operation to be secure, we require that

(a) $(sub, nc) \in authlist$

(b) if $nc \notin OC$, then $curcls(sub) \geq glb(oc_1, oc_2, ..., oc_j)$, where $objcls(nc) = \{oc_1, oc_2, ...oc_j\}$

or

if $nc \in OC$ then $objcls(nc) \geq subcls(sub)$

After the operation is performed, $(sub, nc) \in connlist'$ and $nc \in subrefobj(sub)$.

Information Manipulation Operations

Having connected to a remote network component, a subject can perform operations which allow manipulation of information objects. In general security constraints related to this phase are defined according to the security policy implemented in the TCB of the remote network component rather than the NTCB. We envisage the information manipulation phase to consist of two stages : a binding stage and a manipulation stage. The binding stage involves a subject linking itself to the information unit object on which the operation is to be performed. At the manipulation stage, typically the operations include those operations defined by the Bell-LaPadula model such as read, append, write and execute. In our model, we will only consider one basic manipulation operation which allows transfer of an object from one network component to another, as this is perhaps the most important operation from the network point of view. This operation causes information to flow from one entity to another over the network. (In fact, this operation will form part of other operations as well. For instance, consider a read operation, whereby a user reads a file stored in a remote entity. This operation must include the transfer of the file from the remote network component to the local network component in which the user resides.)

There are also other operations which modify certain security attributes of objects and subjects. In the usual computer security model, these include operations for assigning and changing security classes to users and information objects and assigning and modifying access sets for information unit objects. In the case of our network security model, we need additional operations such as to assign security classes of network component objects, to set authorization list and operations to assign and change roles of the users.

Note that in general for any operation to be performed, the subject must have authorized access to the connection with the remote entity. That is, the Connect Constraint must be satisfied to begin with.

Let us first consider the bind and unbind operations. (We will use the notation x and x' to refer to x at states s and s').

Bind

The operation $bind(iuobj, nc)$ allows a subject sub to link to an information object $iuobj$ in a network component nc. For this operation to be secure, we require that

(a) $(sub, iuobj) \in accset(iuobj)$

(b) $curcls(sub) \geq objcls(iuobj)$

(c) for any $sb \in Sub_s$, $iuobj \notin subrefobj(sb)$

After the operation is performed, $iuobj \in subrefobj'(sub)$

Note that we have included a simple access control based on $accset$ at the remote network component. In practice, a comprehensive access control mechanism is likely to be provided by the TCB of the remote entity. Note that we could have defined the bind operation as part of the connect operation, thereby making the connection to a particular information object at the connect stage rather than to a network component. We have deliberately separated these two operations to differentiate the roles played by the NTCB and the TCB.

Unbind

The operation *unbind(sub,iuobj)* allows a subject *sub* to release its link to an information object *iuobj*. That is, before this operation $iuobj \in subrefobj(sub)$. After this operation, we have $iuobj \notin subrefobj(sub)$.

Let us now consider the information transfer operation and other operations which affect the security attributes of subjects and objects.

Transfer

The operation *transfer(iuobj1, nc1, iuobj2, nc2)* allows a subject *sub* to append the contents of an information unit object *iuobj1* in a network component object *nc1* to the contents of another information unit object *iuobj2* in a network component object *nc2*. For this operation to be secure, we require that

(a) $objcls(iuobj2) \geq objcls(iuobj1)$

(b) $curcls(sub) \geq objcls(iuobj1)$

(c) $(sub, iuobj1) \in accset(iuobj1)$

(d) $(sub, iuobj2) \in accset(iuobj2)$

Further both *iuobj1* and *iuobj2* referenced by the subject *sub* must not be referenced by any other subject. That is,

(c) for any $sb \in Sub_s$, $sb \neq sub$, $iuobj1, iuobj2 \notin subrefobj(sb)$
and $iuobj1, iuobj2 \in subrefobj(sub)$

After the operation is performed, the security classes of the objects *iuobj1* and *iuobj2* remain unchanged. That is,

(d) $objcls'(iuobj1) = objcls(iuobj1)$

(e) $objcls'(iuobj2) = objcls(iuobj2)$

where *objcls'* refers to the new state s'.

Set-sclass-ncobj

The operation *set-sclass-ncobj(nc,scls)* allows a subject *sub* to set the security class of a network component object, *nc*, to *scls*. That is, $objcls'(nc) = \{scls\}$. This operation can be performed only when the component is not being used. Further, only a Network Security Officer (NSO) has the authority to set the security class of a network component object. That is, if this operation is to be performed at state s then the following must be true.

If there exists any $nc \in NC$ such that $objcls(nc) \neq objcls'(nc)$ then

- for any subject $sb \in Sub_s$ ($sb \neq sub$), $nc \notin subrefobj(sb)$ and $(sb, nc) \notin connlist$

(that is, no other subject either referencing or connected to that component)

- $NSO \in role(sub)$ and $currole(sub) = NSO$.

Set-sclass-user

The operation *set-sclass-user(usr,scls)* allows a subject sub to set the security class of a user, usr, to $scls$. That is, $subcls'(usr) = scls$. Typically the conditions we require for this operation to be secure are :

If there exists any $usr \in Users$ such that $subcls(usr) \neq subcls'(usr)$ then

- $NSO \in role(sub)$ and $currole(sub) = NSO$

- if the user is logged in at state s (i.e. $usr \in Users_s$), then $subcls'(usr) \geq curcls(usr)$. (note : $curcls'(usr) = curcls(usr)$).

Set-curclass-user

The operation *set-curclass-user(usr,scls)* allows a subject sub to set the current security class of a user, usr, to $scls$. That is, $curcls'(usr) = scls$. The conditions required for this operation to be secure can be described as follows :

If there exists any $usr \in Users$ such that $curcls(usr) \neq curcls'(usr)$ then

- $NSO \in role(sub)$ and $currole(sub) = NSO$ or $usr = sub$.

- $subcls(usr) \geq curcls'(usr)$

- if the user is logged onto a terminal at state s, then $curcls'(usr) \geq objcls(term(usr))$.

- if the user is connected to a network component at state s which is not an output device, that is, $(usr, nc) \in connlist$ and $nc \notin OC$, then $curcls'(usr) \geq glb(oc_1, ..., oc_n)$ where $objcls(nc) = \{oc_1, oc_2, ..., oc_n\}$.

- if the user is logged in and is connected to an output device, that is, $(usr, nc) \in connlist$ and $nc \in OC$, then $objcls(nc) \geq curcls'(usr)$.

Set-role-user

The operation *set-role-user(usr,rlset)* allows a subject sub to assign a role set $rlset$ to a user, usr. That is, $role'(usr) = \{rlset\}$. For this operation to be secure, we need the following conditions to hold :

If there exists any $usr \in Users$ such that $role(usr) \neq role'(usr)$ then

- $NSO \in role(sub)$ and $currole(sub) = NSO$

- if the user is logged in at state s (i.e. $usr \in Users_s$), then $currole(usr) \in role'(usr)$.

Set-currole-user

The operation *set-currole-user(usr,rl)* allows a subject *sub* to change the current role of a user, *usr*, to *rl*. That is, *currole'(usr)* = *rl*. The security requirements of this operation are :

If there exists any *usr* ∈ *Users* such that *currole(usr)* ≠ *currole'(usr)* then

- Only the user himself or a subject whose current role is a NSO has the authority to change the current role of the user. That is, $NSO \in role(sub)$ and $currole(sub) = NSO$ or *usr = sub*.

- The new role *rl* must be in the set of authorized roles of the user. That is, $currole'(usr) \in role(usr)$.

Set-authlist

The operation *set-authlist(al)* allows a subject to set the authorization list. The *authlist* is of the form (sb, nc), where $sb \in Sub$ and $nc \in NC$.

Again, this operation can only be performed by a subject who can act as a NSO. That is, if *al* ∉ *authlist* and *al* ∈ *authlist'* then $NSO \in role(sub)$ and $currole(sub) = NSO$ where *sub* is the subject performing this operation.

The above list provides just a sample of operations in such a model. It is necessary to extend this list to include other operations (e.g. create and remove network components) and derive appropriate conditions for these operations to be secure. For instance, note that the create operation also needs to set the range of security classes of the new component.

4.2.5 Action

The Action function describes the transition of the system from one state to another by applying one or a sequence of operations described above. That is,

$A : Sub \times O \times S \rightarrow S$

Hence, $s' = A(sub, op, s)$ is the resulting state due to an operation $op \in O$ performed by a subject $sub \in Sub$ starting at state $s \in S$.

4.3 Verification

In order to verify the security of the network system described by the above model, we need to go through the functions performed by the system at each stage and ensure that the system satisfies all the required security properties. We will briefly describe the verification process by going through each of the three phases identified earlier. Here we only state the theorems; refer to [12] for their proofs.

4.3.1　Login Phase

By definition, it is clear that the Login Constraint is satisfied by the system at the initial state s_0. We need to show that no operation exists which can take the system to a state in which the criterion is not satisfied. We can formally state this as follows :

Theorem 1 : The network security model $M = < S, T, A, s_0 >$ satisfies the Login Constraint if the initial state s_0 satisfies the criterion.

4.3.2　Connect Phase

Having logged onto the network, a user may wish to establish a connection with another network component. In determining whether such a connection is to be permitted, the Connect Constraint in Proposition 2 must be taken into account.

The system satisfies the Connect Constraint if and only if starting from the initial state, s_0, every state in the sequence $(s_0, s_1, ...)$ satisfies it.

This leads to the following Theorem.

Theorem 2 : The security model $M = < S, T, A, s_0 >$ satisfies the Connect Constraint if the initial state s_0 satisfies the Connect Constraint.

4.3.3　Operation Phase

Having obtained the connection, the user can perform operations which cause information to flow from one entity to another. These operations in general can cause the system to move from a secure state to an insecure state. *Note that for any of these operations to be performed, the subject must have authorized access to the connection with the remote entity.* That is, the Connect Constraint at that state must be satisfied.

We considered in Section 4.2.4 the conditions required for the operations to be secure. In general, we need to show that a *sequence* of these *secure* operations do not cause the system to go from a secure state to an insecure one.

Definition : The model $< S, O, A, s_0 >$ is **action secure** if

(i) the actions are *bind secure, transfer secure* and *set secure*.
(if there are more operations, we need to include them as well)

(ii) any sequence of these actions will not result in an insecure state, starting from a secure state.

Theorem 3 : The model $< S, O, A, s_0 >$ is action secure.

4.3.4 Basic Security Theorem

Finally it may be useful to give the analog of the Basic Security Theorem for this model. In fact, this is not critical to our model as the concept of a secure system is largely defined in terms of secure transformations.

In the following definition, X_s, $X_{s'}$ refer to the set X at states s and s', and x_s, $x_{s'}$ refer to the element x at states s and s' respectively.

The states s and s' are given as follows :

$$s = < Sub_s, Obj_s, authlist, connlist, accset, subcls, objcls, curcls, subrefobj,$$
$$role, currole, term, contents >$$

$$s' = < Sub_{s'}, Obj_{s'}, authlist', connlist', accset', subcls', objcls', curcls', subrefobj',$$
$$role', currole', term', contents' >$$

Definition :

The system described by the above model $< S, O, A, s_0 >$ is secure, if

- the initial state s_0 is a secure state

- any action A defined by $A(sub, op, s) = s'$ satisfies

  (a) Theorem 1

  (b) Theorem 2

  (c) Theorem 3

  (d) $\forall x' \in IU_{s'}, \forall x \in IU_s, \forall z' \in (IOC_{s'} \cup OC_{s'}), \forall z \in (IOC_s \cup OC_s)$,
  if $(x, contents(x)) \notin view(z)$ and $(x', contents(x')) \in view(z')$ then $objcls(z') \geq objcls(x')$.

  (e) for all users $usr \in Users$ such that $role(usr) \neq role'(usr)$ or $currole(usr) \neq currole'(usr)$ then $currole'(usr) \in role'(usr)$.

Theorem 4 : The model $< S, O, A, s_0 >$ is secure.

5 Discussion

We have considered above the design of "just one possible" network security policy model. In fact, as mentioned earlier, whether a security policy makes sense or not depends on the layer at which security in the network is to be provided. In this section, we illustrate this issue of layering security policy by briefly outlining the design of another network security policy model.

5.1   Security Policy Model 2

System Description

Consider the situation where we have a set of hosts communicating with each other over a network by sending and receiving messages. As in the previous security policy model, we will assume that each host communicates through a trusted network interface unit in which the NTCB partition resides. In fact, as mentioned earlier, the trusted interface unit can also be used to encrypt messages before transferring the message over the network. (The keys required for this process can be handled using a Key Distribution Centre. We will not be considering such issues here).

Network Security Policy

In this model, let us define the set of network subjects and objects to be as follows : Let the *hosts* in the network represent the subjects and let the *connections* between pairs of hosts represent the network objects. A connection indicates the potential for two hosts to communicate with each other by sending and receiving messages.

This might be typically applicable to the transport/network layer in the OSI model.

We will assume that a security class is defined exactly in the same way as before. However in this case, we associate with each subject (host) a range of security classes within which the host can communicate. Each object (connection) is to have a single security class. Hence connections must be established for each security class at which a pair of hosts wishes to communicate.

The network security policy is to control the establishment and the use of connections between pairs of hosts.

A discretionary access control policy can be formulated as follows :
A host can have access to another host via a connection if and only if it has the appropriate authorization.
This policy can be specified using authorization lists as before. The authorization list is of the form $(host_i, host_j)$, which indicates that $host_i$ is authorized to connect to $host_j$.

A mandatory access control policy could be defined as follows :
A host can have access to a connection if and only if the security class of that connection falls within the range of security classes of that host.

For the sake of simplicity, let us again just consider one operation, namely that of transfer of messages. The transfer operation includes both sending and receiving of messages. Sending of messages correspond to a host (subject) *writing* on to a connection (object) whereas receiving of messages corresponds to *reading* from a connection. As in the previous model, we have an additional condition that a host can only transfer messages if it has an authorized access to a connection. However note that in this model there is no information flow control based on the security classes of messages.

As before, we can define a special set of operations which can only be performed by the Network Security Officer.

In this model, we can define a state to be secure if all the *connections* between pairs of hosts at that state are secure. A connection is said to be secure if and only if the security policy is satisfied in the establishment of the connection.

We can again give a state machine representation of the model and use the "inductive nature" of security to show that the system model is secure and that the system progresses from a secure state to another secure state under secure transitions.

6 Conclusions

In this paper, we have considered the development of some multilevel network security policy models. Recently, the question of the design of a security policy model has come under intense scrutiny, largely because of the publication of the Department of Defense Trusted Computer System Evaluation Criteria (TCSEC) and The Trusted Network Interpretation (TNI). In this paper, we have described an abstract network security policy model which addresses some of the access control and information flow control requirements of a multilevel secure network. The network access control policy determines the requirements for establishing connections between network components and information flow policy regulates the flow of information between the network components. The model is used to formally prove that the access control requirements are not violated and that information does not flow from higher security classes to lower ones as a result of network operations.

In developing network security policies, it is necessary to relate them to specific layers of the network architecture. In particular, whether a security policy makes sense or not depends on the layer to which it is applied to. This has been discussed by looking at an other security policy model with different network subjects and objects.

In addition to these access control and information flow requirements, we also need to protect the communications. Hence appropriate cryptographic techniques to provide confidentiality, integrity and authentication of communications are necessary. We have not considered such cryptographic aspects in the description of our model.

Finally, it is worth emphasizing that a major advantage of a formal model is that it enables a precise and unambiguous representation of the security policy. In this respect, we feel that a formal representation of the security policy proposed by Clark and Wilson in their seminal paper ([13]) will be useful and it will help to clarify some of the ambiguities apparently present in that model. At present, a preliminary analysis of the model is being carried out ([14]).

7 References

[1] "DoD Trusted Computer System Evaluation Criteria", DoD 5200.28-STD, Dec.1985.

[2] "Trusted Network Interpretation of the TCSEC", National Computer Security centre, NCSC-TG-005, Vers.1, July 1987.

[3] S.T.Walker, "Network Security Overview", Proc. of IEEE Symposium on Security and Privacy, 1985, pp 62-76.

[4] J.P.Anderson, "A unification of Computer and Network Security Concepts", Proc. of IEEE Symposium on Security and Privacy, 1985, pp 77-87.

[5] D.E.Bell and L.J.LaPadula, "Secure Computer Systems", Vols.I,II and III, The Mitre Corporation, MTR-2547, ESD-TR-73-278.

[6] C.E.Landwehr, "Formal Models for Computer Security", ACM Computing Surveys, Vol.13, No.3, pp 247-278, Sept.1981.

[7] C.E.Landwehr, C.L.Heitmeyer and J.McLean, "A Security Model for Military Message Systems", ACM Transactions on Computer Systems, Vol.2, No.3, pp 198-222, Aug.1984.

[8] J.K.Millen, "A Network Security Perspective", Proceedings 9th National Computer Security Conference, NCSC, pp 7-15, 1986.

[9] J.Rushby and B.Randall, "A Distributed Secure System", Computer, Vol.16., No.7, pp 65-67.

[10] D.P.Sidhu and M.Gasser, "A Multilevel Secure Local Area Network", Proc. of IEEE Symposium on Security and Privacy, 1982, pp 137-143.

[11] D.Denning, *Cryptography and Data Security*, Addison Wesley Pub., 1982.

[12] V.Varadharajan, "Developing Network Security Policy Models", Hewlett-Packard Labs., HPL-ISC-TM-89-089.

[13] D.Clark and D.Wilson, "A Comparison of Commercial and Military Computer Security Policies", IEEE Symposium on Security and Privacy, April 1987.

[14] V.Varadharajan, "Discussion of the Clark-Wilson Integrity Model for Computer Information Systems", Hewlett-Packard Labs., HPL-ISC-TM-88-061.

KEYMEX: An expert system for the design of key management schemes

J.C.A. van der Lubbe   D.E. Boekee

Delft University of Technology
Department of Electrical Engineering
Information Theory Group
P.O. Box 5031
2600 GA Delft, The Netherlands

Abstract

This paper deals with the development of KEYMEX; an expert system for the design of key management schemes. The main task of KEYMEX is the detection of possible weaknesses in key management schemes. An outline of the structure of the system, which is based on semantical networks, is presented. The way in which the knowledge concerning the key management schemes to be tested is incorporated in the knowledge base of the expert system is discussed. Finally, the performance of the system is assessed on the basis of results which are obtained by application of the system to some well-known key management schemes and protocols.

1   Introduction

Considering the increase of digital communication networks as well as the present and forth-coming applications of telecommunication services (videophone, telebanking, tv, etc.) there is an increasing need for protection of the transmitted data against eavesdroppers. In addition to the development of secure cryptographic algorithms the security of key management and key distribution plays an important role. In order to guarantee a frequent change of keys and an adequate authentication and signature scheme very complex key management schemes are demanded. Due to this need for key management schemes with high complexity the designer has the problem to assess a priori all possible loopholes in order to obtain iteratively a system concept which is as secure as possible.

In this paper after a brief introduction related to expert systems and cryptography, the expert system KEYMEX is outlined, which assists the designer with respect to the development and testing of key management schemes and by which so-called loopholes can be detected. With the help of the system some known key management schemes have been tested, which leads to some new insight concerning their weaknesses.

2 Expert systems and cryptography

Recent years have shown an increasing interest in the use of artificial intelligence techniques within the field of cryptography, especially with respect to automatic cryptanalysis and key management testing.

Automatic cryptanalysis

In [4] and especially in [5] and [6] expert systems have been introduced for the automatic cryptanalysis of some classical cipher systems. However, an expert system which can solve more complex ciphers seems impossible. More success may be expected when expert systems are used as a toolbox for the performance assessment of cryptographic algorithms. In that case, although the cipher system is not solved, the results of the various tests as well as the reasoning process performed by the expert system give insight into the possible weaknesses of the algorithms.

Key management testing

The main question in testing key management schemes is whether it is possible to obtain some secret key by some sequence of manipulations. In [2] and [3] an expert systems approach of the key management and distribution problem is given, based on OPS5 and Prolog respectively.

3 The structure of KEYMEX

The KEYMEX project is a joint project of the Faculties of Electrical Engineering and Mathematics and Informatics of the Delft University of Technology. The main objective of the KEYMEX project is to obtain insight into the possibilities to develop an expert system of which the main task is the detection of so-called loopholes and weaknesses in key management schemes and which as such can be an interesting toolbox for the designer of key management schemes. In addition to this we have special interest in the usefulness of expert system shells, which are based on semantical networks, for the development of such systems.

In order to make the expert system a useful toolbox for the designer of key management schemes, flexibility is an important requirement. This means that the system should be useful for testing a large number of key management schemes and protocols and that information about new key management schemes can easily be implemented in the system. During the development of the system, which is still in progress, this is kept in mind constantly since it determines the advantages of an expert system approach in comparison with standard software systems.

Another aspect which has been of importance for the definition of the structure of the expert system was the characteristics of the reasoning process. There is an essential difference between the use of expert systems for cryptographic analysis or for key management schemes. This is related to the question to what extent the reasoning process should deal with uncertainty. With respect to the analysis of cryptrahic algorithms one frequently has to do with a reasoning process which is highly influenced by the uncertainty by which hypotheses are supported, e.g. by outcomes of statistical tests. In the case of key management scheme testing one only searches for certain facts. With the help of some information a secret key can either be obtained or not, there is no middle course like in algorithm testing.

The KEYMEX system as developed is implemented on a SUN 3-60 work station. The applied expert system shell is DELFI-3; an expert system shell based on semantic networks, the development of which started a few years ago at the Delft University of Technology and is still in progress. The choice of DELFI-3 meets the possibilities outlined above. Furthermore, DELFI-3 is extremely useful for the development of complex expert systems and it is characterized by a more powerful data representation than is the case with e.g. Prolog and OPS5.

The functional structure of KEYMEX as developed is illustrated in Figure 1. The relevant parts of the expert system are the global database, the knowledge base and the control mechanism. These three main constituents are now discussed in more detail.

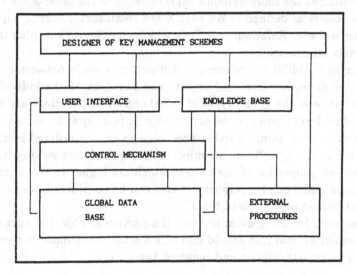

Figure 1: Functional structure of KEYMEX

3.1 The global database

The global database includes all the information with respect to a priori knowledge, e.g. which keys and cipher texts are accessible. Furthermore, it contains information about the ultimate goal and the problem it has to solve; e.g. finding some secret key. Information like this can be entered in the global database at the beginning of the consultation by the designer. During the consultation interim results, such as obtained and required subgoals, are also incorporated in the global database. Furthermore, all used information, inferred knowledge, etc. as well as the order in which they are obtained during the session are also stored in the global database. The global database can be consulted by the designer at the end of the session in order to trace the reasoning process which has been followed. The information in the global database is represented in the form of lists.

3.2   The knowledge base

In the knowledge base all the relevant knowledge with respect to the present problem-domain and the key management scheme to be tested are implemented. This means that it contains a functional description of the key management scheme in terms of input and output models as well as some housekeeping rules.

The language used for the representation of knowledge is LORE (Language of Objects and RElations), of which the data structures are that of the data model DAMOR (DAta Model of Objects and Relations). It can be considered as a type of semantical network. LORE is based on descriptions of objects and relations. Objects express knowledge about entities that are relevant within the problem-domain. In relations the more dynamic aspects between the objects are expressed. Examples of objects as defined in KEYMEX are "plaintext", "key", "ciphertext", "random number" etc. Relations specify in which way the various plaintexts, keys and ciphertexts are connected.

According to DAMOR a fundamental distinction is made between objects and relations at the definition level and objects and relations at the individual level. At the definition level the global types of objects and their relations are described, at the individual level there are the actual objects (e.g. specific keys, ciphertexts etc.) and their interrelations. Furthermore, objects can be ordered hierarchically in the knowledge base by defining an object as "father" of another object, i.e. each object inherits all properties of the objects which are higher in the hierarchy. As such the objects plaintext, key, ciphertext etc. can be considered as examples of a more general defined object "data block".

Since it is essential for the system to know if a particular object is already known or that the expert system just should find it, a distinction is made between known and required plaintexts, known and required keys, etc.

The relations in the knowledge base represent the relations between objects like plaintext, key, ciphertext etc. on an abstract level as well as on a level which is more specific for the key management scheme to be tested. The three main types of relations which are defined in the system are as follows.

- Relations describing the general characteristics of encipherment and decipherment

  An example of such a general relation is that if one has a ciphertext as well as the key with which it has been enciphered, that then the plaintext can be obtained. In fact this is trivial, however this does not hold for the expert system. It should learn that the plaintext can be obtained from both ciphertext and key. Without this knowledge it does not enable the system to put together plaintext, ciphertext and key.

- Relations representing general functions
  In addition to the relations mentioned above there are relations which describe some general functions. In key management schemes use is made of one-way functions, two-way functions, modulo-2 additions etc. All these functions determine relations between inputs and outputs, i.e. between plaintexts, ciphertexts etc. The expert system is taught that for example in the case of a one-way function the output can be obtained on the basis of the input but not vice versa and that in the case of a two-way function with two input variables and one output variable, knowledge about both inputs gives all the information about the output. Also these types of relations are general in the sense that they hold for all key management schemes in which such functions play a role.

- Relations describing specific cryptograhic modules
  Specific key management scheme dependent relations are relations which describe the specific cryptograhic modules of a key management schemes to be tested. E.g. in the case of the IBM key management scheme the Re-encipher From Master-keys (RFM) module should be modelled in the knowledge base. This means it should be described for the system in what way its ouput depends on the two inputs, the master keys KM0 and KM1 and the present encipherment and decipherments within the module.

Functionally seen, the relations defined in the knowledge base perform as if-then rules, the conclusion part of which determines possible actions with respect to the global database. If for example during the consultation an output of some cryptograhic module is required, whereas only one of the two inputs is known, then the corresponding relation which describes this module, acts in order to insert the required input on the list of required data as stored in the global database.

It is clear that the performance of the expert system largely depends on the adequacy of the way in which the knowledge of the key management scheme to be evaluated is formalized by the designer in terms of the semantical network. For instance if some plaintext is needed, in principle every combination of a ciphertext and key can be considered as a subgoal and a required item. As a matter of fact this will increase the computational complexity dramatically. The designer should avoid this by using a more appropriate formulation of the scheme.

3.3 The control mechanism and consultation sessions

The control mechanism performs inferences on the basis of this knowledge base in order to solve a problem requested by the user. It contains the relevant information concerning which part of the knowledge base and what relations should be addressed. The reasoning process is top-down or goal driven. The control mechanism checks in the global database what information is known and what is the ultimate goal (e.g. finding a secret key). With the help of the knowledge base it tries to achieve subgoals, which are relevant and conditional in order to achieve the ultimate goal and which are stored into the global database. Then it starts again in order to obtain these new subgoals by inspecting the knowledge base etc. This process continues untill one of the two following possibilities occur.

The first possibility is clear: the expert system stops if it achieves the goal. This means that the ultimate goal defined by the designer (e.g. a secret session key) can be achieved. The way in which this goal can be obtained can be analyzed by the designer on the basis of the information concerning the followed reasoning process as stored in the global database.

The second possibility is that the ultimate goal will not be achieved, wheras all available knowledge in the knowledge base was applied by the expert system. Clearly, in the case of a well-designed key management scheme this situation should occur. Nevertheless, consultation of the global database, especially with respect to the required subgoals, is important since this information represents all the possible weaknesses of the key management scheme to the designer. It gives a survey of conditions to be satisfied in order that the ultimate goal would be reached.

4 Applications

The expert system has been tested on some well-known key management schemes like the IBM key management scheme, the Needham-Schroeder protocol, the EFTPOS key management scheme etc., in order to assess the quality of the expert system and in order to guarantee that it can obtain at least the same results as mentioned in literature. The results obtained with respect to the well-known key management schemes mentioned above are compatible with those mentioned in [2] and [3]. The average running time was 20 seconds, where it should be taken into account that until now no attention has been paid to speed up the process.

With respect to the IBM key management scheme it is mentioned that contrary to [2] 5 pairs of ciphertexts were generated as possible weaknesses which if available can lead to the session key ks. An example of such a pair is $KM1(N)$ and $KM0(N)$, which are the encipherments of an arbitrary data block N, which is encrypted by the master keys KM0 and KM1, respectively. Another pair is $N(KT)$ and $KM0(N)$, where KT is the transaction key. On the basis of these pairs the session key KS can be obtained. In fact these results can be considered as generalizations of the results obtained in [2] as follows by reading KS or KT instead of N. Although from a cryptanalytical point of view deduction of $KM1(N)$ from an arbitrary $KM0(N)$ or vice versa as needed in the case of the first pair) is equally difficult as deduction of $KM1(KS)$ from public data block $KM0(KS)$ or deduction of $KM0(KT)$ from public data block $KM1(KT)$, the cryptanalist is in principle in a more comfortable position. Due to the fact that some required pairs are independent of KS, which may have a rather short life-time, more time is available to determine $KM1(N)$ from $KM0(N)$ than to determine $KM1(KS)$ from $KM0(KS)$.

The system was also applied to the one-way identity-based key distribution system depending on physical security of IC cards, as introduced in [1] . A personal computer network is assumed which is accessible for the user by means of IC-cards. In the scheme of [1] a master key MK is stored in all IC-cards. The result of a modulo-2 addition of this master key and the two identity numbers IDi and IDj, which are related to user i and user j is a key-encrypting key Kij ,which is used for encryption of some work key WK. Within the terminal the work key WK is available in plaintext, which enables analysis of the random generator which generates the work keys. Among others it can be concluded from the analysis by the expert system that if WK has certain specific values (e.g. WK equal to Kij, which can be tested by comparing the encrypted work key WK(WK) with Kij(WK)) the master key MK can be obtained. Of more interest is the situation where the Okamoto scheme is modified in such a way that the work key WK is kept secret, e.g. similar to the way in which the session key KS in the IBM scheme is kept secret. Actual results on the analysis of the (modified) Okamoto scheme by the expert system will be published in a forth-coming paper; it is beyond the scope of the present paper which is more related to the structure of KEYMEX.

5 Conclusions

One of the main objectives of the KEYMEX project is to obtain insight into the possibilities to develop an expert system which can assist the designer of key management schemes. Moreover, we are interested in the usefulness of especially semantical networks for knowledge representation within such an expert system. Untill now the results with KEYMEX are promising.
With the help of the expert system possible weaknesses of key management schemes can easily be detected.

With respect to the flexibility of the system, which was one of the major requirements to the system design, it seems that key management schemes can easily be tested by the system. Furthermore, it is concluded that in general it does not take much effort to translate key management schemes to be tested into a form suitable to the expert system. However, implementation of a suitable description of the Needham-Schroeder protocol, which was one of the protocols that was tested, required more effort than e.g. the IBM key management scheme. This is due to the fact that within that protocol messages are consisting of a varying sequence of subfields encrypted with a key. The present system is insufficiently adjusted to data blocks of variable length. However, the main reason for this increased effort is that in the Needham-Schroeder protocol the aspect of time plays an important role. The messages which can be transmitted or received at a certain moment are conditioned by the dimension of time. At this moment methods are under development which can simplify the incorporation of such protocols in the expert system.

Acknowledgement

The authors would like to thank ir. H. de Swaan Arons and P.J.L. van Beek (both with Delft University of Technology) for their assistance with respect to the development of the KEYMEX system.

References

[1] Identity-based information security management for personal computer networks, E. Okamoto, K. Tanaka, IEEE Selected Areas in Communications, Vol. 7, no 2, February 1989, pp. 290-294

[2] Expert Systems applied to the analysis of key management schemes, D. Longley, Computers and Security, 6, 1987, pp. 54-67

[3] The Interrogator: protocol security analysis, J.K. Millen, S.C. Clark, S.B. Freedman, IEEE Software Eng.,Vol. SE-13, no 2, February 1987, pp. 274-288

[4] Automated analysis of cryptograms, B.R. Schatz, Cryptologia, April 1977, pp. 116-142

[5] The automated analysis of substitution ciphers, J.M. Caroll, L. Robbins, Cryptologia, Vol. 10, no 4, October 1986, pp. 193-209

[6] The automated cryptanalysis of polyalphabetic ciphers, J.M. Caroll, L. Robbins, Cryptologia, Vol. 11, no 4, October 1987, pp. 193-205

SECTION 4

AUTHENTICATION

On The Formal Analysis of PKCS Authentication Protocols *

Klaus Gaarder   Einar Snekkenes [†]

Abstract

In the quest for open systems, standardisation of security mechanisms, framework and protocols are becoming increasingly important. This puts high demands on the correctness of the standards.

In this paper we use a formal logic based approach to protocol analysis introduced by by Burrows, Abadi and Needham in their paper "Authentication: A Practical Study in Belief and Action" [1]. We extend this logic to deal with protocols using public key cryptography, and with the notion of "duration" to capture some time related aspects. The extended logic is used to analyse an important CCITT standard, the X.509 Authentication Framework. Two claims relating to the assumptions necessary and the goals achieved using strong two-way authentication are proved.

We conclude that protocol analysis can benefit from the use of the notation and that it highlights important aspects of the protocol analysed. Some aspects of the formalism need further study.

*Research sponsored by Royal Norwegian Council for Scientific and Industrial Research under Grant IT 0333.22222

†Both Authors are with Alcatel STK Research Centre, Secure Information Systems, P.O. Box 60 Økern, N-0508 Oslo 5, Norway, Tel. +47 2 63-88-00, e-mail: ...!mcvax!ndosl!stku01!snekkene, ...!mcvax!ndosl!stku01!gaarder

1 Introduction

Michael Burrows, Martín Abadi and Roger Needham recently published a paper suggesting a notation for the formal analysis of protocols [1]. The notation given in [1] is restricted to Symmetric Key Crypto Systems(SKCS), and excludes the explicit mentioning of time.

Public Key Crypto Systems (PKCS) [6], such as RSA [12] seem to be appropriate when considering authentication in open systems. This is recognized by ISO and CCITT [3]. Thus, one would expect that a formalism for analysing authentication protocols not to be restricted to SKCS.

We have extended [1] to cater for PKCS and a slightly generalized notion of timestamp. The use of the extended formalism is demonstrated on a non-trivial example, namely a version of the CCITT X.509, The directory - Authentication Framework, strong two-way authentication (STA).

Our first attempt to prove some properties of X.509 using [1] extended with notation to cater for PKCS failed. To prove some reasonable goals of X.509, we had to make unreasonable assumptions. Our problem occured as a result of limitations in [1] making us unable to capture the timestamp related mechanisms of X.509.

The rest of the section gives a summary of the formalism introduced in [1], presenting the notation for stating assumptions, goals and messages, the rules for proving properties of assumptions and goals, and the rules for constructing proofs.

1.1 Goals, Assumptions and Messages

We may express the goal of our protocol analysis as some theorem of the form $\{X\}\ S\ \{Y\}$ where S is the list of protocol steps, X is the collection of initial assumptions and Y is the goal of the protocol. Goals, assumptions and messages consist of formulae constructed by means of the symbols listed below.

Let X denote some formula and for all i, let U_i denote some communicating entity. Below we list the well-formed formulae together with their informal semantics.

$U_i \models X$ U_i *"believes"* X. If $U_i \models X$ holds, then U_i will behave as if X is true in her interpretation.

$U_i \mid\sim X$ U_i *"once said"* X. $U_i \mid\sim X$ typically holds if U_i has sent a message Y having X as a subformula.

$U_i \Rightarrow X$ U_i *"has jurisdiction over"* X. If you believe U_i has jurisdiction over X and that U_i believes X, then you ought to believe X.

$U_i \lhd X$ U_i *"sees"* X. Typically U_i sees X if she has received a message Y, where X is a subformula of Y.

$\#(X)$ X *"is fresh"*. A formula is fresh if it has not previously occured as the subformula of some sent formula. A formula generated for the purpose of being fresh is usually called a *nonce*.

$U_i \xleftrightarrow{K} U_j$ U_i and U_j *"share the good crypto key"* K, where K is a symmetric crypto key.

$\{X\}_K$ *signed* U_i X *"encrypted under the symmetric key "* K *"by "* U_i. Notice that although both parties sharing the key may produce a signed message, the formula only holds if U_i was in fact the one which signed the message.

1.2 Axioms and Inference Rules

Below we list the inference rules of the original logic. Assume $X_1, \ldots, X_n, Y$ are formulae as above, then the rules below have the following format:

$$\text{Rm} \quad \frac{X_1, \ldots, X_n}{Y}$$

read as "if X_1 and ... and X_n then Y". Basically, the rules make explicit the intuitive consequences of the above given semantics.

It is worth noting that one assumes the existence of an implicit rule:

$$\text{R0} \quad \frac{\Psi}{\Psi^L_{L'}}$$

where L and L' are permuted lists of formulae, and Ψ is some formula containing the formula list L, and $\Psi^L_{L'}$ is Ψ with L' substituted for L.

Let U_i, U_j denote communicating entities, X, Y denote formulae, and K a symmetric cryptokey. Assuming you see a message encrypted under some good key of a SKCS, then you ought to believe that the unencrypted message was uttered by the only other entity knowing the key. This is formalized as:

$$\text{R1} \quad \frac{U_i \models U_j \xleftrightarrow{K} U_i, U_i \vartriangleleft (\{X\}_K \text{ signed } U_j)}{U_i \models U_j \hspace{-0.2em}\mid\hspace{-0.5em}\sim X}$$

Note that entities are not allowed to change their beliefs during a run (R2).

$$\text{R2} \quad \frac{U_i \models \#(X), U_i \models U_j \hspace{-0.2em}\mid\hspace{-0.5em}\sim X}{U_i \models U_j \models X}$$

If you believe U_j has jurisdiction over X, and you believe that U_j believes X, then you believe X (R3).

$$\text{R3} \quad \frac{U_i \models U_j \Rightarrow X, \; U_i \models U_j \models X}{U_i \models X}$$

The rules R4 ...R7, R10 make it possible to break up and aggregate formulae. Their intuitive justification should be obvious.

$$\text{R4} \quad \frac{U_i \models X, U_i \models Y}{U_i \models X, Y} \qquad \text{R5} \quad \frac{U_i \models (X, Y)}{U_i \models X}$$

$$\text{R6} \quad \frac{U_i \models U_j \models (X, Y)}{U_i \models U_j \models X} \qquad \text{R7} \quad \frac{U_i \models U_j \hspace{-0.2em}\mid\hspace{-0.5em}\sim (X, Y)}{U_i \models U_j \hspace{-0.2em}\mid\hspace{-0.5em}\sim X} \qquad \text{R10} \quad \frac{U_i \vartriangleleft (X, Y)}{U_i \vartriangleleft X}$$

The order in which the communicating entities of the *"share the good crypto key"* triple are listed is immaterial (R8,R9), reflecting the symmetry of a SKCS.

$$\text{R8} \quad \frac{U_i \models R \xleftrightarrow{K} R'}{U_i \models R' \xleftrightarrow{K} R} \qquad \text{R9} \quad \frac{U_i \models U_j \models R \xleftrightarrow{K} R'}{U_i \models U_j \models R' \xleftrightarrow{K} R}$$

If you share a key with someone you may read her messages if they are encrypted using the shared key (R11).

$$\text{R11} \quad \frac{U_i \models U_j \xleftrightarrow{K} U_i, U_i \vartriangleleft \{X\}_K \text{ signed } R}{U_i \vartriangleleft X}$$

Clearly if you haven't seen any message X, you cannot have seen any messages where X occured as a subformula. In particular we have:

$$\text{R12} \quad \frac{U_i \models \#(X)}{U_i \models \#(X,Y)}$$

There are other rules, but the above are the most commonly used ones.

1.3 Protocol Annotation

When analysing a protocol, one first derives an *idealized* version of the protocol S, then state assumptions A and goals B, and finally attempt (and hopefully succeed) to prove $\{A\}S\{B\}$ as a theorem using the annotation rules below. An idealized protocol consists of a sequence of steps, each on the form $(U_1 \rightarrow U_2 : X)$, where U_1, U_2, X is sender, recipient and message respectively. The message X is expressed as a formula in the above notation.

There are four annotation rules (A1 ... A4). The annotation rules are presented in the familiar Hoare style [10]. The first rule states that the effect of executing a protocol step is that the sent message becomes visible to the recipient.

$$\text{A1} \quad \vdash \{Y\}(U_i \rightarrow U_j : X)\{Y, (U_j \lhd X)\}$$

The collection of annotated protocols is in a certain sense transitive, as shown below.

$$\text{A2} \quad \frac{\vdash \{X\}S_1 \ldots \{Y\}, \vdash \{Y\}S_1' \ldots \{Z\}}{\vdash \{X\}S_1 \ldots \{Y\}S_1' \ldots \{Z\}}$$

Intermediate assertions and the conclusion may be weakened(A3).

$$\text{A3} \quad \frac{\vdash \{W\}S_1 \ldots \{X_1\} \ldots \{X_n\}, X_i \vdash X_i'}{\vdash \{W\}S_1 \ldots \{X_1\} \ldots \{X_i'\} \ldots \{X_n\}}$$

We may always make consequences of the assumption explicit.

$$\text{A4} \quad \frac{\vdash \{X\}S \ldots \{Y\}, X \vdash X'}{\vdash \{X, X'\}S \ldots \{Y\}}$$

2 The Extensions

In this section we describe the modifications needed in the logic to be able to reason about PKCS and certain aspects of time. For each, we give the necessary symbols, their informal semantics and the corresponding inference rules.

2.1 Public Key Crypto Systems

In a PKCS, we assume the existence of a nonempty collection of communicating entities. Each entity has associated one public and one private key. We only consider signature properties, leaving secrecy issues aside. To check that a message m was signed by U, it is sufficient to know U's public key. When using this approach, it is paramount that U's public key is genuine. However, to sign some message with U's signature, one must be in possession of U's private key. Usually, only U knows U's private key. Below we list the symbols together with their informal semantics.

$\mathcal{PK}(K,U)$ The entity U *"has associated the good public key"* K. Thus, there exists some unique key corresponding to K.

$\Pi(U)$ The entity U *"has associated some good private key"*. Consequently, the key value is only known by U.

$\sigma(X, U)$ The formula X *"signed with the private key belonging to"* U.

The above informal semantics lead to the rule below

$$R13 \quad \frac{U_i \models \mathcal{PK}(p_j, U_j), U_i \models \Pi(U_j), U_i \lhd \sigma(X, U_j)}{U_i \models U_j \hspace{-0.3em}\sim X}$$

Thus, to believe that U_j has once said X, it is sufficient to believe that we have U_j's public key, that U_j's secret key is good and we must see X signed with U_j's private key.

The content of a signed message can always be made visible.

$$R14 \quad \frac{U_i \lhd \sigma(X, U_j)}{U_i \lhd X}$$

2.2 Time

Using the notion of freshness described above, we may well have a fresh formula which was generated at some arbitrary time in the past. Thus, a fresh message might still be very old. Similarly, a very recent message is only fresh if it hasn't been sent before. Thus, we stipulate that recency and freshness are different concepts. The authors of [1] claim that their logic benefits from avoiding the explicit notion of time. As previously stated, one has to include the notion of time to be able to prove certain properties of X.509. Thus, insisting on using [1] to certify protocols removes an important mechanism from the protocol designers repertoire, with the effect that protocols may become unnecessary complex and hard to design.

A timestamp added to a message usually states the time the message was generated [5]. It seems reasonable to assume that a timestamped message was in some sense *good* at the time of generation. From this we develop the notion of a durationstamp. A message might thus be tagged with the durationstamp denoting the time interval during which its creator claims the message is good.

The symbols needed are listed below together with their informal semantics.

$(\Theta(t_1, t_2), X)$ X *"holds in the interval"* t_1, t_2. Thus, the creator which uttered the timestamped message X, claims that X is, or was good in the time interval between t_1 and t_2.

$\Delta(t_1, t_2)$ t_1, t_2 *"denotes a good interval"* Thus, the local unique Real Time Clock (RTC) shows a time in the interval between t_1 and t_2.

Obviously we may model a timestamp, viz. $(\Theta(t, t), X)$. Our semantics above is based on the following assumptions:

1. A process making the explicit reference to some RTC in some message using a durationstamp, does so with respect to his own unique local RTC.

2. The duration of any protocol run must be short.

3. When tagging a message subsequently uttered in a protocol step using a durationstamp, e.g. $(\Theta(t_1, t_2))$, one commits oneself to believe the message X to be OK in the time interval specified by t_1 and t_2.

Recall that formulae of [1] are only assumed to stay true for the duration of a single run of the protocol. However, using the above, one is able to commit oneself for future runs. Later, it is shown that this is exactly what is needed when reasoning about the certificates of X.509.

Based on the assumptions above, we give a rule for reasoning about durationstamps.

$$R15 \quad \frac{P \models Q \models \Delta(t_1, t_2), P \models Q \hspace{0.2em}\sim\hspace{-0.9em}\mid\hspace{0.2em} (\Theta(t_1, t_2), X)}{P \models Q \models X}$$

The rule states that when uttering a durationstamped message, one commits oneself to believe the message for the interval specified by the durationstamp. If one uses Coordinated Universal Time (UTC time) [1], or synchronized clocks, the first premise of R15 may be established by inspecting the local RTC.

3   An Informal Description of X.509

The central items in the X.509 messages are the data structures known as *certificate* and *token* which are the basis of authentication together with *distingushed name*. The X.500 directory is a database.

Certification Authority (CA):   A trusted entity maintaining security information in the directory.

Distinguished name: Each entitiy has a unique *distinguished name* which is allocated by the *naming authority*. The naming authority will usually equal the *Certification Authority*.

Signature (as defined in the X.509 standard) :   Signing a message M is performed as encrypting a hashed version of M under the signer's *private* key and appending it to the message M,

$$\langle M \rangle_U \stackrel{def}{=} M \cdot E(Z_U, h(M)) \tag{1}$$

where '.' denotes concatenation, E is the encryption function, h is a hash function and Z_U is the private key of the signer U. [2]

Certificate: The certificate is a data structure signed by the *certification authority*, connecting a distinguished name to a *public key* of some PKCS. Its structure is as follows :

let A be an algorithm identifier, I be the name of the *issuer*, δ be a *duration* period consisting of two dates (from , to), U the name of the *owner* and P_U the *public key* of U, then the certificate of U is

$$C_U = \langle (A, I, \delta, U, P_U) \rangle_I \tag{2}$$

---

[1]See CCITT Recommendations X.208 and X.209.

[2]The hash function suggested for use in [3] has recently been shown to contain weaknesses [4].

Token: The token is a signed data structure of the following form where t_i, U_i, R_i, U_j, R_j denotes the *generation time* and *expiry date* in UTC time, name of the sender, a random number generated by U_i, name of the receiver and a random number generated by U_j respectively. In addition to these data elements, the token in its most general form contain two further pieces of data: S and $E(p_j, \mathcal{E})$ ($\mathcal{E}$ encrypted with the receiver's public key). S is called 'sgnData' and is a digital signature of some message, and $\mathcal{E}$, called 'encData' may be a secret key for some SKCS, typically a DES key. The token has two forms:

$$T_{i,j} = \langle (t_i, R_i, U_j, S, E(p_j, \mathcal{E})) \rangle_{U_i} \tag{3}$$

(cfr. X.509 9.2.2) or

$$T'_{i,j} = \langle (t_i, R_i, U_j, R_j, S, E(p_j, \mathcal{E})) \rangle_{U_i} \tag{4}$$

The form (4) of the token is used in the reply to the first form (3), X.509-9.3.5.

Basic Idea of X.509 By possessing the public key of the certification authority, the entities in question will be able to verify the signature on the certificates and thus be convinced that it contains the good public key of the desired communication partner. By using this good public key each entity will verify the token received from the other, which was produced using the private key of the sender, and thus authenticating the sender.

The precise role of the certificates is thus to supply trusted information to the authenticating parties (*their respective public keys*) which in turn will make them able to do the signature verification which constitutes the authentication proper. When analysing a protocol such as X.509 we might argue that the certificate is not an integral part of the *authentication* protocol, since the process of authentication consists of the token exchange and verification.

Some workers have chosen this approach [2], but important aspects of X.509 are then lost.

We consider only what is known as "two-way strong authentication". The other modes are "one-way" and "three-way".

Without loss of generality the certification path is considered to be of length one. This may be done since a certification path is simply a sequence of certificates [3]. Then each user need only one certificate, and this is supplied to both by the same certification authority (CA).

3.1 Strong Two-Way Authentication

In order to facilitate later notation we "rename" the participants U_1, U_2 and U_3, where U_1 and U_2 are the parties seeking to authenticate (A and B) and U_3 the Certification Authority (C).

The following messages are passed:

Message 1: $U_1 \longrightarrow U_3 : (U_1, U_2)$ This is the *certificate request*.

Message 2: $U_3 \longrightarrow U_1 : (C_1, C_2)$ The CA forwards U_1's and U_2's certificates to U_1. At this point the only interesting thing to happen is that U_1 checks that the signature of U_3 on C_2 is OK, by using the public key of U_3. If the signature check succeeds then U_1 proceeds.

---

[3] A certification path is defined in [3], to establish a path of trusted points in the DIT, cfr. [3], sect. 7.5.

Message 3: $U_1 \longrightarrow U_2 : (C_1, T_{1,2})$ U_1 forwards her certificate to U_2 together with a *token* of the form (3) above. On receipt of this message U_2 will check the signature on C_1, and if it is OK proceed to check the signature on $T_{1,2}$. If U_2 successfully applies U_1's public key (received in C_1) to verify the signature on T_1, he will proceed to check the validity of the contents.

Message 4: $U_2 \longrightarrow U_1 : (T'_{2,1})$ This is U_2's reply, a token of the form (4). U_1 performs the same check as U_2 did above (also checking the returned part of her own token).

4 The Logic of X.509

The analysis of a particular protocol may be considered to give rise to a theory of its own in the following sense. A logical theory consists of a language, a set of *logical* axioms, a set of inference rules and a set of *non-logical axioms*, see any textbook on logic for details (e.g. [7]). If we consider the assumptions α_i as our non-logical axioms, then we have a theory for each set of such assumptions. Let $\mathcal{P}$ denote the protocol, α the assumptions and γ the goals of the protocol. The analysis of a protocol then consists of proving the theorem Φ

$$\Phi = \{\alpha\}\mathcal{P}\{\gamma\} \tag{5}$$

in the Hoare-style notation. In the formalism of [1] this is easily seen to reduce to proving

$$\alpha, Y \vdash \gamma \tag{6}$$

where Y is the set consisting of the following formulae

$$Y_{il} = U_i \lhd X_l \tag{7}$$

where X_l is the formula sent to U_i in the l'th message during the run of the protocol (or rather *checking* if Φ is a theorem). That the formulae Y_{il} are on the form above is easily seen from the basic annotation axiom.

4.1 The Formalised Assumptions

In a protocol one often include *names* of the principals in certain messages, like in

$$U_i \longrightarrow U_j : (U_j, X) \tag{8}$$

where *the semantics of the informal protocol itself* is such that the inclusion of the name U_j in the message is interpreted as "U_j is the intended recipient of X".(In the X.509 protocol this is explicitly stated as one of the goals achieved by the protocol.)

To handle this we introduce a special symbol for the notion of "recipient" of a message. We write

$$\mathcal{R}(U_i, X) \tag{9}$$

to say that "U_i *is the intended recipient of the message* X". From the X.509 text it is clear that each principal must be assumed to have some kind of jurisdiction over statements involving this construction, at least in the following sense. If U_i believes that U_j has said $\mathcal{R}(U_i, X)$ this will lead U_i to believe that X is indeed meant for her. However we will not allow U_i to conclude anything if what U_j said was $\mathcal{R}(U_k, X), k \neq i$. That is, you only accept as meant for you the messages which contains your name in the recipient field. Thus, you cannot conclude that a message isn't meant for you. [4]

$\mathcal{R}$ may be thought of as a "tag" on a parcel. The reasons for introducing $\mathcal{R}$ can be illustrated with an example from every day life. If you receive an invoice in your mailbox (be it electronic or otherwise) your further action will clearly depend upon wether you were the intended recipient of this invoice or not! Thus, it is a clear need to formalise this notion.

Adding R to the logic, we may still use the inference rules described above. In particular the propagation of freshness also applies to $\mathcal{R}$.

4.1.1 The X.509 Assumptions

We choose to introduce all assumptions needed in the proof at once rather than introducing them along the way.

We recognize 7 assumptions in the X.509 protocol. The assumptions are stated based on the text in the standard [3]. Some are clearly expressed and some have to be read between the lines, and from the intuitive semantics of the protocol design. The assumptions in formal guise are denoted by $\alpha_1, \ldots, \alpha_7$. Recall that $U_1 = A, U_2 = B$ and $U_3 = C$. Let $i, j \in \{1, 2, 3\}$, then we have:

All participants believe that *everyone keeps their private key private*, including their own, X.509-6.4.

$$\alpha_1 : U_i \models \Pi(U_j) \tag{10}$$

All U_i trust all U_j if they say 'U_i is the recipient of X'. i.e. if I receive something with my name on it *is really meant for me* (X.509-9.1.2.a.1, 9.1.2.b.1),

$$\alpha_2 : \{U_1 \models U_2 \Rightarrow \mathcal{R}(U_1, X), \ U_2 \models U_1 \Rightarrow \mathcal{R}(U_2, X)\} \tag{11}$$

Both U_1 and U_2 believe they have the correct public key of U_3, the authority (X.509-7.1.).

NOTE: We suppress the fact of how U_1 and U_2 got hold of p_3.

$$\alpha_3 : \{U_1 \models \mathcal{PK}(p_3, U_3), \ U_2 \models \mathcal{PK}(p_3, U_3)\} \tag{12}$$

U_1 and U_2 trust the certification authority on (public-key , name) pairs (X.509-7.1,X.509-7.2).

$$\alpha_4 : \{U_1 \models U_3 \Rightarrow \mathcal{PK}(p_2, U_2), U_2 \models U_3 \Rightarrow \mathcal{PK}(p_1, U_1)\} \tag{13}$$

$U_1(U_2)$ believes the existence of a nonce (X.509-9.3.6.c, 9.3.3.d respectively).

$$\alpha_5 : U_1 \models \#(N_2) \tag{14}$$

$$\alpha_6 : U_2 \models \#(N_1) \tag{15}$$

Both U_1 and U_2 believe that the certification authority believes that the duration δ_j of U_j's certificate is still good, i.e. that U_3 will not deliver certificates with invalid duration periods, X.509-7.2. This is a critical assumption.

$$\alpha_7 : U_i \models U_3 \models \Delta(\delta_j) \tag{16}$$

---

[4]Note that if the logic is complete in the usual sense (i. e. $T \nvdash \phi \Rightarrow T \vdash \neg\phi$) then we could conclude from the failure to prove beliefs to the negation of these beliefs.

Comments on the assumption α_7 Since the certificate does not contain any *timestamp* or random number making it a *nonce* (i.e. having the "fresh" property) we will not be able to use the rule of jurisdiction to deduce

$$U_i \models \mathcal{PK}(p_j, U_j)$$

from the assumptions α_1–α_6. Thus, the X.509 protocol would seem not to achieve its primary goals. There are at least two ways to remedy this.

One way is to insert a timestamp in the certificate in addition to the duration, making it necessary to fetch the certificates from the CA, with a fresh time stamp, each time they are needed. Essentially this means generating a new certificate each time a request is made. This is exactly the problem one seeks to avoid for performance reasons. It is also possible that this would not solve the problem, the CA might stamp a bad certificate due to lack of knowledge. This is the "rotten fish in todays newspaper" problem, it is still not fresh! To go for this solution would be to say that *the formalism* as it stands captures every aspect of the protocol, and claim that there is a deficiency in *the protocol*.

Another possible solution is the one we have outlined. We argue that the duration period certainly increases the security of the protocol, and that the formalism could not capture this notion of time dependence. Therefore we chose to include the assumption α_7, and the concept of a "good duration period", $\Delta(\delta)$. Since all participants are assumed to use UTC time, it is easy to check $\Delta(\delta)$. Note that *clocks have to be trusted if this is to provide additional security*.

When trying to prove $\Gamma_1, \ldots, \Gamma_8$ without the Δ relation, one might be misled to assume that certificates have the "fresh" property. This is fundamentally wrong since certificates are expected to be used in a large number of protocol runs.

This also reflects the fact that the "fresh" notion is not basically a measure of time, just a stating of the fact that something has not been observed in any previous run of the protocol.

4.2 The Idealized Protocol

The formalisation of the certificate and token are critical aspects of the analysis. The key to the formalisation is the implicit meaning of each message and its contents.

Certificate The important features of the certificate are

- the signature of the certification authority
- the duration
- the name of the owner
- the public key of the owner

Since δ_i is the duration of C_i we write $\Theta(\delta_i)$ for $\Theta(t_i, t_i')$. Then we give the certificate the following formalisation (where $\mapsto$ is read "maps to"):

$$C_i \mapsto \sigma((\Theta(\delta_i), \mathcal{PK}(p_i, U_i)), U_3), i = 1, 2 \tag{17}$$

Token As with the certificate the important features are reflected in the formalisation, these are

- the signature of the issuer
- the timestamp (or random number), expiry date
- the name of the intended recipient

The timestamp and random number should give the token the properties of a nonce. The X.509 document [3] is unclear on this point. Our interpretation is that the token will have the properties of a nonce.

Let N_i be the entire datastructure in the token as defined in (3) or (4) above.

Then we formalise the tokens as :

$$T_{i,j} \mapsto \sigma((N_i, \mathcal{R}(U_j, N_i)), U_i) \tag{18}$$

or

$$T'_{i,j} \mapsto \sigma((N_i, N_j, \mathcal{R}(U_j, (N_i, N_j))), U_i) \tag{19}$$

Note that we have formalised the *name* of the recipient of a token as $\mathcal{R}(U_j, N_i)$ and $\mathcal{R}(U_j, (N_i, N_j))$, to signify the meaning of this name in the token.

4.2.1 Idealizing The Protocol

Exactly how we formalise message 1 (X.509-7.7) is of little importance to the subsequent analysis. The formalisation of messages 2 (X.509-7.7) and 3 (X.509-9.3.1, 9.3.2) are given by (17) and (18) above, message 4 (X.509-9.3.4, 9.3.5) is given by the form of the token in (19).

Message 1

$$U_1 \longrightarrow U_3 : (U_1 \hspace{0.5mm}\vdash\hspace{-1.5mm}\sim (U_1, U_2)) \tag{20}$$

Message 2

$$U_3 \longrightarrow U_1 : \sigma((\Theta(\delta_1), \mathcal{PK}(p_1, U_1)), U_3), \sigma((\Theta(\delta_2), \mathcal{PK}(p_2, U_2)), U_3) \tag{21}$$

Message 3 The idealization of this message is critical, and requires some explaining. Concerning checking of certificates, [3] states that U_2 will check the validity of U_1's certificate. Although [3] says nothing explicitly, we find it reasonable to assume that this applies to U_1 as well, i.e. U_1 checks the validity of U_2's certificate before proceeding with message 3. She would *not* proceed if U_2's certificate was invalid, so if she proceeds this is an implicit statement of the following

$$U_1 \models \mathcal{PK}(p_2, U_2)$$

which is in fact conveyed to U_2 via U_1's token if and when U_1 proceeds in the protocol. Arguing this way leads us to the following formalisation of message number 3:

$$U_1 \longrightarrow U_2 : \sigma((\Theta(\delta_1), \mathcal{PK}(p_1, U_1)), U_3), \sigma((\mathcal{R}(U_2, N_1), \mathcal{PK}(p_2, U_2)), U_1) \tag{22}$$

Message 4 Arguing as above for message 4 gives:

$$U_2 \longrightarrow U_1 : \sigma((\mathcal{R}(U_1, (N_2, N_1)), \mathcal{PK}(p_1, U_1)), U_2) \tag{23}$$

4.3 The Goals of Authentication in X.509

Burrows, Abadi and Needham suggest a set of goals for authentication which any protocol is evaluated against. This might lead one to believe that it is possible to formulate general goals of authentication regardless of the mechanisms used to achieve it. The goals presented by [1] are strictly related to SKCS, since they are formulated as statements about *shared symmetric crypto keys*. Such a notion neither exists in a PKCS nor in a Zero Knowledge Interactive Proof system (ZKIP), see e.g. [8], [9]. In fact, in a ZKIP based protocol no key in the usual sense need exist at all.

We have taken a slightly different approach. Every protocol will be designed to achieve certain goals, and the user of any protocol should be aware of this. So if a protocol design claims to achieve only this much then we can't expect it to achieve more, but we do expect it to achieve what it claims! Until we have an agreed version of the meaning of *authentication*, authentication protocols may be expected to achieve slightly differing goals. We have approached the analysis of X.509 by formalising the goals which [3] claims to achieve, and trying to prove these.

Formal goals of X.509 [5]

These are the goals of the X.509 strong two-way authentication formalised:

$$\Gamma_1 : U_1 \models \mathcal{PK}(p_2, U_2) \tag{24}$$

$$\Gamma_2 : U_2 \models \mathcal{PK}(p_1, U_1) \tag{25}$$

$$\Gamma_3 : U_1 \models U_2 \models \mathcal{PK}(p_1, U_1) \tag{26}$$

$$\Gamma_4 : U_2 \models U_1 \models \mathcal{PK}(p_2, U_2) \tag{27}$$

$$\Gamma_5 : U_1 \models ((U_2 \hspace{1mm}\mid\sim (\mathcal{R}(U_1, N_2), N_2)), \#(\mathcal{R}(U_1, N_2))) \tag{28}$$

$$\Gamma_6 : U_2 \models ((U_1 \hspace{1mm}\mid\sim (\mathcal{R}(U_2, N_1), N_1)), \#(\mathcal{R}(U_2, N_1))) \tag{29}$$

$$\Gamma_7 : U_1 \models \mathcal{R}(U_1, N_2) \tag{30}$$

$$\Gamma_8 : U_2 \models \mathcal{R}(U_2, N_1) \tag{31}$$

Informal semantics of Γ_1–Γ_8

Γ_1–Γ_2: After a run of the protocol each of the parties should believe that they have a valid public key belonging to the other.

Γ_3–Γ_4: The beliefs of the previous goals should be mutual, i.e. not only should you be sure about your communication partner, but you should be sure that he is sure that you're sure ... .

---

[5]The formulation of these goals may be found in the sections of [3] as follows: $\Gamma_1 - \Gamma_4$: 7.1, Γ_5: 9.1.2.b.1, 9.1.2.b.2, Γ_6: 9.1.2.a.1, 9.1.2.a.3, Γ_7: 9.1.2.b.1, Γ_8: 9.1.2.a.2.

Γ_5–Γ_6: These goals are less intuitive. They state the following fact: Each user U_i should end up believing that U_j recently said that U_i is the intended recipient of N_j(viz. the token of U_j), i.e. that the tokens received were produced for U_i by U_j during the current run.

Γ_7–Γ_8: Both should end up believing they are the legitimate receivers of the respective tokens.

We notice the similarity of the goals Γ_1–Γ_4 with the goals presented in [1].

4.4 Proof outline

Conducting the proof itself is rather easy once the assumptions, goals and the idealized protocol are established. At each step only a few inference rules are applicable, making it easy to decide which rules to apply.

The all important thing to note in the proof outline is where and how the different goals appear, and the use of the assumptions. It turns out that the assumptions concerning time aspects are critical. As we have seen, the use of time in X.509 is one of the primary causes for the need to extend the logic.

We will state the following claims:

Claim 1 *With the assumptions $\alpha_1, \dots, \alpha_7$ the above idealized protocol attains the goals $\Gamma_1, \dots, \Gamma_8$.*

Claim 2 *The set $\alpha = \{\alpha_1, \dots, \alpha_7\}$ is minimal in the sense that, with no set $\beta \subseteq \alpha$ will the idealized protocol attain all goals $\Gamma_1, \dots, \Gamma_8$.*

The proof outline will concentrate on claim 1, the establishment of the goals Γ_1–Γ_8.

Claim 2 can be justified by noting that all assumptions are eventually used in the proof of claim 1, and that the assumptions are logically independent.

Goals Γ_1–Γ_2

$\Gamma_1 : U_1 \models \mathcal{PK}(p_2, U_2)$

$\Gamma_2 : U_2 \models \mathcal{PK}(p_1, U_1)$

These goals are proven using the result of messages 2 and 3, which are

$$U_1 \lhd \sigma((\Theta(\delta_2), \mathcal{PK}(p_2, U_2)), U_3) \qquad (32)$$

$$U_2 \lhd \sigma((\Theta(\delta_1), \mathcal{PK}(p_1, U_1)), U_3) \qquad (33)$$

by using the assumptions $\alpha_1, \alpha_3, \alpha_4, \alpha_5$ and α_7 with the proper instances of U_i, δ_j, p_j.

Without the rule R15 and the assumption that the certification authority does not hand out certificates with bad duration (α_7) we could not have proved this.

Goals Γ_3–Γ_4

$\Gamma_3 : U_1 \models U_2 \models \mathcal{PK}(p_1, U_1)$

$\Gamma_4 : U_2 \models U_1 \models \mathcal{PK}(p_2, U_2)$

Obviously the arguments for Γ_3 proceeding from the result of message 4, and for Γ_4 proceeding from message 3 are completely similar, so we use a general form.

Using assumptions $\alpha_1, \alpha_5, \alpha_6$ we first prove

$$U_i \models (U_j \mathrel{\vdash} (\mathcal{R}(.), \mathcal{PK}(p_i, U_i))) \tag{34}$$

by rule R13. Using R12 and α_1 we get

$$U_i \models \#((\mathcal{R}(.), \mathcal{PK}(p_i, U_i))) \tag{35}$$

and further by using R2 we finally get

$$U_i \models U_j \models (\mathcal{R}(.), \mathcal{PK}(p_i, U_i)) \tag{36}$$

Now using projection of beliefs (R6) we get

$$U_i \models U_j \models \mathcal{PK}(p_i, U_i) \tag{37}$$

this proves Γ_3 and Γ_4, by setting $i = 2, j = 1$ for message number 3, and $i = 1, j = 2$ for message 4.

Goals Γ_5–Γ_6

$\Gamma_5 : U_1 \models ((U_2 \mathrel{\vdash} (\mathcal{R}(U_1, N_2), N_2)), \#(\mathcal{R}(U_1, N_2)))$

$\Gamma_6 : U_2 \models ((U_1 \mathrel{\vdash} (\mathcal{R}(U_2, N_1), N_1)), \#(\mathcal{R}(U_2, N_1)))$

From message 4, using the signature verification rule we obtain

$$U_1 \models ((U_2 \mathrel{\vdash} (\mathcal{R}(U_1, N_2), N_2)), \mathcal{R}(U_1, N_2)) \tag{38}$$

The fact that the token has the 'nonce' property of being 'fresh', gives us

$$U_1 \models \#(\mathcal{R}(U_1, N_2)) \tag{39}$$

which together with (38) form goal Γ_5 by the belief aggregation rule. Similarily for Γ_6.

Goals Γ_7–Γ_8

$\Gamma_7 : U_1 \models \mathcal{R}(U_1, N_2)$

$\Gamma_8 : U_2 \models \mathcal{R}(U_2, N_1)$

These are the goals formulated in [3] as the fact that each user should believe he is the intended recipient of the token of the other.

Message number 3 and 4 result in the following formulae being available:

$$U_2 \lhd \sigma(\mathcal{R}(U_2, N_1), U_1) \tag{40}$$

$$U_1 \lhd \sigma(\mathcal{R}(U_1, N_2), U_2) \tag{41}$$

By using the assumptions α_1, α_2 and α_6, it is easy to prove Γ_7 and Γ_8 by using the rules of signature verification and jurisdiction (R13 and R3). Note that it is the PKCS version of signature verification which must be used. This concludes the proof outline for claim 1.

From above we may easily check that all assumptions $\alpha_1 \dots \alpha_7$ have been used. One may show that no assumption may be proven from any subset of the remaining assumptions. From this we conclude claim 2.

5   Conclusions

We have shown that formal proofs in protocol analysis using the formalism presented is feasable, even without machine support. Also, the proofs has given additional insight into the workings of the protocol.

Using a formal notation, we have clarified very important assumptions for X.509 to succed. In particular, the requirement to have knowledge about current time of the certification authority. Also our idealization of the certificates emphasize the fact that the certification authority *must* commit himself when signing the certificates. The suggestion to use blacklists in [3] reflects the view that this commitment might be too strong.

The crucial issue of formalizing goals, assumptions and idealizing the protocol was not always intuitively obvious. This has also been noted by [11]. However, the formalization forced us to consider details which would otherwise have been ignored.

The approach taken is rather abstract, thus one does not increase the confidence in the actual mechanisms. Consequently we have *not* shown that the mechanisms used to implement X.509 cannot be compromised. Thus, our results are not inconsistent with [4].

In our presentation, we have focused on the use of the notation rather than the notation itself. The formal semantics of the notation is the topic for further study.

We hope our extensions have widened the scope of the notation, making its application an interesting approach for the protocol analyst and designer.

6   Acknowledgements

We would like to thank the anonymous referees and our colleagues at Alcatel STK Research Centre, Secure Information Systems for helpful suggestions. We would also like to thank Michael Burrows and Martín Abadi for helpful comments and for making us aware of their recent work [2]. A special thanks to Mr. Kåre Presttun at the research centre for his encouragement and continued support.

References

[1] Michael Burrows, Martín Abadi, and Roger Needham. Authentication: A practical study in belief and action. Technical Report 138, University of Cambridge Computer Laboratory, 1988.

[2] Michael Burrows, Martín Abadi, and Roger Needham. A logic of authentication. Technical Report 39, DEC Systems Research Center, Palo Alto, 1989.

[3] CCITT. *CCITT blue book, Recommendation X.509 and ISO 9594-8, Information Processing Systems - Open Systems Interconnection - The Directory - Authentication Framework.* Geneva, March 1988.

[4] Don Coppersmith. Analysis of ISO/CCITT document X.509 annex D. IBM Thomas J. Watson Research Center, Yorktown Heights, June 1989.

[5] D.E. Denning and G.M. Sacco. Timestamps in key distribution protocols. *CACM*, 24(28):533–536, 1981.

[6] W. Diffie and M.E.Helleman. New directions in cryptography. *IEEE Transactions on Information Theory*, IT-22(6), 1976.

[7] H.D. Ebbinghaus, J. Flum, and W. Thomas. *Mathematical Logic.* Springer-Verlag, 1984.

[8] U. Feige, A.Fiat, and A. Shamir. Zero-knowledge proofs of identity. *Journal of Cryptology*, 1(2):77–94, 1988.

[9] S. Goldwasser, S. Micali, and C. Rackoff. Knowledge complexity of interactive proof systems. *SIAM Journal of Computing*, 18(1):186–208, 1989.

[10] C.A.R. Hoare. An axiomatic basis for computer programming. *CACM*, 12(10):576–580, 1969.

[11] Catherine Meadows. Using narrowing in the analysis of key management protocols. In *IEEE Computer Society Symposium on Security and Privacy*, pages 138–147, 1989.

[12] R.L. Rivest, A. Shamir, and L. Adleman. A method for obtaining digital signatures and public key crypto systems. *Communications of the ACM*, 21(2):120–126, 1978.

Some Remarks on Authentication Systems

Martin H.G. Anthony* Keith M. Martin*,
Jennifer Seberry** Peter Wild*

Abstract

Brickell, Simmons and others have discussed doubly perfect authentication systems in which an opponent's chance of deceiving the receiver is a minimum for a given number of encoding rules. Brickell has shown that in some instances to achieve this minimum the system needs to have splitting. Such a system uses a larger message space. Motivated by Brickell's ideas we consider authentication systems with splitting and the problems of reducing the message space.

1 Authentication

We use the model of authentication described by Simmons [9, 10] and Brickell [1]. There are three participants involved in this model; a transmitter T, a receiver R and an opponent O. T wants to communicate some information to R. It is not necessary that the information be kept secret, but R wants to be sure that the information did indeed come from T.

An item of information that the transmitter might want to send to the receiver is called a source state, and we denote by S the set of source states. We assume that there is some fixed probability distribution P_S on S ($P_S(s)$ is the probability that $s \in S$ is to be communicated on any given occasion).

In order to relay a source state $s \in S$ to R, T encodes it (using some encoding rule chosen from a set I of encoding rules) as a message m and sends m to R.

In order for R to be able to determine which source state is being relayed it is necessary that for any given encoding rule a message m can relay at most once source state under that rule. T and R agree on which encoding rule they will use before communication starts.

Let M be the set of messages that T can send to R. Let 0 be an element not belonging to S. Associated with an encoding rule i is a mapping $f_i : M \to S \cup \{0\}$ given by $f_i(m) = s$ if T can encode $s \in S$ as m under encoding rule i and $f_i(m) = 0$ if no source state can be encoded as m under i. R accepts a message m as authentic (relaying source state s) if $f_i(m) = s$. R rejects m if $f_i(m) = 0$.

We call the triple (I, M, S) an authentication system and if $|I| = b$, $|M| = v$ and $|S| = k$ we denote it by $AS(b, v, k)$. It may be represented by a matrix whose rows are indexed by the set I of encoding rules and whose columns are indexed by the set M of messages with entry $f_i(m)$ in row i, column m. Alternatively it may be represented by a $b \times k$ array $A = (a_{is})$ where $a_{is} = \{m \in M | f_i(m) = s\}$ for $i \in I$, $s \in S$. We call this $b \times k$ array A an authentication array corresponding to the authentication system.

Example 1: Authentication array for $AS(9, 9, 3)$ with $M = \{a, b, c, d, e, f, g, h\}$, is

$$
\begin{array}{ccc}
a & d & g \\
a & e & h \\
a & f & i \\
b & d & h \\
b & e & i \\
b & f & g \\
c & d & i \\
c & e & g \\
c & f & h \\
\end{array}
$$

The opponent O attempts to get R to accept some information that did not come from T. If O knows which encoding rule T and R have agreed upon then O may succeed with probability 1. We assume that T and R share an encoding rule in secret for each transmission and that the encoding rule is chosen according to a probability distribution P_I on the set I of encoding rules. O may deceive R by impersonation or substitution. O impersonates T by sending a message when in fact T has not sent a message. O is successful if R accepts the message as authentic. If T sends a message m, relaying source state s, then O may intercept it and substitute a different message m'. O is sucessful if R accepts the substituted message m' and this message relays a source state different form s.

If $i \in I$ and there exists $m_1 \neq m_2$ such that $f_i(m_1) = f_i(m_2) \neq 0$ then we say splitting occurs in encoding rule i. If splitting occurs then two or more messages may relay the same source state for some encoding rule. In this case T also chooses a splitting strategy. Given that encoding rule $i \in I$ and the source state $s \in S$ are used, a splitting strategy determines the probability that T sends message m for each message m with may relay s under i.

An optimal strategy for T is a probability distribution P_I on the set I of encoding rules and a splitting strategy which minimizes the probability that O may successfully deceive R. This probability is denoted by V_G and is a measure of the security afforded by the authentication system.

2   Cartesian Doubly Perfect Authentication Systems

Simmons and Brickell [1] have given a bound on V_G in terms of the size of the set I of encoding rules. They show that $V_G \leq b^{-\frac{1}{2}}$. This result was also obtained by

Gilbert, MacWilliams and Sloane [5] for a slightly different situation.

An authentication system for which $V_G = b^{-\frac{1}{2}}$ is called doubly perfect.

In an authentication system (I, M, S), for each $m \in M$, let $I(m)$ denote the set $\{i \in I | f_i(m) \in S\}$. The proof of the following result is contained in the proof of Theorem 6 of Brickell [1].

Lemma 1 *Let* (I, M, S) *be a doubly perfect authentication system* $AS(b, v, k)$ *with* $V_G = \alpha$. *Then* $n = 1/\alpha$ *is an integer,* $b = n^2$ *and* $|I(m)| = n$ *for all* $m \in M$.

The bound of the following lemma is given by Simmons [11]. Simmons also shows that if equality holds then splitting does not occur in any encoding rule of an optimal strategy. In an optimal strategy for a doubly perfect authentication system all encoding rules are equally likely.

Hence we have

Lemma 2 *Let* (I, M, S) *be an authentication system* $AS(b, v, k)$. *Write* $n = 1/V_G$. *Then* $v \geq kn$. *If the system is doubly perfect then equality holds if and only if there is no splitting.*

An authentication system is called cartesian if whenever $f_i(m) \neq 0$ and $f_j(m) \neq 0$ for $i, j \in I$ and $m \in M$ then $f_i(m) = f_j(m)$. In a cartesian authentication system a message relays the same source state whichever encoding rule is being used. The sets $M(s) = \{m \in M | f_i(m) = s$ for some $i \in I\}$ for $s \in S$ then partition M. In the $b \times k$ array A representing a cartesian authentication system the entry a_{is} is a subset of $M(s)$ which is the set of messages relaying source state s.

Suppose that, for each $s \in S$, ϕ_s is a bijection from $M(s)$ to the set of integers $\{1, 2, \ldots, |M(s)|\}$. Thus ϕ_s labels the messages of $M(s)$ with the integers 1 to

$|M(s)|$. Then $A' = (a_{is})$ where $a_{is} = \phi_s(a_{is})$ is a $b \times k$ array with integer entries. We refer to A' as a cartesian authentication array.

Example 2: Cartesian authentication array corresponding to $AS(9,9,3)$ of Example 1.

$$
\begin{array}{ccc}
\text{Source States} & & \\
1 & 1 & 1 \\
1 & 2 & 2 \\
1 & 3 & 3 \\
2 & 1 & 2 \\
2 & 2 & 3 \\
2 & 3 & 1 \\
3 & 1 & 3 \\
3 & 2 & 1 \\
3 & 3 & 2
\end{array}
$$

(Encoding Rules, to the right of rows 4 and 5.)

Brickell [1] has constructed cartesian doubly perfect authentication systems using cartesian authentication arrays which he has called orthogonal multi-arrays. An orthogonal multi-array $OMA(k, n; r_1, \ldots, r_k)$ is a $n^2 \times k$ array $A = (a_{ij})$ satisfying (i) a_{ij} is a r_j-subset of the set $\{1, 2, \ldots, r_j\}$ and (ii) given integers x, y with $1 \le x < y \le k$ and integers m_1, m_2 with $1 \le m_1 \le r_x n$ and $1 \le m_2 \le r_y n$ there exists exactly one i such that $m_1 \in a_{ix}$ and $m_2 \in a_{iy}$. An $OMA(k, n; r_1, \ldots, r_k)$ corresponds to a cartesian doubly perfect authentication system with $b = n^2$, $v = n \sum r_j$ and $V_G = 1/n$. This system has splitting if and only if $r_j > 1$ for some j.

An $OMA(k, n; 1, \ldots, 1)$ is called an orthogonal array and denoted $OA(k, n)$. An $OA(k, n)$ is equivalent to a set of $k - 2$ mutually orthogonal latin squares of order n. The maximum number of mutually orthogonal latin squares of order n is $n - 1$. A set of $n - 1$ mutually orthogonal latin squares of order n is called a complete set. Complete sets of mutually orthogonal latin squares are known to exist when n is a prime power.

$$A' = (\phi_s(a_{is})) = \begin{array}{|cccc|}
\hline
1 & 1 & 1 & 1,7 \\
1 & 2 & 2 & 2 \\
1 & 3 & 3 & 5 \\
1 & 4 & 4 & 6 \\
1 & 5 & 5 & 3 \\
1 & 6 & 6 & 4,8 \\
2 & 1 & 2 & 6 \\
2 & 2 & 3 & 1,8 \\
2 & 3 & 6 & 3 \\
2 & 4 & 1 & 2 \\
2 & 5 & 4 & 4,7 \\
2 & 6 & 5 & 5 \\
3 & 1 & 3 & 3 \\
3 & 2 & 6 & 6 \\
3 & 3 & 2 & 4,7 \\
3 & 4 & 5 & 1,8 \\
3 & 5 & 1 & 5 \\
3 & 6 & 4 & 2 \\
4 & 1 & 4 & 8 \\
4 & 2 & 1 & 4 \\
4 & 3 & 5 & 2,6 \\
4 & 4 & 2 & 3,5 \\
4 & 5 & 6 & 1 \\
4 & 6 & 3 & 7 \\
5 & 1 & 5 & 4 \\
5 & 2 & 4 & 3,5 \\
5 & 3 & 1 & 8 \\
5 & 4 & 6 & 7 \\
5 & 5 & 3 & 2,6 \\
5 & 6 & 2 & 1 \\
6 & 1 & 6 & 2,5 \\
6 & 2 & 5 & 7 \\
6 & 3 & 4 & 1 \\
6 & 4 & 3 & 4 \\
6 & 5 & 2 & 8 \\
6 & 6 & 1 & 3,6 \\
\hline
\end{array}$$ Encoding Rules

(the numbers represent messages)

Table 1.

If there do not exist $k - 2$ mutual orthogonal latin squares of order n then a cartesian doubly perfect authentication system with $V_G = 1/n$ and $|S| = k$ must have splitting. For example, since there does not exist a pair of orthogonal latin squares of order 6, a cartesian doubly perfect authentication system with $V_G = 1/6$ and $|S| = 4$ must have splitting. In such a case $v > 24$. Brickell [1] gives an example of an $OMA(4, 6; 1, 1, 1, 2)$. This example corresponds to an authentication system with $v = 30$. This is the minimum size of M that such a system arising from an OMA can have. The following example shows that it is possible for a cartesian doubly perfect authentication system with $V_G = 1/6$ and $|S| = 4$ to have fewer than 30 messages.

This example is a cartesian authentication array corresponding to a cartesian doubly perfect authentication system with $V_G = 1/6$, $|S| = 4$ and $v = 26$.

Stinson [11] has used transversal designs to construct a cartesian authentication system with $V_G = 1/6$, $|S| = 7$ and $v = 42$. (This system has a subsystem with 4 source states and 24 messages). However this example has $b = 72$ and is not doubly perfect. In the light of the above example we may state a result in a slightly more general form that that given in theorems 5 and 6 of Brickell [1].

Theorem 1 *Let $S = \{s_1, \ldots, s_k\}$ and let $M(s_1), \ldots, M(s_k)$ be disjoint sets. Put $M = M(s_1) \cup \ldots \cup M(s_k)$. An $n^2 \times k$ array $A = (a_{is})$ where $a_{is} \subseteq M(s)$ for $1 \leq i \leq n^2$, $s \in S$ is an authentication array corresponding to a cartesian doubly perfect authentication system with $V_G = 1/n$ and $|S| = k$ if and only if*

(i) $a_{is} \neq \phi$ *for all* $1 \leq i \leq n^2$, $s \in S$

(ii) *for all* $s \in S$ *and* $m \in M(s)$, $I(m) = \{i | m \in A_{is}\}$ *has n elements*

(iii) *for any* $s_i, s_j \in S$, $s_i \neq s_j$, $|I(m_1) \cap I(m_2)| \leq 1$ *for all* $m_1 \in M(s_1)$ *and* $m_2 \in M(s_2)$.

3   Incidence Structures

An incidence structure is a triple $(\mathbf{P}, \mathbf{B}, I)$ where $\mathbf{P}$ and $\mathbf{B}$ are non-empty disjoints sets and $\mathbf{I} \subseteq \mathbf{P} \times \mathbf{B}$. The elements of $\mathbf{P}$ are called points and the elements of B are called blocks. We say $P \in \mathbf{P}$ is incident with $x \in B$ if and only if $(P, x) \in \mathbf{I}$.

Let (I, M, S) be a cartesian doubly perfect authentication system $AS(n^2, v, k)$. We may define an incidence structure $(I, M, \mathbf{I})$ by $(i, m) \in \mathbf{I}$ if and only if $f_i(m) \neq 0$. We note that each block $m \in M$ is incident with n points. We also note that if two blocks $m, m' \in M$ are incident with the same set of points then $(I, M \backslash \{m'\}, S)$ would also be a cartesian doubly perfect authentication system. We therefore assume throughout that no two blocks of $(I, M, \mathbf{I})$ are incident with the same set of points. An incidence structure with these two properties is called a design.

The design $(I, M, \mathbf{I})$ has the property that there is a partition of blocks into classes $M(s_1), \ldots M(s_k)$ such that

(i) every point belongs to at least one block of every class,

(ii) any two blocks, belonging to distinct classes, have at most one point in common.

Indeed the existence of a cartesian doubly perfect authentication system with $V_G = 1/n$ and $|S| = k$ is equivalent to the existence of a design with n^2 points, block size n and such a partition.

If the authentication system (I, M, S) has been constructed from an orthogonal multi-array then the partition $M(s_1), \ldots, M(S_k)$ of the design $(I, M, \mathbf{I})$ has the property that each point is incident with a constant number of blocks from each class. Such a partition is called a resolution. Moreover any two blocks from distinct classes are incident with exactly one common point. Such a resolution is called an *outer resolution*. If the orthogonal multi-array is in fact an orthogonal array, so that there is no splitting in the authentication system, then the blocks of a class are disjoint and the design is a net (See Hughes and Piper [6]). A net with k classes is equivalent to $k - 2$ mutually orthogonal Latin squares.

There are two problems:

(1) for a given n find the largest integer k such that there exists a cartesian doubly perfect authentication system with $V_G = 1/n$ and $|S| = k$.

(2) for given n and k find a cartesian doubly perfect authentication system with the minimum number of messages.

These problems correspond to constructing designs admitting partitions of the blocks having the properties described above with the maximum number of blocks. The largest value of k is $n + 1$ and the minimum number of blocks is kn. These solutions correspond to orthogonal arrays. See Stinson [11] for a description of these systems in terms of transversal designs.

Theorem 2 *Let (I, M, S) be a cartesian doubly perfect authentication system with $V_G = 1/n$. Let A be a cartesian authentication array for it.*

Then

(i) $|S| \leq n + 1$ with equality if and only if A is an orthogonal array

(ii) $|M| \geq n|S|$ with equality if and only if A is an orthogonal array.

Proof: (i) Let $|S| = k$. Let $i \in I$. There exist $m_1, \ldots, m_k \in M$ such that $f_i(m_1)$, $\ldots, f_i(m_k)$ are the k elements of S. Then $I(m_1), \ldots I(m_k)$ are k n-subsets of I which intersect pairwise in $\{i\}$. Thus $1 + k(n-1) \leq |I| = n^2$, Hence $k \leq n + 1$.

If $k = n + 1$ then $I(m_1), \ldots, I(m_k)$ cover the $n^2 - 1$ points of I distinct from i exactly once. Suppose $m \in M$ and $f_i(m) \neq 0$. Then $f_i(m) = f_i(m_j)$ for some j. Now $|I(m) \cap I(m_h)| \leq 1$ for $h \neq j$, so $I(m) \cap I(m_h) = \{i\}$ and we must have $I(m) = I(m_j)$. Hence $m = m_j$.

It follows that there is no splitting, blocks within a class are disjoint and blocks from distinct classes meet in exactly one point. Thus $(I, M, \mathbf{I})$ is a net, and A is an orthogonal array, (ii) follows by lemma 2 and Brickell [1] theorem 6.

When n is a prime power constructions of appropriate nets in the case of equality in theorem 2 are well known for any $k = |S| \leq n + 1$.

4 Mutually Orthogonal F-squares

By lemmas 1 and 2 an authentication system (I, M, S) with $|S| = k$, $|I| = n^2$ and $|M| < nk$ cannot be doubly perfect. That is if $|M| \leq nk$ then $V_G \geq 1/n$.

However, for some applications, it may be important that the size of the message space is kept small. In this section we use F-squares which are generalizations of latin squares to construct cartesian authentication systems $AS(n^2, v, k)$ with $v < nk$. By theorem 1 of Simmons [9], which asserts that $V_G \geq |S|/|M|$ it follows that for these systems $V_G > 1/n$, and they are not doubly perfect.

Let n be a positive integer and let $(\lambda_1, \ldots, \lambda_m)$ be a vector of positive integers such that $\lambda_1 + \ldots + \lambda_m = n$. An F-square of order n with frequency vector $(\lambda_1, \ldots, \lambda_m)$ is an $n \times n$ matrix $X = (x_{ij})$ with entries x_{ij} from a set

$U = \{u_1, \ldots, u_n\}$ of m symbols such that each element $u_i \in U$ appears exactly λ_i times in each row and in each column of X. The F-square X is denoted $F(n; \lambda_1, \lambda_2, \ldots, \lambda_m)$. An $F(n; 1, \ldots, 1)$ (also denoted $F(n; 1^n)$) is a latin square. The integers $\lambda_1, \ldots, \lambda_m$ are called the frequencies of the symbols in the square. Two F-squares $F_1(n; \lambda_1, \ldots, \lambda_m)$ and $F_2(n; \mu_1, \ldots \mu_p)$ are called orthogonal if for all $1 \le i \le m$ and $1 \le j \le p$ the pair (i, j) occurs exactly $\lambda_i \mu_j$ times when F_1 and F_2 are superimposed. (See Denes and Keedwell [4]).

Theorem 3 *Suppose $F_1 = (x_{ij}^1) = F_1(n; \lambda_{11}, \ldots, \lambda_{1m}), \ldots, F_{k-2} = (x_{ij}^{k-2}) = F_{k-2}(n; \lambda_{(k-2),1}, \ldots, \lambda_{(k-2),m_{(k-2)}})$ are a set of $k-2$ mutually orthogonal F-squares. Put $\lambda = max(\lambda_{ij})$ and $t = min(m_i)$. Then there is a cartesian authentication system $AS(b, v, k)$ with $b = n^2$, $v = 2n + \sum m_i$ and $1/n \le V_G \le \lambda/n$.*

Proof: The $k-2$ mutually orthogonal F-squares determine a $b \times k$ array, $A' = (a_{ij})$ where, for $1 \le i, j \le n$,

$$a_{n(i-1)+j,h} = x_{ij}^{h-2} \quad h = 3, \ldots, k$$
$$a_{n(i-1)+j,1} = i$$
$$a_{n(i-1)+j,2} = j$$

Corresponding to A' is a cartesian authentication system $(I, M, S) = AS(n^2, v, k)$ where $v = 2n + \sum M_i$. Let $S = \{s_i, \ldots, s_k\}$ and let the partition of M be

$$M_{s_i} = \{m_{i1}, \ldots, m_{i_{p_i}}\}.$$

Suppose $p_h = t$, and suppose O impersonates T using message m_h with probability $1/t$ for $j = 1, \ldots, t$. Since there is, for each encoding rule, exactly one message in M_s which is valid under that rule the probability that O is successful is $\sum p(i)1/t = 1/t$. Hence $V_G \ge 1/t$.

Now suppose that T uses a strategy with the uniform probability distribution on the encoding rules. Suppose O impersonates T by sending m_{ij}. Then there are $n\lambda_{ij}$ encoding rules under which m_{ij} is valid and the probability that O succeeds is

$$\frac{n\lambda_{ij}}{n^2} = \frac{\lambda_{ij}}{n} \le \frac{\lambda}{n}.$$

Suppose O observes the message m_{ij} and substitutes m_{ab} for it. There are $n\lambda_{ij}$ encoding rules under which m_{ij} is valid and of these there are $\lambda_{ij}\lambda_{ab}$ under which m_{ab} is also valid. Hence O succeeds in deceiving R with a substitution attack with probability

$$\frac{\lambda_{ij}\lambda_{ab}}{n\lambda_{ij}} = \frac{\lambda_{ab}}{n} \le \frac{\lambda}{n}.$$

It follows that $V_G \le \lambda/n$.

Remark: If we identify some symbols in any column of A' so that the total number of occurrences of this combined symbol is at most λn then we obtain an $AS(n^2, v', k)$ which still satisfies $(1/n) \le V_G \le \lambda/n$ but for which $v' \le v$. By making suitable identifications of symbols in an array arising from mutually orthogonal F-squares, we obtain the following result.

Corollary 1 *Suppose there exist $k-2$ mutually orthogonal $F(n, \lambda^t)$-squares (necessarily $n = \lambda t$). Then there exists a cartesian authentication system $AS(n^2, kn/\lambda, k)$ with $V_G = \lambda/n$.*

Example 3: Table 2 is a cartesian authentication array corresponding to an $AS(36, 16, 6)$ with $1/2 \le V_G \le 2/3$ constructed as above from four mutually orthogonal F-squares: $F_1(6; 1^6)$, $F_2(6; 1^4, 2)$; $F_3(6; 2^3)$, $F_4(6; 2^1, 4^1)$,

$$
\begin{array}{|cccccc|}
\hline
1 & 2 & 3 & 4 & 5 & 6 \\
2 & 3 & 6 & 1 & 4 & 5 \\
3 & 6 & 2 & 5 & 1 & 4 \\
4 & 1 & 5 & 2 & 6 & 3 \\
5 & 4 & 1 & 6 & 3 & 2 \\
6 & 5 & 4 & 3 & 2 & 1 \\
\hline
\end{array}
\qquad
\begin{array}{|cccccc|}
\hline
1 & 2 & 5 & 5 & 3 & 4 \\
5 & 1 & 3 & 2 & 4 & 5 \\
3 & 5 & 4 & 1 & 5 & 2 \\
5 & 4 & 2 & 3 & 1 & 5 \\
4 & 3 & 5 & 5 & 2 & 1 \\
2 & 4 & 1 & 4 & 5 & 3 \\
\hline
\end{array}
$$
$$
F_1(6; 1^6) \qquad\qquad F_2(6; 1^4 2)
$$

$$
\begin{array}{|cccccc|}
\hline
1 & 2 & 3 & 3 & 2 & 1 \\
1 & 2 & 3 & 3 & 2 & 1 \\
3 & 1 & 2 & 2 & 1 & 3 \\
2 & 3 & 1 & 1 & 3 & 2 \\
3 & 1 & 2 & 2 & 1 & 3 \\
2 & 3 & 1 & 1 & 3 & 2 \\
\hline
\end{array}
\qquad
\begin{array}{|cccccc|}
\hline
1 & 2 & 2 & 2 & 2 & 1 \\
2 & 1 & 2 & 2 & 1 & 2 \\
1 & 2 & 2 & 2 & 2 & 1 \\
2 & 2 & 1 & 1 & 2 & 2 \\
2 & 2 & 1 & 1 & 2 & 2 \\
2 & 1 & 2 & 2 & 1 & 2 \\
\hline
\end{array}
$$
$$
F_3(6; 2^3) \qquad\qquad F_4(6; 2^1, 4^1)
$$

Seberry [8] has shown how to construct a set of $n - 1$ mutually orthogonal $F(n; \lambda^t)$-squares using a generalized Hadamard matrix of size $n = \lambda t$ with entries from a group G of order t.

Several families of generalized Hadamard matrices $GH(n; G)$ of size n with entries from G are known to exist including the families: $n = 2p^\alpha$, $G = Z_p^\alpha$ (Jungnickel [7], Street [12]); $n = 4p^\alpha$, $G = Z_p^\alpha$ (Dawson [2]; and $n = (p^\alpha - 1)p^\alpha$, $G = Z_p^\alpha$ (Seberry [8]) (where p is a prime and α is a positive integer). These give families of cartesian authentication systems $AS(b, v, k)$ with $b = n^2$, $v = p^\alpha(n-1)$, $k = n - 1$ and $V_G = 1/p^\alpha$.

5 Cyclotomy and Mutually Orthogonal F-squares

In this section we use sets of mutually orthogonal F-squares and cyclotomy to construct authentication systems. This construction is based on a method of Parker (see [4]) for constructing sets of mutually orthogonal latin squares. It produces authentication schemes $AS(b, v, k)$ with similar properties to those of the previous section: $b = (q + f)^2$, $v < (q + f)k$ and $V_G \leq \lambda/(q + f)$ where q is a prime power and f is the order of the F-squares of some set of $k - 2$ mutually orthogonal F-squares.

Source States

	1	1	1	1	1	1
	1	2	2	2	2	2
	1	3	3	5	3	2
	1	4	4	5	3	1
	1	5	5	3	2	2
	1	6	6	4	1	2
	2	1	2	5	1	2
	2	2	3	1	2	2
	2	3	6	3	3	1
	2	4	1	2	3	2
	2	5	4	4	2	2
	2	6	5	5	1	1
	3	1	3	3	3	2
	3	2	6	5	1	2
	3	3	2	4	2	1
Encoding Rules	3	4	5	1	2	1
	3	5	1	5	1	2
	3	6	4	2	3	2
	4	1	4	5	2	2
	4	2	1	4	3	1
	4	3	5	2	1	2
	4	4	2	3	1	2
	4	5	6	1	3	2
	4	6	3	5	2	1
	5	1	5	4	3	2
	5	2	4	3	1	1
	5	3	1	5	2	2
	5	4	6	5	2	2
	5	5	3	2	1	1
	5	6	2	1	3	2
	6	1	6	2	2	1
	6	2	5	5	3	2
	6	3	4	1	1	2
	6	4	3	4	1	2
	6	5	2	5	3	1
	6	6	1	3	2	2

(the numbers represent messages)

Table 2.

Let $q = mf + 1$ be a prime power. The multiplicative group of $GF(q)$ is cyclic of order $q - 1$ and has a unique subgroup H of order f. The cyclotomic classes of index m of $GF(q)$ are the cosets of H in the multiplicative group of $GF(q)$.

Suppose $F_1 = (x_{ij}^1)$, $F_2 = (x_{ij}^2)$, ..., $F_m = (x_{ij}^m)$ are m mutually orthogonal F-squares of order f. Suppose that the rows and columns for these F-squares are indexed by an F-set U. Suppose that U_i is the set of symbols appearing in F_i, $i = 1, \ldots, m$. Let $\psi : H \to U$ be a bijection and, for $i = 1, \ldots, m$ let $\phi_i : H \to U_i$ be a function such that for each $u \in U$ there are exactly λ elements $h \in H$ with $\phi_i(h) = u$ where u appears λ times and each row (and column) of F_i.

Let $D = \{(a_1, 1), (a_2, 2), \ldots, (a_{m+1}, m+1)\} \subseteq GF(q) \times Z_{m+2}$ such that $a_i \neq a_j$ for $i \neq j$. We define a $(q + f)^2 \times m + 2$ array A in the following way:

(1) the f^2 rows of the $f^2 \times (m+2)$ array B obtained as in the previous section from $F_1, \ldots, F_m$ are rows of A

(2) for each $a \in GF(q)$ the row $aa \ldots a$ is a row of A

(3) for each $(h, j, a) \in H \times Z_{m+2} \times GF(q)$ there is a row whose dth entry $(1 \leq d \leq m + 2)$ is

$$ha_i + a \text{ where } i + j \equiv d \pmod{m+2} \text{ if } j \not\equiv d \pmod{m+2}$$

$$\phi_{d-2}(h) \text{ if } j \equiv d \pmod{m+2}, \quad d \neq 1, 2$$

$$\psi(h) \text{ if } j \equiv d \pmod{m+2}, \quad d = 1, 2$$

We note that the array A has the following properties: each symbol appearing in columns 1 and 2 appears $q + f$ times; each element of $GF(q)$ appears $q + f$

times in each of the columns 3 to $m+2$; if a symbol appears λ times in each row (and column) of F_i then it appears $\lambda(q+f)$ times in column $i+2$ of A.

Two symbols, neither of which belongs to $GF(q)$, appear together in the same row only in rows of B. The number of times this happens is given by the product of the frequencies of the two symbols (symbols in columns 1 and 2 have frequency 1). A symbol not in $GF(q)$ and an element of $GF(q)$ occur together in λ rows where λ is the frequency of the symbol. Two elements $a, b \in GF(q)$, $a \neq b$, occur together in the same row if and only if there exist $a_i, a_j \in D$ such that $a - b$ or $b - a$ and $a_i - a_j$ belong to the same cyclotomic class.

Put $D_k = \{a_i - a_j | a_i, a_j \in D, i - j = k(\mathrm{mod}\, m+2)\}$, $k = 1, \ldots, m+1$.

Suppose μ_k is the maximum number of representatives of any one cyclotomic class which belong to D_k. Then for $i - j \equiv k(\mathrm{mod}(m+2))$, μ_k is the maximum number of rows of A in which any pair $a, b \in GF(q)$ appear together in columns i, j (respectively). Let μ be the maximum of $\mu_1, \ldots, \mu_{m+1}$. The following theorem may be proved in a similar fashion to theorem 3.

Theorem 4 *Let $q = mf+1$ be a prime-power. Suppose $F_1, \ldots F_m$ are mutually orthogonal F-squares of order f. Let λ be the maximum of the frequencies occurring in $F_1, \ldots F_m$. Suppose $D = \{(a_1, 1), \ldots, (a_{m+1}, m+1)\} \subseteq GF(q) \times Z_{m+2}$ is such that $a_i \neq a_j$ for $i \neq j$. Let μ be as described above. Then the array A as defined above is a cartesian authentication array corresponding to a cartesian authentication system $AS(b, v, k)$ with $b = (q+f)^2$, $v \leq (q+f)k$ and $V_G \leq max(\lambda, \mu)/(q+f)$.*

By making identifications of symbols we may obtain other cartesian authentication systems $AS(b, v', k)$ with $V_G \leq max(\lambda, \mu)/(q+f)$ and $v' < v$.

We remark that the construction may be generalized by using a cartesian authentication array of size $f^2 \times (m+2)$ in place of B in D above.

Example 4: Consider $m = 4$. Let $q = 4f + 1$ be a prime power. Suppose there exist four mutually orthogonal F-squares of order f such that each symbol in these squares has frequency 1 or 2. Let α be a primitive element of $GF(q)$ and put $D = \{ (0,1), (1,2), (1+\alpha, 3), (1+\alpha+\alpha^2, 4), (1+\alpha+\alpha^2+\alpha^3, 5) \}$. Then if f is odd D_k contains at most 2 elements from any cyclotomic class of index 4 for $k = 1, 2, 3, 4, 5$. Hence in this case the above construction yields an $AS((q+f)^2, 6(q+5), 6)$ with $V_G \leq 2/(q+f)$. For example with $f = 9$, $q = 37$, since there exist four mutually orthogonal latin orders of order 9 we obtain an $AS(46^2, 6.46, 6)$ with $V_G \leq 2/46$.

Ackowledgement: Professor Jennifer Seberry received support under SERC Visiting Fellow Research Grant GR/E83870. The work of Martin H.G. Anthony was supported by an SERC Studentship and the work of Keith M. Martin was supported by a SERC CASE award.

References

[1] Brickell, E.F. A few results in message authentication. *Congressus Numerantium*, **43** (1984), 141-154.

[2] Dawson, J.E. A construction for generalized hadamard matrices $GH(4q, EA(q))$, *J. Stat. Planning and Inference*, **11** (1985), 103-110.

[3] de Launey, W. A survey of generalized Hadamard matrices and difference matrices $D(k, \lambda, G)$ with large k, *Utilitas Math.*, **30** (1986), 5-29.

[4] Denes, J. and Keedwell, A.D. *Latin Squares and their Applications*, English Universities Press, London, 1974.

[5] Gilbert, E.N., MacWilliams, F.J. and Slone, N.J.A. Codes which detect deception. *The Bell Sys. Tech Journal*, **53** (1974), 405-414.

[6] Hughes, D.R. and Piper, F.C. *Design Theory*, Cambridge University Press, Cambridge, 1985.

[7] Jungnickel, D. On difference matrices, resolvable TD's and generalized Hadamard matrices, *Math. Z*, **167** (1979), 49-60.

[8] Seberry, J. A construction for generalized Hadamard matrices, *J. Stat. Planning and Inference*, **4** (1980), 365-368.

[9] Simmons, G.J. Message authentication: a game on hypergraphs, *Congressus Numerantium*, **45** (1984), 161-192.

[10] Simmons, G.J. Authentication theory / coding theory, in *"Advances in Cryptology: Proceedings of Cryto '84"*, Lecture Notes in Computer Science, vol. 196, 411-432, Springer Verlag, Berlin, 1985.

[11] Stinson, D.R. Some constructions and bounds for authentication codes, *J. of Cryptology*, **1** (1988), 37-51.

[12] Street, D.J. Generalized Hadamard matrices, orthogonal arrays and F-squares, *Ars Combinatoria*, **8** (1979), 131-141.

*Department of Mathematics
RHBNC
University of London
Egham, Surrey TW20 0EX,
U.K.

**Department of Computer Science
University College
University of New South Wales
Australian Defence Force Academy
Canberra, A.C.T. 2600
Australia

Meet-in-the-Middle Attack
on Digital Signature Schemes

Kazuo Ohta   Kenji Koyama[†]

NTT Communications and Information Processing Laboratories

Nippon Telegraph and Telephone Corporation

1-2356, Take, Yokosuka-shi, Kanagawa, 238-03 Japan

[†]NTT Basic Research Laboratories

Nippon Telegraph and Telephone Corporation

3-9-11, Midori-cho, Musashino-shi, Tokyo, 180 Japan

Abstract : The meet-in-the-middle attack can be used for forging signatures on mixed-type digital signature schemes, and takes less time than an exhaustive attack. This paper formulates a meet-in-the-middle attack on mixed-type digital signature schemes, shows the necessary conditions for success, and discusses the relationships between computational and space complexities as well as success probability during the attack. We also analyze an optimal strategy for forgers to apply this attack, pointing out that an intermediate value of 64 bit length is not secure for any mixed-type digital signature scheme. Finally, we show how to design secure mixed-type digital signature schemes.

Key words. Authentication, Digital signature, Hash function, One-way function, Meet-in-the-middle attack.

1. Introduction

Authentication, which certifies the message contents and the originator, becomes an important technique, because transactions of valuable information for electronic funds transfer, business contract and other functions are widely spread by use of telecommunications or computer networks. To ensure authenticity, a digital signature is effective in place of a handwritten signature [1,2]. Digital signature schemes based on a public-key cryptosystem are realized in the following way; after a sender signs a message using a secret key and sends it, a receiver verifies whether the contents of message and their originator are valid using the sender's public key.

Recently, various cryptosystems and digital signature schemes have been proposed and their efficiency and security have been discussed [3,4,5].

The meet-in-the-middle attack is an attack which can be used for cryptanalysis and signature forgery with less computational time and memory size than the exhaustive attack. The meet-in-the-middle attack has the following two negative applications:

(i) secret key guessing,

(ii) digital signature forgery.

The type (i) attack finds a secret key k which satisfies the relationship $y_0 = f(k, x_0)$ for a given encryption function f and a pair (x_0, y_0). This means, if the attack succeeds, the cryptosystem f is totally broken. For type (i), several cryptanalyses are discussed; a known plaintext attack on a double usage of DES [8], a chosen plaintext attack on a triple usage of DES with two secret keys [16] and a general cascade type of DES [17]. The type (ii) attack finds a pair (x, y) which satisfies the relationship $y = f(x)$ for a given signature generating function f. This attack can partially break a signature scheme and is a kind of ad-hoc forgery. As an example of type (ii) example, the schemes by Rabin[18] and by Davies and Price[7] are analyzed in [20, 15]. There are also other references [6, 13, 14, 19, 21] related to meet-in-the-middle attacks.

These two kinds of meet-in-the-middle attack are different in definition of success and in attacking strategy. In comparison with the type (i) attack, there have not been sufficient studies of the type (ii) attack. For purposes of application, the conditions for success of meet-in-the-middle attacks and the relationships between the success probability and the computational and space complexities of the attacks have not been clearly analyzed. The security of mixed-type signature schemes (see Section 2), which are secure and effective against a meet-in-the-middle attack, has not yet been analyzed.

In this paper, we first generally formulate the meet-in-the-middle attack on mixed-type digital signature schemes, show the necessary conditions of success, and clarify the relationship among the various parameters to design the mixed-type digital signature schemes. Next, we will analyze an optimal strategy for a forger to apply this attack. Finally, we show how to design secure mixed-type digital signature schemes.

2. Mixed-Type Digital Signature Schemes

The mixed-type digital signature scheme [5,7] authenticates data using a hash function and a public-key cryptosystem for a digital signature of the hashed value (Fig. 1).

Step 1: The signer calculates signature $y = f_2(f_1(x))$, where x is a plaintext, f_1 is a *public* data compression function, and f_2 is a *secret* signature function known to the sender, and sends both x and y to a verifier.

Step 2: The verifier checks whether the relationship $f_2^{-1}(y) = f_1(x)$ holds using f_1 and a *public* validation function f_2^{-1}. If this equation holds, the message is authenticated.

Public data compression function f_1 is often called a "hash function" and it generates a digest compressed from a plaintext of any length. Therefore, f_1 is not injective. For all x, $f_1(x)$ should be random numbers that depend on all the bits of the plaintext x. Moreover, it should be a one-way function so that it is difficult to guess any partial information of x from $f_1(x)$. As examples of f_1, repeated transformations in cipher block chaining (CBC) modes [11] using DES [9] and Rabin's public-key cryptosystem [10] are known.

Function f_2 is the signature function of a public-key cryptosystem. Since $f_2^{-1}(f_2(x)) = x$ holds for any x, function f_2 is injective and f_2^{-1} is surjective. Moreover, f_2 is calculated by the sender, and f_2^{-1} can be calculated by anyone. Basically, f_2^{-1} is a trapdoor one-way function. The RSA cryptosystem [12] is a typical example of f_2. Here, note that the signature or validation function is bijective (i.e., injective and surjective).

3. Formulation of Meet-in-the-Middle Attacks

The meet-in-the-middle attack for forgery of a digital signature is defined as finding a pair (x, y) that satisfies $y = f(x)$ for a given signature generation function f.

The necessary conditions to apply the meet-in-the-middle attack are:

(i) the signature generating function f can be decomposed into

$f(x) = f_2(f_1(x))$ and f_1 and f_2^{-1} can be calculated independently.

(ii) f_1 is not injective.

The meet-in-the-middle attack procedure is as follows:

Step 1: An attacker selects the set of attacked signatures, S, calculates $f_2^{-1}(y)$ for each $y \in S$, sorts the calculated values, and stores all pairs of $(f_2^{-1}(y), y)$ in a database as signature candidates.

Step 2: The attacker selects the set of candidates of plaintext, M, calculates $f_1(x)$ for each $x \in M$, and checks $f_1(x) \in f_2^{-1}(S)$. Finally the attacker finds $x \in M$ and $y \in S$ that satisfy $f_1(x) = f_2^{-1}(y)$.

Let s be the number of elements of tested signature set S. Let m be the number of elements of tested message set M. It is clear that memory size needed in the meet-in-the-middle attack is s blocks, where the block length depends on the cryptosystem f_2. In the attack, f_1 is applied m times, and f_2^{-1} is applied s times. Sorting of s elements requires $s \log s$ basic operations, and coincidence checks between random m elements and sorted s elements require $m \log s$ basic operations. Thus, the time T required in the attack is

$$T = mT_1 + sT_2 + s \log s + m \log s$$

where T_1 and T_2 denote the numbers of basic operations of one-time applications of f_1 and f_2^{-1}, respectively. Here, the processing time of basic operation is supposed to be one. Since the terms of mT_1 and sT_2 are sufficiently greater than the terms of $s \log s$ and $m \log s$, we have

$$T = mT_1 + sT_2$$

Furthermore, assume that $T_1 = T_2$. By rescaling the time unit, we have $T = m + s$.

Let signature generating function f be mapping from the set of plaintexts $\mathbf{M}$ to the set of signatures $\mathbf{S}$ and decomposed into $f = f_2 \cdot f_1$. Let R_1 be the range of f_1 and D_2 be the domain of f_2, the intermediate value space C can be denoted as $C = R_1 \cap D_2$. The signature schemes are constructed as $R_1 \subseteq D_2$, thus $C = R_1$. Let c be the number of elements of C (see Figure 2). Note that if f_2 is an RSA signature function, $f_2(D_2) = \mathbf{S}$ holds, because f_2 is bijective.

Again for simplicity, we will discuss the case where a set of known used signatures S is included in $f_2(f_1(M))$. Here, because f_2 is injective, $|f_2^{-1}(S)| = s$. The case where S is randomly selected will be discussed in 4.2.2.

If $f_1(M) \cap f_2^{-1}(S) \neq \phi$, that is, there is at least one pair (x, y) satisfying $x \in M$, $y \in S$ and $f_1(x) = f_2^{-1}(y)$, the meet-in-the-middle attack succeeds. Inversely, if $f_1(M) \cap f_2^{-1}(S) = \phi$, that is, $f_1(x) \notin f_2^{-1}(S)$ for every $x \in M$, the meet-in-the-middle attack never succeeds.

Let S be a set of known used signatures, $S \subseteq f_2(f_1(M)) \subseteq f_2(D_2)$ holds. Because f_2^{-1} is injective on S, the ratio of $f_2^{-1}(S)$ to C is s/c. The probability that $f_1(x)$ for every $x \in M$ is not in $f_2^{-1}(S)$ is $(1-s/c)$. Thus, the probability that $f_1(x) \notin f_2^{-1}(S)$ for all $x \in M$ is $(1-s/c)^m$, where $m = |M|$. Therefore, the probability P that $f_1(x) \in f_2^{-1}(S)$ holds for at least one $x \in M$ is

$$P = 1 - (1 - s/c)^m. \tag{1}$$

Hereafter, we call P as "the success probability of the meet-in-the-middle attack."

4. Analysis of Meet-in-the-Middle Attacks
4.1 Success Probability

The properties of the success probability P are clarified as follows. Since $s \leq c$ and

$$\frac{\partial P}{\partial m} = -(1 - \frac{s}{c})^m \log(1 - \frac{s}{c}) \geq 0, \tag{2}$$

P increases as m increases, when both s and c are constant and $s \neq c$. Since

$$\frac{\partial P}{\partial c} = -\frac{ms(1 - \frac{s}{c})^m}{c^2(1 - \frac{s}{c})} \leq 0, \tag{3}$$

P decreases as c increases, when both s and m are constant and $s \neq c$. Since

$$\frac{\partial P}{\partial s} = \frac{m}{c}(1 - \frac{s}{c})^{m-1} \geq 0, \tag{4}$$

P increases as s increases, when both m and c are constant and $s \neq c$.

Let

$$r = \frac{ms}{c}, \tag{5}$$

then the success probability P is rewritten as

$$P = 1 - (1 - \frac{r}{m})^m. \tag{6}$$

Since $s/c \leq 1$ implies $m \geq r$ and

$$\frac{\partial P}{\partial r} = (1 - \frac{r}{m})^{m-1} \geq 0, \tag{7}$$

P increases as r increases, when m is constant and $m \neq r$.
Since

$$\frac{\partial P}{\partial m} = -(1 - \frac{r}{m})^m \{\log(1 - \frac{r}{m}) + \frac{r}{m(1 - r/m)}\} \leq 0, \tag{8}$$

P decreases as m increases (i.e. s decreases), when r is constant and $m \neq r$.

Note that

$$(1 - \frac{r}{m})^m = 1 - r + \frac{1}{2}\frac{(m-1)}{m}r^2 - \frac{1}{6}\frac{(m-1)(m-2)}{m^2}r^3$$
$$+ \frac{1}{24}\frac{(m-1)(m-2)(m-3)}{m^3}r^4$$
$$- \frac{1}{120}\frac{(m-1)(m-2)(m-3)(m-4)}{m^4}r^5$$
$$+ O(r^6)$$

and

$$e^{-r} = 1 - r + \frac{1}{2}r^2 - \frac{1}{6}r^3 + \frac{1}{24}r^4 - \frac{1}{120}r^5 + O(r^6)$$

imply

$$(1 - r/m)^m \approx e^{-r},$$

and

$$P \approx 1 - e^{-r}, \tag{9}$$

where $m \gg 1$.

The relationship between P and r is shown in Table 1. Table 1 shows that the probability P will be saturated if $r > 10$.

Consider a special case of the meet-in-the-middle attack with $|S| = 1$, the exhaustive attack, where an attacker finds $x (\neq x_0)$ satisfying $y_0 = f(x)$ with a real pair (x_0, y_0) composed of a plaintext and a signature. The success probability P is

$$P = 1 - (1 - \frac{1}{c})^m.$$

Note that an attack with m calculations for f_1 and s calculations for f_2^{-1} has the same success probability as an attack with $m \times s$ calculations for f_1 and *one* calculation for f_2^{-1}. Because the number of the intermediate elements and $r = ms/c$ are constant, the success probability is constant.

4.2 Optimal Strategy

Assume that the success probability P and the number of intermediate elements c are given. Optimal strategy for the meet-in-the-middle attacks is defined to minimize the computational and space complexities of the attack by optimally choosing m and s, where m is the number of tested plaintexts and s is the number of tested signatures.

4.2.1 Attack with Known Used Signatures

(1) Minimization of computational complexity

From the discussion in section 3 and equation (5), the computational complexity U is

$$U = m + s = \frac{rc}{s} + s,$$

where the value r is chosen by the attacker, and the value c is chosen by the system designer. Because $\partial U/\partial s = 0$ implies $s = \sqrt{rc}$ and $\partial^2 U/\partial^2 s > 0$ with $s = \sqrt{rc}$, the minimal value of U is $2\sqrt{rc}$ with $s = \sqrt{rc}$ and $m = \sqrt{rc}$. Choosing s and m in this way minimizes computational complexity. Note that the probability of success is $1 - e^{-r}$. For example, if $r = 1$, success probability is about 63% and the minimal value of U is $2\sqrt{c}$ with $s = \sqrt{c}$ and $m = \sqrt{c}$.

The above discussion describes a case where there is no memory space constraints. Strategy in a case where there is memory constraint is as follows: when $s \leq s_{max} < \sqrt{rc}$, where s_{max} is the maximal memory size used by the attacker, minimal value U is $(s_{max} + rc/s_{max})$ with $s = s_{max}$ and $m = rc/s_{max}$.

(2) Minimization of the computational and space complexities product

Similarly, the the computational and space complexities product V is

$$V = (m + s)s = (\frac{rc}{s} + s)s = rc + s^2.$$

The minimal value of V is $(rc + 1)$ with $s = 1$ and $m = rc$, because $s > 0$. It is remarkable that this optimal strategy is the exhaustive attack.

(3) Minimization of the computational and space complexities sum

The computer costs depend on computational time and memory size respectively. Therefore, estimation of the computational and space complexities sum are meaningful. Here, for simplicity, suppose that the units cost for time and memory, respectively, are unities.

The computational and space complexities sum W is

$$W = (m + s) + s = \frac{rc}{s} + 2s.$$

Because $\partial W/\partial s = 0$ implies $s = \sqrt{rc/2}$ and $\partial^2 W/\partial^2 s > 0$ with $s = \sqrt{rc/2}$, the minimal value of W is $2\sqrt{2rc}$ with $s = \sqrt{rc/2}$ and $m = \sqrt{2rc}$.

4.2.2 Attack with Randomly Chosen Signatures

In this subsection, we will formulate the-meet-in-the-middle attack strategy where an attacker randomly chooses signature candidates from the set of signatures $\mathbf{S}$ without using known signatures.

Let c_1 be the number of elements of the range of f_1, that is, $c_1 = |R_1|$. Let c_2 be the number of elements of the domain of f_2, that is, $c_2 = |D_2|$. Suppose $c_1 \leq c_2$ as described in section 3. Figure 3 shows the difference between known used signatures and randomly chosen signatures. Let S' be a set of signatures randomly chosen from $\mathbf{S}$. The probability that $f_2^{-1}(y)$ for any $y \in S'$ is included in the intermediate value set $C(= R_1)$ is c_1/c_2. Thus, $s'(= sc_2/c_1)$ calculations for the f_2^{-1} operation is required to get the set of intermediate values of C where s is the number of elements in the set C. Therefore, the computational complexity needed for the meet-in-the-middle attack in this case, is $(m + s')$ and the space complexity needed for the attack is s. The success probability of this attack is presented in equation (1) as well as in the attack with known used signatures.

(1) Minimization of computational complexity

The computational complexity U' is

$$U' = m + s' = \frac{rc_1}{s} + \frac{sc_2}{c_1} = \frac{c_2}{c_1}(\frac{c_1^2 r}{sc_2} + s).$$

The minimal value of U' is $2\sqrt{rc_2}$ with $s = c_1\sqrt{r/c_2}$, $s' = \sqrt{rc_2}$ and $m = \sqrt{rc_2}$. Note that the memory size needed to store the intermediate values is $s = c_1\sqrt{r/c_2}$.

(2) Minimization of the computational and space complexities product

The computational and space complexities product V' is

$$V' = (m + s')s = (\frac{rc_1}{s} + \frac{sc_2}{c_1})s = rc_1 + \frac{c_2}{c_1}s^2.$$

The minimal value of V' is $(rc_1 + c_2/c_1)$ when $s = 1$, $s' \approx c_2/c_1$ and $m = rc_1$, because $s > 0$.

(3) Minimization of the computational and space complexities sum

The computational and space complexities sum W' is

$$W' = (m + s') + s = \frac{rc_1}{s} + (\frac{c_2}{c_1} + 1)s.$$

Because $\partial W'/\partial s = 0$ implies $s = c_1\sqrt{r/(c_1 + c_2)}$ and $\partial^2 W'/\partial^2 s > 0$ with $s = c_1\sqrt{r/(c_1 + c_2)}$, the minimal value of W' is $2\sqrt{r(c_1 + c_2)}$ with $s = c_1\sqrt{r/(c_1 + c_2)}$, $s' = c_2\sqrt{r/(c_1 + c_2)}$ and $m = \sqrt{r(c_1 + c_2)}$.

5. Countermeasures to the Meet-in-the-Middle Attack

5.1 Range Size of Data Compression Function

As described above, minimum values of the computational complexity $U(U')$, the computational and space complexities product $V(V')$ and the computational and space complexities sum $W(W')$ are

$$U_0 = 2\sqrt{rc_1}, \ V_0 = rc_1 + 1, \ W_0 = 2\sqrt{2rc_1},$$

$$U_0' = 2\sqrt{rc_2}, \ V_0' = rc_1 + c_2/c_1, \ W_0' = 2\sqrt{r(c_1 + c_2)}.$$

Here, $U_0' \geq U_0, V_0' \geq V_0$, and $W_0' \geq W_0$, because $c_2 \geq c_1$. Thus forgery with known used signatures is easier than that with randomly chosen signatures. The larger U_0, V_0 and W_0 make the scheme more secure. Therefore, the security of the mixed-type digital signature is dependent on U_0, V_0 and W_0. These values are proportional to $\sqrt{c_1}$ or c_1. Because parameter c_1 is selected by the system designer, c_1 should be so large that the meet-in-the-middle attack is impossible due to the enormous mount of computational and space complexities required. The following example shows that c_1 should be much larger than 2^{64}.

[Example] Suppose f_1 is implemented by the CBC mode of DES with $c_1 = 2^{64}$, and f_2 is implemented by typical RSA cryptosystem with $c_2 = 2^{512}$. Assume the operational time of f_1 and f_2^{-1} is $1\mu \ sec/operation$ and $1m \ sec/operation$, respectively. Let $ms = 2^{64}$, $r = 1$ so that the success probability be 63%.
(i) The minimal operational time of the sum of m calculations for f_1 and s calculations for f_2^{-1} is obtained when $s = 2^{27}$ and $m = 2^{37}$. The required time for this optimal attack is

$$2^{37} \times 10^{-6} + 2^{27} \times 10^{-3} \, sec \approx 2^{18} \, sec \approx 73 hours \approx 3 days.$$

Here, the required memory size for the intermediate values is

$$2 \times 512 \times 2^{27} bits \approx 16 \ Gigabytes.$$

(ii) Let us consider the situation where the number of known used signatures is s and parallel processing is available. Suppose $s = 2^{10} = 1024$, then $m = 2^{54}$. Suppose $2^{13} (= 8192)$ parallel processors are available. Here, the required time for this attack is

$$2^{54} \times 10^{-6} + 2^{10} \times 10^{-3} sec/2^{13} processor$$
$$\approx 2^{21} sec/processor \approx 24 days/processor.$$

Here, the required memory size for the intermediate values is

$$2 \times 512 \times 2^{10} bits \approx 128 Kbytes.$$

The memory size and the number of parallel processing computers required in (i) and (ii) are easily implemented with the present computer technology. Therefore, it is clear that $c_1 = 2^{64}$ is not secure.

5.2 Domain Sizes of Signature Function

Domain size c_2 is determined to secure f_2. For example, when f_2 is the RSA cryptosystem, c_2 is recommended to be amount of about 500 bits to make the factoring difficult. The computational complexity of factoring the composite c_2 is on the order of $\exp \sqrt{\log c_2 \times \log(\log c_2)}$. Note that

$$\sqrt{c_2} \gg \exp \sqrt{\log c_2 \times \log(\log c_2)},$$

when c_2 is of about 500 bits. Thus the minimal computational complexity U_0' of meet-in-the-middle attack is much larger than the complexity of factoring. Therefore, if c_2 is of about 500 bits, security of the mixed-type digital signature scheme depends mainly on the complexity of factoring rather than the complexity of the meet-in-the-middle attack.

6. Conclusion

The meet-in-the-middle attack can reduce the time to do fraud in mixed-type digital signature schemes or to find secret keys in multiple encryptions much further than the exhaustive attack. In this paper, we have formulated the meet-in-the-middle attack on mixed-type digital signature schemes, have shown the necessary success conditions, and have discussed the relationships between computational and space complexities as well as success probability and the intermediate size needed in the attack. We also have clarified the relationships between the meet-in-the-middle attack and the exhaustive attack.

The meet-in-the-middle attack can be always used in any mixed-type digital signature scheme. We have analyzed an optimal strategy for a forger to apply this attack from various standpoints, and we have shown how to design a secure mixed-type digital signature scheme. We especially have pointed out that 64 bits of intermediate value space with hashing function using DES is too small to counter the meet-in-the-middle attack.

References

[1] Diffie, W., and Hellman, M.: "New direction in cryptography", IEEE Trans. Inf. Theory. IT-22, 6, pp.644-654 (Nov. 1976)

[2] Akl, S. G.: "Digital signatures : A tutorial survey", IEEE Computer, 16, 2, pp.15-24 (Feb. 1983)

[3] Denning, D. E.: "Protecting public keys and signature keys", IEEE Computer, 16, 2, pp.27-35 (Feb. 1983)

[4] Davies, D. W.: "Applying the RSA digital signature to electronic mail", IEEE Computer, 16, 2, pp.55-62 (Feb. 1983)

[5] Koyama, K.: "Fast and Secure Digital Signature Using Public-Key Cryptosystems", Trans. of IECE of Japan, J67-D, 3, pp.305-312 (Mar. 1984)

[6] Ohta, K., and Koyama, K.: "A meet-in-the-middle attack against digital signature methods", Trans. of IECE of Japan, J70-D, 2, pp.415-422 (Feb. 1987)

[7] Davies, D. W., and Price, W. L.: "The application of digital signatures based on public key cryptosystems", Proc of ICC, pp.525-530 (Oct. 1980)

[8] Diffie, W., and Hellman, M.: "Exhaustive cryptanalysis of the NBS data encryption standard", IEEE Computer. 10, 6, pp.74-84 (June. 1977)

[9] National Bureau of Standards: "Data Encryption Standard", FIPS PUB 46, NBS (Jan. 1977)

[10] Rabin, M. O.: "Digitalized signatures and public-key functions as intractable as factorization", Tech. Rep. MIT/LCS/TR MIT Lab. Comput. Sci. (1979)

[11] International Organization for Standardization: "Modes of operation for a 64bit block cipher algorithm, ISO8372 (1987)

[12] Rivest, R. L., Shamir, A., and Adlemen, L.: "A method of obtaining digital signature and public-key cryptosystems", Commun. ACM, 21, 2, pp.120-126 (Feb. 1978)

[13] Winternitz, R. S.: "Producing a one-way hash function from DES", Advances in Cryptology Proceedings of Crypto 83, Plenum Press, pp.203-207, New York (1984)

[14] Akl, S. G.: "On the security of compressed encoding", Advances in Cryptology Proceedings of Crypto 83, Plenum Press, pp.209-230, New York (1984)

[15] Coppersmith, D.: "Another birthday attack (Draft)", Proceedings of Crypto 85, Lecture Note in Computer Science, 218, Springer-Verlag, pp.14-17 (1986)

[16] Merkle, R. C.: "On the Security of Multiple Encryption", Commun. ACM, 24, 7, pp.465-467 (1981)

[17] Even, S., and Goldreich, O.: "On the power of cascade cipher", Advances in Cryptology Proceedings of Crypto 83, Plenum Press, pp.43-50, New York (1984)

[18] Rabin, M. O,: "Digital signatures", Foundation of Secure Computation, Academic Press (1978)

[19] Mueller-Schloer, C.: "DES-generated checksum for electronic signatures", Cryptologia, pp.257-273 (July 1983)

[20] Yuval, G.: "How to swindle Rabin", Cryptologia, 3, 3, pp.187-190 (July 1979)

[21] Girault, M., Cohen, R., and Campana, M.: "A Generalized Birthday Attack", Proceedings of Eurocrypt 88, Lecture Note in Computer Science, 330, Springer-Verlag, pp.129-156 (1988)

Table 1.   Success Probability P and r

r = m s / c	P (%)
0 . 0 1	1 . 0
0 . 0 5	4 . 9
0 . 1	9 . 5
0 . 5	3 9 . 4
1	6 3 . 2
5	9 9 . 3
1 0	9 9 . 9 9 5

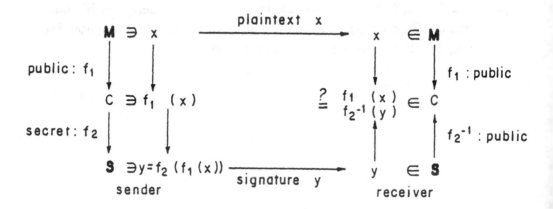

Fig. 1   Mixed-type digital signature scheme

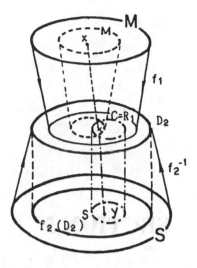

Fig. 2 A meet-in-the-middle attack model

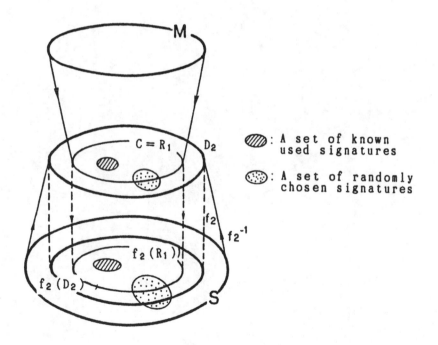

Fig. 3   Known used signatures
and randomly chosen signatures

SECTION 5

PSEUDORANDOMNESS AND SEQUENCES II

A Binary Sequence Generator Based on Ziv-Lempel Source Coding

CEES J.A. JANSEN
Philips USFA B.V.
P.O. Box 218, 5600 MD Eindhoven
The Netherlands

DICK E. BOEKEE
Technical University of Delft
P.O. Box 5031, 2600 GA Delft
The Netherlands

Summary

A new binary sequence generator is proposed, which is based on Ziv-Lempel source coding. In particular the Ziv-Lempel decoding algorithm is applied to codewords generated by linear feedback shift registers. It is shown that the sequences generated in this way have a high linear complexity and good statistical properties. This sequence generator can also be viewed at as an LFSR which is 'filtered' by a nonlinear feedforward function containing memory. In the simplest case the sequence generator reduces to an LFSR filtered by a memoryless nonlinear feedforward function, thereby demonstrating that our construction can be seen to produce very efficient feedforward functions from a circuit complexity point of view.

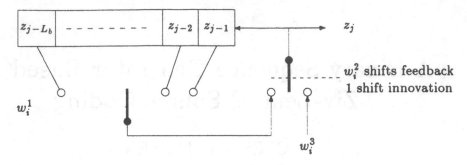

Figure 1: Ziv-Lempel decoding algorithm as a FSR

1 The Ziv-Lempel Data Compression Algorithm

In [Ziv 77] a universal algorithm for sequential data compression has been presented. The algorithm is based on the complexity considerations of Lempel and Ziv as described in [Lemp 76] and uses maximum length copying from a buffer containing a fixed number of most recent output characters. The idea is that a sequence can be parsed into successive components (subsequences). Each component is a copy of some subsequence occurring earlier in the sequence, except for its last character which is called the innovation. So each component is fully determined by a pointer to the starting position of the subsequence to be copied, the length of this subsequence and the innovation character. The compression is generally obtained by only transmitting this pointer value, the length and the innovation character.

In the coding algorithm a buffer of length L_b is employed, which contains the last L_b characters of some sequence z. Initially the buffer is filled with some fixed sequence. Successive subsequences of z are encoded by searching for the longest subsequence in the buffer which can be copied to obtain the sequence z. The length of the subsequence to be copied from the buffer is also limited to some prescribed value L_s. Hence, the codewords consist of a starting point in the buffer (called the pointer), the copy length and the innovation character.

The decoding algorithm works in a similar way. It also uses a buffer of length L_b, which contains the last L_b characters of the sequence z to copy from. The decoding algorithm can very elegantly be implemented with a shift register as depicted in Figure 1. If $W_i = w_i^1 w_i^2 w_i^3$ denotes the ith codeword, then w_i^1 denotes the position of the tap of the shift register

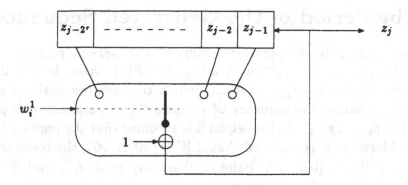

Figure 2: A switch controlled feedback shift register

which is fed back to the input, w_i^2 denotes the number of shifts the register performs in this feedback mode, and w_i^3 denotes the innovation character which is shifted into the register without feedback.

2 Generating Sequences with the Ziv-Lempel Decoding Algorithm

The Ziv-Lempel decoding algorithm is simplified if we restrict the character alphabet to be $GF(2)$. In this case the innovation character does not have to encoded, as it is the complement of the character which follows the subsequence that has been copied from the register. So now w_i^2 shifts are performed in the feedback mode, followed by one shift with the feedback complemented.

Our approach is to use the binary Ziv-Lempel decoding algorithm as a sequence generator. To achieve this, the codewords comprising w_i^1 and w_i^2 are generated by linear feedback shiftregisters. In doing so, we aim at obtaining an output sequence wich resembles a real random sequence very well. It seems, however, that this sequence generator is rather complex to analyze in great detail and therefore we restrict ourselves to the special case where $w_i^2 = 0$ for all i, as depicted in Figure 2. We call this FSR sequence generator a *Switch Controlled FSR*. In the next sections various properties of this sequence generator will be considered.

3 The Period of the Generated Sequence

Let $\underline{s}_0, \underline{s}_1, \ldots, \underline{s}_{r-1}$ denote periodic sequences with periods $p_0, p_1, \ldots, p_{r-1}$, i.e. $\underline{s}_i = (s_{0,i}, s_{1,i}, \ldots, s_{p_i-1,i})^\infty$ with all $s_{j,i} \in GF(2)$. Also, let $\underline{W}_j$ denote the binary r-tuple $(s_{j,0}, s_{j,1}, \ldots, s_{j,r-1})$ which controls the position of the feedback tap, hence, the sequence of r-tuples $\underline{W}_j$ is periodic with period $p = \mathrm{lcm}(p_0, p_1, \ldots, p_{r-1})$. In the sequel it is assumed that the register length $L_b = 2^r$. Moreover it is assumed that if $\underline{W}_j = (0, \ldots, 0)$, the feedback tap is at z_{j-1}, if $\underline{W}_j = (0, \ldots, 0, 1)$ the feedback tap is at z_{j-2} and if $\underline{W}_j = (1, \ldots, 1)$, the feedback tap is at z_{j-2^r}.

The output sequence $\underline{z}$ satisfies the recursion:

$$
\begin{aligned}
z_j &= z_{j-I(\underline{W}_j)} + 1 \\
&= F(\underline{W}_j, z_{j-1}, \ldots, z_{j-2^r}) \\
&= 1 + \sum_{i=0}^{2^r-1} F_i(\underline{W}_j) z_{j-i-1},
\end{aligned}
\tag{1}
$$

where $I(\underline{W}_j)$ denotes the integer representation of the r-tuple $\underline{W}_j$ and the switching functions F_i, $i = 0, 1, \ldots, 2^r - 1$ are mappings from $GF(2)^r$ onto $GF(2)$, defined as:

$$
F_i(\underline{X}) = \begin{cases} 1, & \text{if } I(\underline{X}) = i, \\ 0, & \text{else.} \end{cases}
$$

Equation (1) shows that our sequence generator can be regarded as a FSR with a nonlinear feedback function containing memory.

The generated sequence $\underline{z}$ will ultimately be periodic with some period p_z. To determine the period p_z we first determine the period p_d of the sequence $\underline{d}$, which is obtained from $\underline{z}$ by addition of a one times delayed version, i.e. $d_j = z_j + z_{j-1}$ for all j. By adding z_{j-1} to both sides of equation (1) a recursion for $\underline{d}$ is obtained which resembles that of $\underline{z}$, i.e.:

$$
\begin{aligned}
d_j &= 1 + \sum_{i=0}^{2^r-1} G_i(\underline{W}_j) d_{j-i-1} \tag{2} \\
&= 1 + \sum_{i=1}^{I(\underline{W}_j)} d_{j-i}, \tag{3}
\end{aligned}
$$

where the switching functions G_i, $i = 0, 1, \ldots, 2^r - 1$ are mappings from $GF(2)^r$ onto $GF(2)$, defined as:

$$
G_i(\underline{X}) = 1 + \sum_{k=0}^{i} F_k(\underline{X}) = \begin{cases} 0, & \text{if } I(\underline{X}) \le i, \\ 1, & \text{else.} \end{cases}
$$

The following result is an immediate consequence of the recursion relation (3).

Lemma 1 *Let $\underline{W}_j$ and $I(\cdot)$ be as defined before. If for some $i > 0$, we have that $I(\underline{W}_{i+k}) \leq k$, for $k = 0, 1, \ldots, 2^r - 1$, then $d_j = d_{j+p}$ for all $j \geq i$.*

Proof. Equation (3) implies that by the condition $I(\underline{W}_{i+k}) \leq k$, for $k = 0, 1, \ldots, 2^r - 1$ all d_j, $j \geq i$, are completely determined by the sequence of $\underline{W}_j$'s and not by the initial z-register contents. But because of the periodicity of the sequence of r-tuples the same is true for the d_{j+p}. The result now follows by induction on the d_j. $\qquad\square$

Note that Lemma 1 states a sufficient condition for the period of $\underline{d}$, p_d, to divide p, but this condition is not strictly necessary.

If the switch driving sequence is generated by a linear FSR, the period of $\underline{d}$ is exactly equal to p, as expressed by the next proposition.

Proposition 2 *If the r sequences $\underline{s}_0, \underline{s}_1, \ldots, \underline{s}_{r-1}$ are taken from r different stages of a linear feedback shift register with irreducible connection polynomial having period p, and the condition $\underline{W}_i = \underline{W}_{i-1} = \cdots = \underline{W}_{i+2-2^r} = (0, \ldots, 0)$ holds, then the period p_d of $\underline{d}$ is equal to p.*

Proof. From Lemma 1 it follows that p_d divides p. Due to equation (3) the condition $\underline{W}_i = \underline{W}_{i-1} = \cdots = \underline{W}_{i+2-2^r} = (0, \ldots, 0)$ implies that a run of $2^r - 1$ or more ones occurs at least once in $\underline{d}$. This run, however, can only be terminated if the switch, pointing to an even tap position, is set to point to an odd position. This odd position can only be obtained by having $s_{i,0} = 1$ for some $i > 0$. Consequently, as $d_j = d_{j+p_d}$ for all j, there must at least exist one i such that $s_{i,0} = s_{i+p_d,0} = s_{i+2p_d,0} = \cdots = 1$. As the sequence $\underline{s}_0$ is periodic with period p, it follows from the decimation principle (see e.g. [Ruep 84, Ch. 6]) that $p_d = p$. $\qquad\square$

One can prove that Proposition 2 also holds if the r sequences, which constitute $\underline{W}_j$, are obtained from r separate linear FSR's with primitive connection polynomials which are pairwise relatively prime.

From the relation between $\underline{d}$ and $\underline{z}$ it is easy to see that the period p_z of $\underline{z}$ is equal to p_d if the number of ones in one period of $\underline{d}$ is even. If, however, the number of ones in one period of $\underline{d}$ is odd, $p_z = 2p_d$ and the second half of $\underline{z}$ is the complement of the first half.

Connection polynom. of driving LFSR	Period	$\Lambda(\underline{d})$		
		$r = 1$	$r = 2$	$r = 3$
$x^3 + x + 1$	7	6		
$x^4 + x + 1$	15	12	10	14
$x^5 + x^2 + 1$	31	30	31	31
$x^6 + x + 1$	63	48	56	60
$x^7 + x^3 + 1$	127	126	127	120
$x^8 + x^4 + x^3 + x^2 + 1$	255	250	255	254
$x^9 + x^4 + 1$	511	510	511	510
$x^{10} + x^3 + 1$	1023	1007	1013	1022
$x^{11} + x^2 + 1$	2047	2046	2047	2047

Table 1: Linear complexity of $\underline{d}$

4   Complexity Aspects

The main purpose of the nonlinear switching function is to generate an output sequence with complexity close to that of a real random sequence, whereas the switch driving sequences have low complexities. Also, the good statistical properties should be maintained. The linear complexity of $\underline{d}$ has been determined experimentally for several values of r and the length L of the linear FSR driving the feedback switch. In the experiment it was ensured that d has the same period as the switch driving sequences. The results of these experiments are listed in Table 1. Table 1 shows that the linear complexities have values very close to the corresponding periods, which is obviously what one would expect for periodic random sequences [Ruep 84, pg. 51]. It also appears, however, that for some combinations of linear FSR and r degeneracies occur.

The maximum order complexity as introduced by Jansen [Jans 89] has also been considered and it appears that the maximum order comlexities vary much more than the linear complexities. The results are shown in Table 2. The maximum order complexities as listed in Table 2 were determined using one period of $\underline{z}$ if p_z was equal to p_d, else only half a period of $\underline{z}$ was used. In the latter case the table entries are marked with a * and have $p_z = 2p_d$. From the table it can be seen that with every value of r there tends to be some region in the LFSR order L, for which $c(\underline{z}^1) \approx 2L$. For higher and lower values of L the comlexities can deviate substantially from the expected value of $2L$.

Connection polynom. of driving LFSR	$c(\underline{z}^1)$			
	$r=1$	$r=2$	$r=3$	$r=4$
$x^3 + x + 1$	2			
$x^4 + x + 1$	6	5	8	
$x^5 + x^2 + 1$	12	7*	12*	
$x^6 + x + 1$	11	10	12	
$x^7 + x^3 + 1$	21	17*	15*	12
$x^8 + x^4 + x^3 + x^2 + 1$	16	17*	15	13
$x^9 + x^4 + 1$	28	23*	16	19*
$x^{10} + x^3 + 1$	41	26*	20	20*
$x^{11} + x^2 + 1$	45	31*	33*	20*

Table 2: Maximum order complexity of a single period of $\underline{z}$

5 An Efficient Feedforward Function

For the case $r = 1$ the sequence generator of Figure 2 allows a nice alternative representation. In this case there is only one switch driving sequence, which determines whether z_{j-1} or z_{j-2} is fed back. We find the following recursion relations:

$$z_j = (s_j + 1)z_{j-1} + s_j z_{j-2} + 1, \tag{4}$$
$$d_j = s_j d_{j-1} + 1. \tag{5}$$

The corresponding $\underline{d}$ sequence generator is depicted in Figure 3. The recurrence relation (5) can be written as:

$$d_j = 1 + s_j(1 + s_{j-1}(1 + s_{j-2}(1 + \cdots))). \tag{6}$$

However, since $\underline{s}$ is a linear FSR sequence, the product terms in (6) of degree exceeding the FSR length L will all be zero. Hence, we obtain:

$$d_j = 1 + s_j(1 + s_{j-1}(1 + \cdots (1 + s_{j-L+1}))) \tag{7}$$
$$= 1 + s_j + s_j s_{j-1} + \cdots + s_j s_{j-1} \cdots s_{j-L+1}. \tag{8}$$

So for $r = 1$ the sequence $\underline{d}$ is identical to the sequence obtained directly from the LFSR by a nonlinear feedforward function as given by expression (8). From a circuit complexity point of view the feedforward function of Figure 3 is much more efficient than that given by (8).

For nonlinear feedforward functions, acting on LFSR sequences, Rueppel [Ruep 84] has given a lowerbound on the linear complexity. However,

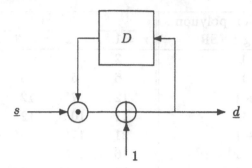

Figure 3: Feedforward equivalent of $\underline{d}$ generator for $r = 1$

because the highest order product term in (8) has degree L, this lower-bound does not explain the experimental results as listed in Table 1. In connection with this table we conjecture that for $r = 1$ and L prime the linear complexity of $\underline{d}$ is equal to $2^L - 2$.

6 Statistical Properties

The recurrence relation (1) provides for a balanced number of ones and zeroes to be stored in the register. This may be regarded as a form of 'negative feedback', i.e. an unbalance in the distribution of ones and zeroes will eventually be corrected.

If we assume that the switch is driven by real random sequences, then the expectation of an output character is fully determined by the distribution of the previous characters stored in the register. Consequently, there exists a correlation between consecutive characters of $\underline{z}$. This correlation depends on the register length L_b. If L_b is small the correlation is rather strong, which is demonstrated by the case $r = 1$. Then equation (4) shows that, if $z_{j-2} = z_{j-1}$, z_j becomes the complement of z_{j-1}, independent of s_j. As a result the maximum run-length of runs in $\underline{z}$ will be 2. In general the maximum run-length of runs in $\underline{z}$ will be limited to L_b. We therefore conclude that the period p of the switch driving sequence should be close to 2^{L_b} and L_b sufficiently large.

There is obviously also correlation between the sequences $\underline{s}$ and $\underline{z}$. The recurrence relation (8) can be obtained by a simplification of the 'flipflop' sequence generator, proposed by Pless [Ples 77]. It can be seen that the $\underline{d}$ sequence can be obtained from the $\underline{s}$ sequence through one JK-flipflop. Generating $\underline{z}$ requires an additional JK-flipflop as depicted in Figure 4.

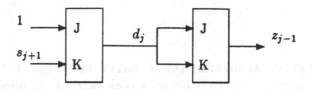

Figure 4: Equivalent JK–flipflop generator for $r = 1$

In [Rubi 79] it was shown that it is computationally feasible to break Pless' cipher by using the correlation between input and output sequences. A similar attack is possible on the generator proposed in this section for larger values of L_b. However, the computational complexity will be highly increased.

References

[Jans 89]   C. J. A. Jansen. *Investigations On Nonlinear Streamcipher Systems: Construction and Evaluation Methods*, PhD. Thesis, Technical University of Delft, Delft, april 1989.

[Lemp 76]   A. Lempel and J. Ziv. "On the Complexity of Finite Sequences", *IEEE Trans. on Info. Theory*, vol. IT–22, no. 1, pp. 75–81, January 1976.

[Ples 77]   V. S. Pless. "Encryption Schemes for Computer Confidentiality", *IEEE Trans. on Comp.*, vol. C–26, pp. 1133–1136, November 1977.

[Rubi 79]   F. Rubin. "Decrypting a Stream Cipher Based on JK-flipflops", *IEEE Trans. on Comp.*, vol. C–28, pp. 483–487, July 1979.

[Ruep 84]   R. A. Rueppel. *New Approaches to Stream Ciphers*, PhD. Thesis, Swiss Federal Institute of Technology, Zurich, 1984.

[Ziv 77]   J. Ziv and A. Lempel. "A Universal Algorithm for Sequential Data Compression", *IEEE Trans. on Info. Theory*, vol. IT–23, no. 3, pp. 337–343, May 1977.

A FAST ITERATIVE ALGORITHM FOR A SHIFT REGISTER INITIAL STATE RECONSTRUCTION GIVEN THE NOISY OUTPUT SEQUENCE

Miodrag J. Mihaljević

Jovan Dj. Golić

Institute of Applied Mathematics and Electronics, Belgrade,
Faculty of Electrical Engineering, University of Belgrade,
Bulevar Revolucije 73, 11001 Beograd, Yugoslavia

ABSTRACT: A novel fast algorithm for the correlation attack on a class of stream ciphers is proposed. The algorithm is based on the error correction principle and the finite-state matrix representation of a linear feedback shift register. Some general properties of the algorithm are pointed out and illustrated by some experimental results.

I. INTRODUCTION

A weakness of a class of running key generators for stream ciphers is demonstrated in [1]. It is shown there that if the key stream is correlated to the output sequence of a binary linear feedback shift register (LFSR), a correlation attack can be performed to reconstruct its initial state. This problem of cryptanalysis can be regarded as the problem of a LFSR initial state reconstruction using the noisy output sequence or, equivalently, as a decoding problem of the corresponding binary linear block code. In [2] two algorithms (Algorithms A and B) for efficient realization of this task are presented.

In this paper we consider a class of algorithms to which Algorithm B belongs. In this class the initial state reconstruction is based on the error correction principle. It means that the procedure is iterative: in each step we first calculate the posterior probabilities, bit-by-bit (phase I), and then make a bit-by-bit decision (phase II). Note that the phase I of Algorithm B is based on the same principle as the decoding scheme for low-density parity-check code from [3]. The phase II of Algorithm B is a bit-by-bit decoding typical for an error correction procedure.

Denote by $\{x_n\}_{n=1}^{N}$ the output segment of a LFSR of length L. In a statistical model, a binary noise sequence $\{e_n\}_{n=1}^{N}$ is assumed to be a realization of a sequence of i.i.d. binary variables $\{E_n\}_{n=1}^{N}$ such that $\Pr(E_n=1) = p_n$, n=1,2,...,N . Let $\{z_n\}_{n=1}^{N}$ be a noisy version of $\{x_n\}_{n=1}^{N}$ defined by $z_n = x_n \oplus e_n$, n=1,2,...,N . The problem under consideration is a reconstruction of the LFSR initial state given the segment $\{z_n\}_{n=1}^{N}$, provided that the feed-back function and p $(p_n=p$, n=1,2,...,N) are known.

II. ON ITERATIVE RECALCULATION OF POSTERIOR PROBABILITIES

Suppose that there are M_n orthogonal parity-check sets involving the n-th bit e_n and let $\Pi_{n,m}$ denote the m-th parity-check set of the n-th bit, m=1,2,...,M_n , n=1,2,...,N. In [2], the cardinality of every set $\Pi_{n,m}$ is equal to the number of nonzero coefficients of the LFSR characteristic polynomial.

Using [3, Theorem 1], the following formula for the posterior probabilities can be established:

$$A_n = \frac{\Pr(E_n=0 \mid \{\Pi_{n,m}\}_{m=1}^{M_n})}{\Pr(E_n=1 \mid \{\Pi_{n,m}\}_{m=1}^{M_n})} = \frac{1-p_n}{p_n} \prod_{m=1}^{M_n} \left\{1 + \frac{2}{\left[\prod_{\ell \in \Pi_{n,m}} \left(1 + \frac{2}{\frac{1-\hat{p}_\ell}{\hat{p}_\ell} - 1}\right)\right] - 1}\right\},$$

$$(1)$$

where for every $\Pi_{n,m}$, $m=1,2,\ldots,M_n$, $n=1,2,\ldots,N$, :

$p_n \neq 0.5$,

$\hat{p}_\ell = p_\ell$, for $\ell \in \Pi_{n,m}$, $\ell \neq n$,

$$\hat{p}_\ell = \begin{cases} p_\ell & , \ \ell = n \text{ , when m-th parity-check is satisfied} \\ 1-p_\ell & , \ \ell = n \text{ , when m-th parity-check is not satisfied} \end{cases} .$$

    In [2, Algorithm B, phase I] the posterior probabilities are iteratively recalculated, so that in the i-th iteration step the prior probabilities are replaced with the posterior ones calculated in the (i-1)-th iteration step.

    The main effect of the iterative procedure is to take into account a large number of error bits and thus make a more reliable decision. However in formula (1) the bits are treated as if they were independent, which is not true. Moreover, the interdependence is inherently introduced. If all the error bits were independent in each iteration step, that is, if all the parity-checks involved were orthogonal, which is desirable [3] , [4] , then the recalculation procedure would be exactly the same as in [3]. The possibility to apply formula (1) to the case when the parity checks are not orthogonal was also mentioned in [3].

    Instead of implicit use of a large number of orthogonal [3] or non-orthogonal [2] parity-checks through an iterative recalculation procedure, in this paper we propose an algorithm which in one step explicitly employs a large number of orthogonal parity-checks, based on the finite-state matrix representation of LFSR. As in [2 , Algorithm B, phase II], we also incorporate an appropriately defined error correction principle.

III. A NOVEL ALGORITHM FOR INITIAL STATE RECONSTRUCTION

    Denote by $X_t = [x_{t+\ell}]_{\ell=1}^L$ an L-dimensional binary vector defining the t-th state of LFSR

$$X_t = A^t X_0 \qquad , \tag{2}$$

where X_0 is the initial state, and A^t is the t-th power, in GF(2) , of the LxL-dimensional binary matrix determined by the LFSR's characteristic polynomial. Let the segments of $\{e_n\}_{n=1}^N$ and $\{z_n\}_{n=1}^N$ define the L-dimensional binary vectors $E_t = [e_{t+\ell}]_{\ell=1}^L$ and $Z_t = [z_{t+\ell}]_{\ell=1}^L$. It is clear that

$$E_t = (Z_t \oplus A^{t-k}Z_k) \oplus A^{t-k}E_k$$

or

$$E_t \oplus A^{t-k}E_k = (Z_t \oplus A^{t-k}Z_k) \tag{3}$$

On the basis of (3) a great number of orthogonal parity-checks can be defined. Precisely, we use the following parity-check set.

Definition 1: Π_n is a set of orthogonal parity-checks of the n-th bit, being the union of two sets $\Pi_n^{(*)}$ and $\Pi_n^{(**)}$ that are defined by:

- $\Pi_n^{(*)}$: maximum cardinality set of the standard mutually orthogonal Meier-Staffelbach (M-S) parity-checks of the n-th bit ;
- $\Pi_n^{(**)}$: maximum cardinality set of the mutually orthogonal parity-checks corresponding to (3) that are orthogonal to the parity-checks from $\Pi_n^{(*)}$.

Let $N_n(w)$ be the number of parity-checks in Π_n with w elements and $S_n(w)$ be the number of satisfied parity-checks among them, $w=1,2,\ldots,L+1$, $n=1,2,\ldots,N$.

With $p_n = p < 0.5$, $n=1,2,\ldots,N$, denote by p_w , $w=1,2,\ldots,L+1$, a sequence of real numbers defined by, see [3]:

$$p_w = \frac{1 - (1 - 2p)^w}{2} , \qquad w=1,2,\ldots,L+1 , \tag{4}$$

which is the solution of the corresponding recursion [2] . Then the characteristic ratio of posterior probabilities is:

$$A_n = \frac{1 - P(E_n=1 \mid \Pi_n)}{P(E_n=1 \mid \Pi_n)} = \frac{1 - p}{p} \prod_{w=1}^{L+1} [(1/p_w)-1]^{2S_n(w)-N_n(w)} ,$$

$$n=1,2,\ldots,N \quad . \quad (5)$$

It can be shown that on an infinite sequence, $N = \infty$, formula (5) ensures the distinction between the cases $E_n = 0$ and $E_n = 1$, as desired.

We are now ready to define a <u>basic</u> form of the algorithm. (further improvements of the algorithm are possible)

A L G O R I T H M :

Step 1: Form the set Π_n and determine $N_n(w)$, $w=1,2,\ldots,L+1$, $n=1,2,\ldots,N$. Define the initial value p and adopt the algorithm parameters ϵ_1 , ϵ_2 , $0 < \epsilon_1, \epsilon_2 < 1$. Calculate $No = p (1-\epsilon_1) N$. (typically, ϵ_1 is close to one and ϵ_2 is close to zero)

Step 2: Determine $S_n(w)$, $w=1,2,\ldots,L+1$, $n=1,2,\ldots,N$.

Step 3: Using (4) for the current p calculate the sequence p_w , $w=1,2,\ldots,L+1$.

Step 4: According to (5), calculate the sequence A_n , $n=1,2,\ldots,N$.

Step 5: Determine a set π of No indices n that correspond to the smallest values of A_n .

Step 6: For every $n \in \pi$ complement the current value z_n .

Step 7: Estimate the average number of erroneosly changed digits $\Delta = \sum_{n \in \pi} A_n (1 + A_n)^{-1}$. Estimate the new value of p according to the following:

$$p = p_{new} = p_{old} - (No-2\Delta)/N .$$

Step 8: If $p > \epsilon_2$ repeat the Steps 2 - 7 ;

If $p \leq \epsilon_2$ go to Step 9 .

Step 9: Using the standard Siegenthaler's correlation method [1] on the initial noisy sequence $\{z_n\}_{n=1}^{N}$ test the hypothesis that the shifting block of L succesive bits is correct until the true initial state is reached.

IV. DISCUSSION AND EXPERIMENTAL RESULTS

In this section, we point out some general properties of the proposed algorithm, supported by some experimental results.

(i)   Algorithm is not worse than the Meier-Staffelbach's Algorithm B.

According to Definition 1 (see meaning of the set $\pi_n^{(*)}$) and the structure of the algorithm it is clear that the performance of the proposed algorithm can not be worse than the performance of the Meier-Staffelbach's Algorithm B. The novel algorithm includes all the M-S parity-checks and the additional ones can not degrade the performance. This is due to a pattern recognition principle: more information in general gives rise to a smaller decission error.

An experimental example where both the M-S Algorithm B and our algorithm work is the following:

L = 29 , # feedback tapes = 4 , $N/L = 10^3$, p = 0.275 .

(ii)   <u>The expected number of the low-weight matrix parity-checks is considerable and independent of the number of feed-back tapes.</u>

Define the weight of a parity-check as the number of involved elements. It can be shown that the average relative number of the parity-checks of weight w in $\Pi_n^{(**)}$, on the whole period $N = 2^L - 1$, corresponds to a binomial distribution $B(L, 0.5)$, which is independent of the number of feedback tapes.

Accordingly, the expected number $\overline{N}_k$ of the matrix parity-checks of weight k, given the observed segment of length N, is

$$\overline{N}_k = \frac{N}{L\,2^L} \binom{L}{k} \quad . \tag{6}$$

So, the expected total number of the matrix parity-checks of weight less than or equal to k is

$$\overline{N}_{1,k} = \frac{N}{L\,2^L} \sum_{\ell=1}^{k} \binom{L}{\ell} \quad . \tag{7}$$

On the other hand, the expected number of the Meier-Staffelbach's parity-checks (of weight w) is

$$\overline{M}_w = (w+1)\,\log_2(N/2L) \quad . \tag{8}$$

where w is the number of feedback tapes.

The following tables provide a numerical comparison between $\overline{N}_{1,w}$, $\overline{N}_w$, and $\overline{M}_w$. We can see that the expected number of the matrix parity-checks with the weight equal to the weight of the M-S parity-checks becomes significantly greater than the number of the M-S parity-checks, after some value N/L .

Table 1 a) Expected number of the parity checks as a function of
N/L when shift-register length is 29 and number of
feedback tapes is 4 .

N/L	exp. # of matrix parity checks with the "weight" 1 – 4	exp. # of matrix parity checks with the "weight" 4	exp. # of the Meier-Staffelbach's parity checks (the "weight" = 4)
10	0	0	10
10^2	0	0	25
10^3	0	0	40
10^4	1	0	60
10^5	5	4	75
10^6	52	44	90
10^7	519	442	110
10^8	5186	4424	125
10^9	51856	44240	140
10^{10}	518560	442397	160

Table 1 b) Expected number of the parity checks as a function of
N/L when shift-register length is 60 and number of
feedback tapes is 18 .

N/L	exp. # of matrix parity checks with the "weight" 1 – 18	exp. # of matrix parity checks with the "weight" 18	exp. # of the Meier-Staffelbach's parity checks (the "weight" = 18)
10	0	0	38
10^2	0	0	95
10^3	1	1	152
10^4	13	8	228
10^5	134	80	285
10^6	1335	802	342
10^7	13352	8023	418
10^8	133522	80234	475
10^9	1335218	802335	532
10^{10}	13352181	8023353	608

(iii)   <u>A lower bound on the acceptable noise due to the impact of the low-weight matrix parity-checks can be derived</u>.

The acceptable noise is the noise for which an algorithm is successful in most experiments.

In the Meier-Staffelbach's paper [2], an analysis of the acceptable noise as a function of the ratio N/L and the number of feedback tapes w is presented.

In a simmilar manner, a lower bound on the acceptable noise can be established for the novel algorithm as well. We take into account only the effect of the low-weight matrix parity-checks of weight equal to the weight of the M-S parity-checks.

The following tables give an illustration of the acceptable noise for both the Meier-Staffelbach's and our algorithms. Note that the acceptable noise is a function of:
-     N/L and w in the Meier-Staffelbach's algorithm,
-     N/L , w , and L in the novel algorithm.

Table 2 a) The values of acceptable noise as a function of N/L when shift-register length is   29   and number of feedback tapes is 4 .

N/L	sum of parity checks with "weight" = 4 M-S's + matrix's	M-S Algorithm expected acceptable noise	Novel Algorithm expected lower bound on acceptable noise
10	10 + 0	0.120	0.120
10^2	25 + 0	0.246	0.246
10^3	40 + 0	0.292	0.292
10^4	60 + 0	0.321	0.321
10^5	75 + 4	0.337	0.337
10^6	90 + 44	0.350	0.361
10^7	110 + 442	0.359	0.393
10^8	125 + 4424	0.366	0.427
10^9	140 + 44240	0.372	
10^{10}	160 + 442397	0.379	

Table 2 b) The values of acceptable noise as a function of N/L
          when shift-register length is   40   and number of
          feedback tapes is 16 .

N/L	sum of parity checks with "weight" = 16 M-S's + matrix's	M-S Algorithm expected acceptable noise	Novel Algorithm expected lower bound on acceptable noise
10	34 + 1	0.020	0.020
10^2	85 + 6	0.074	0.075
10^3	136 + 57	0.092	0.096
10^4	204 + 572	0.104	0.122
10^5	255 + 5716	0.111	0.155
10^6	306 + 57164	0.117	
10^7	374 + 571637	0.123	
10^8	425 + 5716365	0.126	
10^9	476 + 57163653	0.130	
10^{10}	544 + 571636534	0.134	

The Tables are self-explanatory. When the number of the
matrix parity-checks of weight equal to the weight of the M-S
parity-checks, is significantly larger than the number of the M-S
parity-checks, the novel algorithm becomes supperior.
    Consequently, we give an experimental example where our
algorithm succeeds and the M-S one fails:
    L=40 , # feedback tapes = 16 , N/L=10^3 , p=0.102 .
    Note that for 16 feedback tapes and p=0.104 , the M-S
algorithm needs N/L=10^4.

(iv)   <u>Our algorithm works for large number of feedback
         tapes, where the M-S one fails.</u>

    Namely, according to (ii) it follows that for a large number
of feedback tapes ($\approx$ or $>$ L/2) the number of the matrix

parity-checks prevails over the number of the M-S ones. Therefore, for sufficiently long an observed sequence our algorithm can work where the M-S one fails.

V. CONCLUSION

Starting from the Meier-Staffelbach's algorithm [2] and taking the decoding theory approach, a novel algorithm for the initial state reconstruction given the noisy output sequence, using the finite-state matrix representation of a linear feed-back shift register and iterative error correction principle, is proposed. Some general properties of the proposed algorithm are pointed out and illustrated by some experimental results. The novel algorithm is not worse than the Meier-Staffelbach's Algorithm B, and in general it works for large number of feed-back tapes and larger noise, assuming a sufficiently long observed sequence, where the M-S one fails.

REFERENCES

[1] T.Siegenthaler, "Decrypting a Class of Stream Ciphers Using Ciphertext Only", IEEE Trans. Comput., vol. C-34, Jan. 1985, pp.81-85.

[2] W.Meier, O.Staffelbach, "Fast Correlation Attacks on Certain Stream Ciphers", Journal of Cryptology, vol.1, 1989., pp.159-176.

[3] R.G.Gallager, "Low-Density Parity-Check Codes", IRE Trans. Inform. Theory, vol. IT-8, Jan. 1962, pp.21-28.

[4] G.Battail, M.C.DeCouvelaere, P.Godlewski, "Replication Decoding", IEEE Trans. Inform. Theory, vol. IT-25, May 1979, pp.332-345.

Parallel Generation of Pseudo-Random Sequences

Reihaneh Safavi-Naini

Department of Computer Science

University College, ADFA

UNSW, Canberra, ACT 2600

Australia

Abstract

Nonlinear filtering of the states of LFSR is proposed to generate parallel pseudo-random (PR) sequences. The result of filtering is a set of sequences over $GF(2^m)$ called exponent sequences. Each exponent sequence can be regarded as m parallel component sequences. Assessment of the parallel generator is discussed and statistical properties of individual component sequences and their mutual statistical dependence are examined. Upper bounds on the linear equivalence of exponent sequences and their associated component sequences are developed and it is shown that in both cases higher linear complexity can only be achieved in higher exponents. It is noticed that exponents that are prime to 2^m-1 produce sequences of maximum period and their component sequences have small cross correlation (equal to -1) and low statistical dependence. Finally directions for further research are proposed.

1. Introduction

Non-linear filtering of the states of a linear feedback shift register (LFSR) is studied in [1]. In this paper we introduce exponentiation function over finite fields as a non-linear filter. The study is limited to the case where the field is determined by the characteristic polynomial of the LFSR. The exponentiation is applied to the stages of a LFSR of length m and primitive characteristic polynomial, the output of which is a sequence over $GF(2^m)$ which can be regarded as m parallel component sequences. The properties of these sequences with regard to their application to secure systems are studied.

Section 2 is devoted to a general study of q-ary sequences, their associated component sequences and their statistical assessment. In section 3, exponent sequences are introduced, and their properties are studied in the section 4. In section 5 and 6 linear equivalence of exponent sequences and component sequences are analysed. Finally, in section 7, conclusions and direction for future research are presented. All the proofs are omitted and more details can be found in [8].

2. Q-ary Sequences and Their Linear Equivalence

Let $s=s_0, s_1,...$ denote a sequence with $s_i \in GF(2^q)$. The sequence s is called a q-ary sequence. Every element of s can be represented by a binary q-tuple :

$$s_i=(s_{i0}, s_{i1}, s_{i2},...s_{1,q-1}) \qquad s_{ij} \in GF(2)$$

Definition: The binary sequence s_{se} obtained from s by replacing s_i with the binary q-tuple representing s_i is called the serial sequence :

$$s_{se}=s_{00},s_{01},...s_{0,q-1},s_{10},s_{11},...$$

Definition: The binary sequence $s(i)$, $0 \leq i \leq q-1$, obtained by taking the ith components of consecutive elements of the sequence s is called the ith component sequence:

$$s(i)=s_0(i), s_1(i),..= s_{0i}, s_{1i}, s_{2i},.. \qquad 0 \leq i \leq q-1$$

a.   Linear Equivalence
Linear equivalence of a sequence is a well established measure of its complexity [1]. A q-ary sequence s satisfies a q-ary linear recurrence of degree m if every element of the sequence can be written as a linear combination over $GF(2^q)$ of m sequence element preceding it:

$$s_j = \sum_{i=1}^{m} a_i s_{j-i} \qquad a_i \in GF(2^q) \qquad j > m$$

The order of the q-ary recurrence of minimum order satisfied by s is called the *q-ary linear equivalence* of s and can be calculated by using the Berlekamp-Massey algorithm [3].

A q-ary sequence s satisfies a *binary linear recurrence* if there exist a linear recurrence with binary coefficients satisfied by s:

$$s_j = \sum_{i=1}^{m} a_i s_{j-i} \qquad a_i \in GF(2)$$

Proposition 1
Let s satisfy a q-ary recurrence of degree m. Every component of s_i can be written as a linear combination of mq components of s_{i-j}, $1 \leq j \leq m$. If s satisfies a binary recurrence of order m every component sequence satisfies the same recurrence.

Corollary 1: Let s denote a q-ary sequence that satisfies a binary recurrence of order m with characteristic polynomial $f(x)$. The serial sequence s_{se} satisfies a recurrence of order mq the characteristic polynomial of which is $f(x^q)$.

This is true because all component sequences satisfy the same recurrence.

Corollary 2 : The order of the binary recurrence satisfied by a q-ary sequence gives an upper bound on the order of the minimum recurrences for component sequences.

In fact every binary recurrence satisfied by a q-ary sequence is satisfied by all its component sequences. It is seen that if the order of a linear recurrence satisfied by a sequence s is known, the elements of the sequence can be used to determine the coefficients of the recurrence. In fact it follows that from the above proposition, for a sequence satisfying a q-ary recurrence of order m, mq linear equations (q binary equations for every q-ary equation) should be solved to find a_{ij}, $0 \leq i \leq q-1$, $1 \leq j \leq m$, while for sequences satisfying a binary recurrence, $m + |m/q|$ consecutive elements are sufficient (because every extra element provides q linear equation).

Definition: Weight of $u \in GF(2^q)$ is the number of non-zero components of u as a binary vector of length q.

Definition: The cyclotomic coset mod 2^m-1 over $GF(2)$ which contains i is denoted by C_i and defined by :

$$C_i = \left\{ i, 2i, i2^2, \ldots i2^{m_i-1} \right\}, \quad i2^{m_i} = i \bmod 2^{m-1}$$

Corollary: The weight of the elements of a cyclotomic coset is the same.

b.   Assessment of q-ary sequences
Let $s = s_0, s_1, s_2, ..$ denote a q-ary sequence with q component sequences. Assessment of s as a q-ary sequence is by using q-ary statistical tests and q-ary linear equivalence. The component sequences of s are binary sequences and their cryptographic assessment as q parallel pseudo-random (PR) generator require not only the binary version of the above tests but also tests to ensure their mutual statistical independence. This would ensure that the knowledge of a subset of $p < q$ component sequences would not deliver extra information about a component sequence not in the subset. Hence the statistical assessment of these sequences includes:

i)   statistical properties of individual component sequences;
ii)   tests to assess the statistical dependence of component sequences.

There are a number of well-known tests to evaluate the statistical performance of a sequence locally and globally [4], [5]. In the following statistical inter-dependence is measured by a set of statistics that is closely related to the mutual information between random variables:

Let $s = s_0, s_1, s_2, ..$ denote a q-ary sequence. Statistical dependence of a component sequence $s(i)$ on a subset of m component sequences $s(i_1), s(i_2), ..s(i_m)$, can be studied by defining $N_h(i / i_1, i_2, ..i_m; \alpha_1 \alpha_2 ..\alpha_m)$, $h=0,1$, to denote the number of j $(0\leq j \leq T-1)$, for which $s_j(i)=h$, $h=0, 1$ while $s_j(i_1)=\alpha_1$, $s_j(i_2)=\alpha_2,..., s_j(i_m)=\alpha_m$ $i.e.$:

$$N_h(i \mid i_1 i_2 \cdot i_m; \alpha_1 \alpha_2 .... \alpha_m) =$$

$$| \{j \mid 0 \leq j \leq T - 1, s_j(i) = h, s_j(i_1) = \alpha_1, s_j(i_2) = \alpha_2 .. s_j(i_m) = \alpha_m \} |, \qquad h = 0, 1$$

where T is the length of the sequences.

Example 1
Consider the following component sequences:

s(0)	1	**1**	0	1	0	**0**	0	1	0	1	0	**0**	1	1
s(1)	0	**1**	0	0	1	**1**	1	0	1	0	1	**1**	0	1
s(2)	0	**0**	1	0	1	**0**	1	1	1	0	1	**0**	1	1

Let $i=0, i_1=1, i_2=2, \alpha_1=1$ and $\alpha_2=0$. Then $N_1(0/12; 10)=1$ and $N_0(0/12; 10)=2$.

QED

Definition: The set S of the q component sequences of a q-ary sequence s is statistically independent of order $m<q$ if for a component sequence $s(i)$, an arbitrary subset of m component sequences and an arbitrary binary m-tuple $\alpha_1\alpha_2...\alpha_m$, we have:

$$N_h(i \mid i_1 i_2 \cdot i_m; \alpha_1 \alpha_2 .... \alpha_m) = \frac{n_h(i)}{2^m} \qquad h = 0,1$$

where $N(.)$ is as defined earlier and $n_h(.)$ is the number of j, $0\leq j\leq T-1$, such that $s_j(i)=h$, $h=0,1$. If each sequence $s(i)$ corresponds to the consecutive value of a random variable σ_i and each element of the sequence corresponds to the outcome of an independent trial, the statistical independence of σ_i from $\sigma_{i_1}, \sigma_{i_2}, \sigma_{i_3},..\sigma_{i_m}$ is equivalent to independent probability distribution i. e.:

$$p(\sigma_i \mid \sigma_{i_1} \sigma_{i_2} \sigma_{i_3}..\sigma_{i_m}) = p(\sigma_i)$$

In fact under the above assumptions the frequency counts tend to probability as the length of the sequence is increased and hence:

$$p(\sigma_i \mid \sigma_{i_1} \sigma_{i_2} ...\sigma_{i_m}) = \alpha_1 \alpha_2 .... \alpha_m) = \frac{N_h(i \mid i_1 i_2 .. i_m; \alpha_1 \alpha_2 ... \alpha_m)}{T} = \frac{n_h(i)}{T \times 2^m}$$

$$\sum_{\sigma_{i_1} \sigma_{i_2} ..\sigma_{i_m}} p(\sigma_i \mid \sigma_{i_1} \sigma_{i_2} ...\sigma_{i_m}) = \frac{n_h(i)}{T} = p(\sigma_i = h) \qquad h = 0,1$$

where the same symbol is used to denote a random variable and its value and the distinction is left to the context. This results in zero mutual information between σ_i and $\sigma_{i_1} \sigma_{i_2} \sigma_{i_3}..\sigma_{i_m}$ i. e.:

$$I(\sigma_i ; \sigma_{i_1}\sigma_{i_2}\sigma_{i_3}..\sigma_{i_m})=0$$

Because:

$$H(\sigma_i|\sigma_{i_1}\sigma_{i_2}..\sigma_{i_m}) = - \sum_{\sigma_i \sigma_{i_1}\sigma_{i_2}..\sigma_{i_m}} p(\sigma_i \sigma_{i_1}\sigma_{i_2}..\sigma_{i_m})\log p(\sigma_i|\sigma_{i_1}\sigma_{i_2}..\sigma_{i_m})$$

$$= - \sum_{\sigma_i \sigma_{i_1}\sigma_{i_2}..\sigma_{i_m}} p(\sigma_i \sigma_{i_1}\sigma_{i_2}..\sigma_{i_m})\log p(\sigma_i)$$

$$= - \sum_{\sigma_i} p(\sigma_i)\log p(\sigma_i) = H(\sigma_i)$$

But:

$$I(\sigma_i;\sigma_{i_1}\sigma_{i_2}..\sigma_{i_m}) = H(\sigma_i) - H(\sigma_i|\sigma_{i_1}\sigma_{i_2}..\sigma_{i_m}) = 0$$

Proposition 2
If the component sequences of a q-ary sequence s are statistically independent of order m $(m<q)$, they are statistically independent of order $m-1$.

However statistical independence of order m does not imply dependence of order $m+1$.

3. Exponent Sequences of LFSR

a.   State Sequence
Let $s_j=(s_{j0}, s_{j1}, s_{j2},..s_{j,m-1})$, $s_{ji}\in GF(2)$, $0\leq i\leq m-1$, denote the jth state of a LFSR with a primitive feedback polynomial $F(D)=1+c_1D+c_2D^2+..+c_mD^m$(or characteristic polynomial $f(x)=x^m+c_1x^{m-1}....+c_m$).

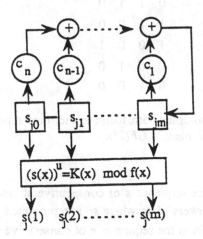

The m-ary sequence $s =s_0, s_1, s_2,...$ of the consecutive states of the LFSR is a sequence over $GF(2^m)$. Obviously s is a periodic sequence of period 2^m-1. This is because the feedback polynomial is primitive and the LFSR goes through all its possible states before returning to its initial state. Moreover s can be regarded as m binary component sequences.

Proposition 3
s satisfies a binary recurrence relation given by:

$$s_{m+k}=c_1 s_{m+k-1}+c_2 s_{m+k-2}+...+s_0 \qquad k>0$$

Let α denote a root of the primitive feedback polynomial $f(x)$. The elements of s correspond to distinct powers of α.

Example 2
Let $m=4$, $f(x)=x^4+x^3+1$ and $s_0=(1\ 0\ 0\ 0)$. Using polynomial representation of the elements of $GF(2^4)$ the sequence s will have the following form:

i	s_i				s_i	s_i
	s_{i0}	s_{i1}	s_{i2}	s_{i3}		
0	1	0	0	0	1	1
1	0	0	0	1	α^3	α^3
2	0	0	1	1	$\alpha^2+\alpha^3$	α^{14}
3	0	1	1	1	$\alpha + \alpha^2 +\alpha^3$	α^8
4	1	1	1	1	$1+\alpha+\alpha^2+\alpha^3$	α^6
5	1	1	1	0	$1+\alpha+\alpha^2$	α^7
6	1	1	0	1	$1+ \alpha+\alpha^3$	α^5
7	1	0	1	0	$1+\alpha^2$	α^9
8	0	1	0	1	$\alpha+\alpha^3$	α^{10}
9	1	0	1	1	$1+ \alpha^2+\alpha^3$	α^{11}
10	0	1	1	0	$\alpha+\alpha^2$	α^{13}
11	1	1	0	0	$1+\alpha$	α^{12}
12	1	0	0	1	$1+\alpha^3$	α^4
13	0	0	1	0	α^2	α^2
14	0	1	0	0	α	α

where the three last columns correspond to the three representations of consecutive states of LFSR as an element of $GF(2^4)$.

It can be seen that the sequence s of consecutive states of the LFSR can also be obtained by taking powers of α, where s_i is represented by α^{n_i}. So the sequence s can be characterised from the sequence n of consecutive powers of α, i.e. $n=n_0, n_1, n_2,...n_{T-1}$, $T=2^m-1$:

$$s=\alpha^{n_0}, \alpha^{n_1}, \alpha^{n_2}, \alpha^{n_3}....\alpha^{n_{T-1}}$$

If the LFSR starts from another initial state given by α^u, the corresponding state sequence r is a shift of s. Again a sequence n^u of powers can be determined such that:

$$n^u = n_0^u, \; n_1^u, \; n_2^u \ldots, \qquad\qquad n_0^u = u$$

$$r = \alpha^{n_0^u}, \; \alpha^{n_1^u}, \; \alpha^{n_3^u} \ldots, \qquad\qquad \alpha^{n_0^u} = \alpha^u$$

and n^u is a cyclic shift of n. In fact since every number between one and T occurs exactly once in n, n^u depends on the initial state u and $n^u \neq n^v$ if $u \neq v$.

Corollary: The sequence of powers corresponding to distinct initial value of a LFSR are distinct and one can be obtained from the other by a proper cyclic shift.

b.   Nonlinear Filtering
One of the well-known methods of generating pseudo-random (PR) sequences with high linear equivalence and good statistical distribution is by applying a nonlinear function to the sequence of states of a LFSR. We study the properties of exponentiation as a nonlinear function on the sequence of states of a LFSR. We restrict ourselves to the case where the exponentiation is modulo the characteristic polynomial of LFSR. This non-linear filter generates parallel sequences the properties of which are studied in the rest of this paper.

Consider a LFSR of length m with primitive characteristic polynomial $f(x)$ and state sequence s:

Definition: An exponent sequence $s^{(j)}$, $1 \leq j < T$, is a m-ary sequence obtained by raising the elements of the sequence of state s to an exponent j, over the field $GF(2^m)$ characterised by $f(x)$.

Definition: An exponent of the form 2^i is called a *basic exponent*, a sequence of the form $s^{(2^i)}$ is called a *basic sequence* and its linear recurrence relation is called a *basic recurrence*.

An exponent sequence $s^{(u)}$ is a periodic sequence with period at most $T = 2^m - 1$, hence it satisfies a recurrence relation of degree at most $2^m - 1$. However there exists a recurrence of minimum degree that $s^{(u)}$ satisfies, the order of which gives the linear equivalence of the sequence.

Let $\Sigma(s)$ denote the set of all exponent sequences of s:

$$\Sigma(s) = \{ s^{(j)} | \; s_i^{(j)} = (s_i)^j \bmod f(\alpha), \; i \geq 0, \; 1 \leq j \leq 2^m - 1 \}$$

Proposition 4
$\Sigma(s)$ can be partitioned into subsets $\Sigma_i(s)$, $i = 1, 2, ..t$ (t is the number of cyclotomic cosets modulo $2^m - 1$) such that every class is closed under exponentiation to basic exponents i.e. if $r \in \Sigma_i(s)$ then $r^{2^j} \in \Sigma_i(s)$ for $j = 1, 2, ..m-2$.

4. Properties of Exponent Sequences

Proposition 5
Consider the exponent sequences of the state sequence s of a LFSR with primitive characteristic polynomial $f(x)$;
i) $GCD(u, 2^m-1)=1$. Every power of α, $(f(\alpha)=0)$, occurs exactly once in a period of $s^{(u)}$;
ii) $GCD(u, 2^m-1)=h$. There are $(2^m-1)/h$ distinct powers of α in $s^{(u)}$ each with the same multiplicity h.

In fact the $(2^m-1)/h$ distinct powers of α correspond to the elements of the cyclic subgroup of the multiplicative group of $GF(2^m)$.

Proposition 6
The exponent sequences with exponents in the same cyclotomic coset are consisted of the same elements (same powers of α), but in different order.

Example 3
Consider a LFSR of length 6 and characteristic polynomial x^6+x+1. Let s denote the output sequence of the LFSR. Since $63=3 \times 21$, there are 21 distinct elements of $GF(2^6)$ in $s^{(3)}$ each with multiplicity three.

QED

This result can be used to find the frequency of ones and zeros in the component sequences and also their cross-correlation. Again the method is explained through an example which can easily be generalized .

Example 4
Consider $s^{(3)}$ in Example 2.

$s^{(3)}$ $\quad$ 1 α^9 α^{12} α^9 α^3 α^6 $\quad$ 1 α^{12} 1 $\quad$ α^3 α^9 α^6 α^{12} α^6 α^3

The elements of the sequence $s^{(3)}$ are the elements of the cyclic subgroup of the multiplicative group of $GF(2^4)$ generated by α^3, i. e. the set $\{1, \alpha^3, \alpha^6, \alpha^9, \alpha^{12}\}$. Moreover each element of the subgroup occurs the same number of times. The binary representation of these elements is:

	1	α^3	α^6	α^9	α^{12}
	1	0	1	1	1
	0	0	1	0	1
	0	0	1	1	0
	0	1	1	0	0

Since the number of zeros and ones in each row and the *Hamming distance* (defined as the number of coordinate places the two rows are not equal) between the two rows does not depend on the actual order of the α^i's in the sequence, the frequency of ones and zeros for the component sequences and their cross-correlation can be found from the above array. Let $n^{(i)}_j(0)$, $0 \leq j \leq 3$ denote the number of zeros in the jth component sequence of $s^{(i)}$, $d^{(i)}_{jk}$ denote the distance between the jth and k^{th} component sequences and $C^{(i)}(j, k)=T-2 \times d^{(i)}_{jk}$ the cross correlation between the same component sequence, of $s^{(i)}$. The above array is used to derive these values for $s^{(3)}$ as: $n^{(3)}_0(0)=3$, $n^{(3)}_1(0)=9$, $n^{(3)}_2(0)=9$ and $n^{(3)}_3(0)=9$. Also $d^{(3)}_{01}=6$, $d^{(3)}_{02}=6$, $d^{(3)}_{03}=12$, $d^{(3)}_{12}=6$, $d^{(3)}_{13}=6$ and $d^{(3)}_{23}=6$ (notice that $d^{(k)}_{ij}=d^{(k)}_{ji}$) and the corresponding cross-correlations are $C^{(3)}_{01}=C^{(3)}_{02}=C^{(3)}_{12}=C^{(3)}_{13}=C^{(3)}_{23}=15-12=3$ and $C^{(3)}_{03}=15-24=-9$.

The same counts are valid for all the sequences of $\Sigma_3(s)$ i.e. exponent 6, 9 and 12 (above proposition).

The same argument can be applied to C_5. Let $t \in C_5$:

$$
\begin{array}{ccc}
1 & \alpha^5 & \alpha^{10} \\
1 & 1 & 0 \\
0 & 1 & 1 \\
0 & 0 & 0 \\
0 & 1 & 1
\end{array}
$$

and $n^{(5)}_0(0)=5$, $n^{(5)}_1(0)=5$, $n^{(5)}_2(0)=15$ and $n^{(5)}_3(0)=5$. Similarly $d^{(5)}_{01}=10$, $d^{(35}_{02}=10$, $d^{(5)}_{03}=10$, $d^{(5)}_{12}=10$, $d^{(5)}_{13}=0$ and $d^{(5)}_{23}=10$ and the corresponding cross-correlations are $C^{(5)}_{01}=C^{(5)}_{02}=C^{(5)}_{12}=C^{(5)}_{03}=C^{(5)}_{23}=15-20=-5$ and $C^{(5)}_{13}=15$.

Finally if the exponent $t \in C_2 \cup C_7$ the q-ary sequence consists of all powers of α^i, $1 \leq i \leq 2^m-1$ and each component sequence has exactly 7 zeros and 8 ones and the distance between any two component sequences is 8 and the cross-correlation is -1.

QED

In general we state the following proposition the proof of which can be easily derived from the above example:

Proposition 7
The frequency of ones and zeros in a component sequence of $s^{(u)}$ and the cross-correlation between two component sequences can be completely determined by the elements of cyclic group generated by α^u. If $GCD(u, 2m-1)=1$, $n^{(u)}_i(0)=2m-1-1$ and $n^{(u)}_i(1)=2m-1$ for $0 \leq i \leq m-1$ and $C^{(u)}_{ij}=-1$ for $0 \leq i,j \leq m-1$.

Let $s(j_1)$, $s(j_2)$,..$s(j_k)$ denote a subset of $k<m$ component sequences and $N(j_1j_2..j_k;$ $\alpha_1\alpha_2..\alpha_k)$ denote the number of times a given k-tuple $\alpha_1\alpha_2..\alpha_k$ occurs as $s(j_1)$, $s(j_2)$,..$s(j_k)$. Then $N(j_1j_2..j_k; \alpha_1\alpha_2..\alpha_k)=2^{m-k}$ for $\alpha_1\alpha_2..\alpha_k \neq 00..0$ and $N(j_1j_2..j_k;$ $\alpha_1\alpha_2..\alpha_k)=2^{m-k}-1$ otherwise. Using this result for $\alpha_1\alpha_2..\alpha_k \neq 00..0$ we have

$$
N_h(j_{k+1} | j_1 j_2..j_k; \alpha_1\alpha_2..\alpha_k) = 2^{m-(k+1)} \qquad h=0,1
$$

where $N_h(.)$ is as defined in section 2 and is independent of $\alpha_1\alpha_2..\alpha_k$. Also

$$
N_0(j_{k+1} | j_1 j_2..j_k; 00..0) = 2^{m-(k+1)} - 1
$$

$$
N_1(j_{k+1} | j_1 j_2..j_k; 00..0) = 2^{m-(k+1)}
$$

Example 5

Consider $j_1=0, j_2=1$ and $j_3=2$ in Example 2. Then:

$$N_1(2|01; 10) = N_0(2|01; 10) = \frac{2}{4} = \frac{1}{2}$$

and $j_1=0, j_2=1, j_3=2$ and $j_4=3$:

$$N_1(1|012; 100) = N_0(3|012; 100) = \frac{1}{2}$$

QED

Proposition 8

Consider an LFSR of length m with primitive characteristic polynomial, state sequence s and let $GCD(u, 2^m-1)=1$. The component sequences of $s^{(u)}$ have very low statistical dependence which reduces with m.

b. Period

The maximum possible value for the period of an exponent sequence is $T=2^m-1$.

Proposition 9

Let $GCD(u,T)=1$. Then

i)   period of $s^{(u)}$ is maximum;

ii)   the difference between the number of ones and zeros in $s^{(u)}{}_{se}$ (the binary sequence obtained from $s^{(u)}$ by replacing each element by its binary representation) is m.

Corollary: If T is a prime number, all the exponent sequences and their component sequences will have maximum period.

c. Generation

An exponent p can be written in terms of powers of 2 (base two representation):

$$p = p_0 + p_1 \times 2 + p_2 \times 2^2 \dots\dots + p_{m-1} \times 2^{m-i}$$

i. e. every exponent p can be written as a linear combination of a subset of basic exponents. This subset is called the *binary components* of p. Since:

$$s_t^{(p)} = (\alpha^p)^n{}_t = (\alpha^{\sum_{i=0}^{m-1} p_i \times 2^i})^n{}_t = \prod_{i=0}^{m-1} (\alpha^{p_i \times 2^i})^n{}_t$$

$$= \prod_{i=0}^{m-1} (s_t^{(2^i)})^{p_i}$$

we have:

Corollary: Elements of an exponent sequence can be obtained as a product (over $GF(2^m)$) of the corresponding elements of at most m basic exponent sequences.

Hence generation of exponent sequences can be done in two steps:

i) generation of basic exponent sequences;
ii) multiplying the subset of basic exponent sequences that corresponds to the base two representation of the exponent.

The first step can be performed as part of pre-calculation. The second step that correspond to a specific exponent can requires at most m multiplication over $GF(2^m)$ for generation of m pseudo-random bits.

5. Linear Recurrence of Exponent Sequences

The minimum binary recurrence relation of an exponent sequence is used to assess the complexity of the sequences.

Proposition 10
Elements of $\Sigma_1(s)$ satisfy a minimum binary linear recurrence of order m, the characteristic polynomial of which is $f(x)$.

It was noted that elements of $s^{(p)}$ can be written as a product of the corresponding elements of a subset of component sequences. Hence the minimum linear recurrence satisfied by $s^{(p)}$ can be obtained by multiplying the basic recurrences of the binary components of p. The coefficients of the recurrence of $s^{(p)}$ can be written as an integral polynomial in the coefficients of the basic recurrences [7] and hence are binary.

Corollary: Every exponent sequence satisfies a minimum binary recurrence.

Proposition 11
All the elements of $\Sigma_v(s)$, $v=1, 2, ..t$, satisfy the same minimum binary recurrence relation.

Corollary: It is sufficient to find the binary recurrence of one exponent sequence $s^{(i)}$ in each $\Sigma_i(s)$, $i=1, 2,..t$. This recurrence can be evaluated as a product of basic recurrences of binary components of i.

The same result applies to the solution of the recurrence of $s^{(p)}$, i. e. the solution to the linear recurrence satisfied by $s^{(p)}$ can be written as the product of the solutions of the recurrences of binary components of p. The procedure is described by the following example, the generalisation of which is straightforward.

Definition: Weight of an integer u is denoted by $w(u)$ and is the number of its non-zero binary components.

Example 6
The minimum recurrence relation satisfied by the state sequences of Example 2 is

$$s_4 = s_3 + s_0$$

with characteristic polynomial $f(x)$:

$$f(x) = x^4 + x^3 + 1.$$

The following table gives the elements $s^{(i)}_j$, $j=0, 1, ..T-1$ and $i=1, 2, ..T$ of the exponent sequences:

i \ j	0	1	2	3	4	5	6	7	8	9	10	11	12	13	14
1	1	α^3	α^{14}	α^8	α^6	α^7	α^5	α^9	α^{10}	α^{11}	α^{13}	α^{12}	α^4	α^2	α
2	1	α^6	α^{13}	α	α^{12}	α^{14}	α^{10}	α^3	α^5	α^7	α	α^9	α^8	α^4	α^2
3	1	α^9	α^{12}	α^9	α^3	α^6	1	α^{12}	1	α^3	α^9	α^6	α^{12}	α^6	α^3
4	1	α^{12}	α^{11}	α^2	α^9	α^{13}	α^5	α^6	α^{10}	α^{14}	α^7	α^3	α	α^8	α^4
5	1	1	α^{10}	α^{10}	1	α^5	α^{10}	1	α^5	α^{10}	α^5	1	α^5	α^{10}	α^5
6	1	α^3	α^9	α^3	α^6	α^{12}	1	α^9	1	α^6	α^3	α^{12}	α^9	α^{12}	α^6
7	1	α^6	α^8	α^{11}	α^{12}	α^4	α^5	α^3	α^{10}	α^2	α	α^9	α^{13}	α^{14}	α^7
8	1	α^9	α^7	α^4	α^3	α^{11}	α^{10}	α^{12}	α^5	α^{13}	α^{14}	α^6	α^2	α	α^8
9	1	α^{12}	α^6	α^{12}	α^9	α^3	1	α^6	1	α^9	α^{12}	α^3	α^6	α^3	α^9
10	1	1	α^5	α^5	1	α^{10}	α^5	1	α^{10}	α^5	α^{10}	1	α^{10}	α^5	α^{10}
11	1	α^3	α^9	α^{13}	α^6	α^2	α^{10}	α^9	α^5	α	α^8	α^{12}	α^{14}	α^7	α^{11}
12	1	α^6	α^3	α^6	α^{12}	α^9	1	α^3	1	α^{12}	α^6	α^9	α^3	α^9	α^{12}
13	1	α^9	α^2	α^{14}	α^3	α	α^5	α^{12}	α^{10}	α^8	α^4	α^6	α^7	α^{11}	α^{13}
14	1	α^{12}	α	α^7	α^9	α^8	α^{10}	α^6	α^5	α^4	α^2	α^3	α^{11}	α^{13}	α^{14}

where α denotes a root of $f(x)$.

The solution to the recurrence relation of s is given by:

$$s_n = A_0(\alpha)^n + A_1(\alpha^2)^n + A_2(\alpha^{2^2})^n + A_3(\alpha^{2^3})^n \qquad A_i = GF(2^4), i = 0, 1, 2, 3$$

The value of A_i, $i=0,1,2,3$ can be obtained by considering the four initial values of s, i. e. the four initial states of the LFSR:

$$s_0 = 1 \qquad s_1 = \alpha^3 \qquad s_2 = \alpha^{14} \qquad s_3 = \alpha^8$$

This results in the following set of linear equation:

$$\begin{bmatrix} 1 \\ \alpha^3 \\ \alpha^{14} \\ \alpha^8 \end{bmatrix} = \begin{bmatrix} 1 & 1 & 1 & 1 \\ \alpha & \alpha^2 & \alpha^4 & \alpha^8 \\ \alpha^2 & \alpha^4 & \alpha^8 & \alpha \\ \alpha^3 & \alpha^6 & \alpha^{12} & \alpha^9 \end{bmatrix} \times \begin{bmatrix} A_0 \\ A_1 \\ A_2 \\ A_3 \end{bmatrix}$$

The matrix of coefficient is non-singular and in fact because $\{1, \alpha, \alpha^2, \alpha^3\}$ is a basis for $GF(2^4)$, its determinant is one ([6], Lemma 18 chapter 4).

It can be noted that since zeros of a binary recurrence relation occur in conjugates, non of the A_i, $0 \le i \le 3$ can be zero.

Similar results can be obtained for $s^{(2)}$, $s^{(4)}$ and $s^{(8)}$ by replacing the four initial elements of each sequence by the following constant vectors respectively:

$$\begin{bmatrix} 1 \\ \alpha^6 \\ \alpha^{13} \\ \alpha \end{bmatrix} \quad \begin{bmatrix} 1 \\ \alpha^{12} \\ \alpha^{11} \\ \alpha^2 \end{bmatrix} \quad \begin{bmatrix} 1 \\ \alpha^9 \\ \alpha^7 \\ \alpha^4 \end{bmatrix}$$

The result of this computation gives the following set of equations:

$$s_t^{(2^i)} = \sum_{j=0}^{3} d_{ij}(\alpha^2)^t \qquad 0 \leq i \leq 3 \qquad t \geq 4$$

where non of the coefficients are zero. It is easy to show that d_{ij}, $0 \leq$, $j \leq 3$, are related as:

$$\begin{bmatrix} s_t \\ s_t^{(2)} \\ s_t^{(4)} \\ s_t^{(8)} \end{bmatrix} = \begin{bmatrix} A_0 & A_1 & A_2 & A_3 \\ A_3^2 & A_0^2 & A_1^2 & A_2^2 \\ A_2^4 & A_3^4 & A_0^4 & A_1^4 \\ A_1^8 & A_2^8 & A_3^8 & A_0^8 \end{bmatrix} \times \begin{bmatrix} (\alpha)^t \\ (\alpha^2)^t \\ (\alpha^4)^t \\ (\alpha^8)^t \end{bmatrix} \qquad (I)$$

Once the solution to the recurrences of the basic sequences is found the results can be used to find those of other exponent sequences:

i)   exponents of weight 2

Suppose we want to find the minimum binary recurrence relation of $s^{(3)}$. Since $3 = 2^0 + 2^1$, we can write:

$$s_t^{(3)} = s_t^{(2)} \times s_t$$

where:

$$s_t = A_0(\alpha)^t + A_1(\alpha^2)^t + A_2(\alpha^4)^t + A_3(\alpha^8)^t$$

$$s_t^{(2)} = A_3^2(\alpha)^t + A_0^2(\alpha^2)^t + A_1^2(\alpha^4)^t + A_2^2(\alpha^8)^t$$

and $s^{(3)}$ can be written as:

$$s_t^{(3)} = B_0(\alpha)^t + B_1(\alpha^2)^t + B_3(\alpha^4)^t + B_6(\alpha^8)^t + B_2(\alpha^3)^t + B_4(\alpha^5)^t + B_7(\alpha^9)^t$$
$$+ B_5(\alpha^6)^t + B_8(\alpha^{10})^t + B_9(\alpha^{12})^t$$

The non-zero B_i's, $0 \leq i \leq 9$ determine zeros of the recurrence of $s^{(3)}$.

The same argument applies to all exponents of weight two, i.e. 3, 5, 6, 9, 12 and 10. These exponents fall into two disjoint cyclotomic cosets. Since the recurrence relation of exponents belonging to one coset is the same (proposition 5), it suffices to find the recurrence of the coset leader by multiplying proper basic recurrences.

A general rule for testing the existence of a possible zero is to evaluate a sub-determinant of order two of the matrix of coefficients in (I) (for $s^{(3)}$ the determinants of order two of the first two rows should be considered) .

ii)   exponents of weight 3

Next we consider an exponent that has weight three i.e., 7, 11, 13 and 14. These exponents all belong to the same cyclotomic coset. The minimum recurrence relation of $s^{(7)}$ (which is the recurrence relation of the other exponents too) can be obtained by multiplying recurrences of s, $s^{(2)}$, $s^{(4)}$.

Possible zeros of the minimum recurrence can be determined by considering all α^{u} where u belongs to the following set:

$$\left\{ u \mid u = 2^{i_1} + 2^{i_2} + 2^{i_3} \bmod 15, \quad 0 \leq i_1, i_2, i_3 \leq 3 \right\}$$

and the coefficient of α^{u} is:

$$\sum_{\substack{0 \leq i_1, i_2, i_3 \leq 3 \\ 2^{i_1} + 2^{i_2} + 2^{i_3} = u \bmod 15}} d_{0 i_1} d_{1 i_2} d_{2 i_3}$$

where d_{ij} denote the ij^{th} element of the coefficient matrix as given above.

iii)   exponents of weight four

The only exponent of weight four is 15. Since $15 = 2^0 + 2^1 + 2^2 + 2^3$ the recurrence relation of $s^{(15)}$ is a product of all basic recurrences and has all powers of α as possible zeros. The coefficient of $(\alpha^{u})^{n}$ is:

$$\sum_{\substack{0 \leq i_1, i_2, i_3, i_4 \leq 3 \\ 2^{i_1} + 2^{i_2} + 2^{i_3} + 2^{i_4} = u \bmod 15}} d_{0 i_1} d_{1 i_2} d_{2 i_3} d_{3 i_4}$$

which can be calculated to test the existence of α^{u} in the set of zeros of $s^{(15)}$.

QED

The result of the above example can be generalised to an arbitrary LFSR with primitive characteristic polynomial.

The following theorem gives an upper bound on the linear equivalence of the exponent sequences:

Theorem 1

Let a LFSR of length m and a primitive characteristic polynomial $f(x)$ generate a sequence of state s and consider the exponent sequence $s^{(u)}$, $0<u<2^m-1$. Let $w(u)=t$. The linear equivalence of $s^{(u)}$ is upper bounded by:

$$\sum_{i=1}^{t} \binom{m}{i}$$

6. Linear Equivalence of Component Sequences

It was noted earlier that the order of the binary recurrence relation of $s^{(u)}$ is an upperbound on the corresponding orders of the component sequences but as the following example shows this upper bound is not achieved in most cases and the actual degree of recurrence for binary component sequences should be calculated individually. The linear equivalence of a component sequence can be calculated by a method similar to the one given for exponent sequences. Again the procedure is best explained through an example:

Example 7:

Let s_0 denote the initial state of the LFSR in Example 2. The element s_i of the state sequence can be represented as a polynomial of degree 3 in α, $f(\alpha)=0$:

$$s_i = a_{i0} + a_{i1}\alpha + a_{i2}\alpha^2 + a_{i3}\alpha^3 \qquad i \geq 0$$

where a_{ij}, $0 \leq j \leq 3$, $i \geq 0$ is the ith element of the jth component sequence of s. Exponentiating s to the exponent u results in another polynomial in α:

$$s_i^{(u)} = a_{i0}^{(u)} + a_{i1}^{(u)}\alpha + a_{i2}^{(u)}\alpha^2 + a_{i3}^{(u)}\alpha^3$$

and $a_{ij}^{(u)}$, $i \geq 0$, $0 \leq j \leq 3$ is the ith element of the jth component sequence of $s^{(u)}$. The coefficient of α^i, $0 \leq i \leq 3$ in $s^{(u)}$ correspond to the ith component sequence of the exponent sequence $s^{(u)}$. Component sequences of $s^{(u)}$ are expressible as the sum of products of the component sequences of s, e.g. $u=2$:

$$(s_i)^2 = s_i^{(2)} = (a_{i0} + a_{i2} + a_{i3}) + a_{i3}\alpha + (a_{i1} + a_{i3})\alpha^2 + (a_{i2} + a_{i3})\alpha^3$$

and gives the component sequences of $s^{(2)}$ as linear combination of component sequences of s. The same results apply to component sequences of all basic exponents.

The component sequences of the basic sequences are given in the matrix of coefficient of the following set of equalities:

$$
\begin{pmatrix} s_i \\ s_i^{(2)} \\ s_i^{(4)} \\ s_i^{(8)} \end{pmatrix} =
\begin{pmatrix}
a_{i0} & a_{i1} & a_{i2} & a_{i3} \\
a_{i0}+a_{i2}+a_{i3} & a_{i3} & a_{i1}+a_{i3} & a_{i2}+a_{i3} \\
a_{i0}+a_{i1}+a_{i3} & a_{i2}+a_{i3} & a_{i2} & a_{i1}+a_{i4} \\
a_{i0}+a_{i3} & a_{i1}+a_{i2} & a_{i1}+a_{i3} & a_{i1}
\end{pmatrix}
\begin{pmatrix} 1 \\ \alpha \\ \alpha^2 \\ \alpha^3 \end{pmatrix}
$$

or:

$$s_i^{(2^j)} = \sum_{j=0}^{3} b_{ij}\alpha^j \qquad j = 0, 1, 2, 3 \qquad i \geq 4$$

QED

Corollary: Component sequences of basic sequences are linear combination of component sequences of s and hence $f(x)$ is the characteristic polynomial of their minimum recurrences.

The results of Example 7 is used to find component sequences of other exponent sequences:

i) Let $w(u)=2$, *i.e.* $u = 2^{u_1} + 2^{u_2}$ and

$$s_i^{(u)} = s_i^{(2^{u_1})} \times s_i^{(2^{u_2})}$$

The product of the polynomial representation of $s_i(2^{u_1})$ and $s_i(2^{u_2})$ results in a polynomial of degree 6 which can be reduced to degree 3 by using the minimum polynomial of α:

$$s_i^{(u)} = A_{i0} + A_{i1}\alpha + A_{i2}\alpha^2 + A_{i3}\alpha^3 + A_{i4}\alpha^4 + A_{i5}\alpha^5 + A_{i6}\alpha^6$$

$$\alpha^4 = \alpha^3 + 1$$

$$s_i^{(u)} = (A_{i0} + A_{i4} + A_{i5} + A_{i6}) +$$

$$+ (A_{i1} + A_{i5} + A_{i6})\alpha + (A_{i2} + A_{i6})\alpha^2 + (A_{i3} + A_{i4} + A_{i5} + A_{i6})\alpha^3$$

for all i. A_{ij} is a sum of the terms of the form $a_{mn}a_{pq}$, $0 \leq m, n, p, q \leq 3$ and a_{ij} is as defined above hence the coefficient of α^j, $0 \leq j \leq 3$, expresses the jth component sequence as a sum of products of at most two component sequences of s. The results of [1] applies directly in his case and the linear complexity of each component sequence of $s^{(2)}$ is upper bounded by :

$$\sum_{i=1}^{2} \binom{4}{i} = 10$$

ii) $w(u)=3$ and $u = 2^{u_1} + 2^{u_2} + 2^{u_3}$. Describing $s^{(u)}$ as:

$$s^{(u)} = s^{(2^{u_1})} \times s^{(2^{u_2})} \times s^{(2^{u_3})}$$

results in a polynomial of degree at most 9 in α which can be reduced modulo $f(\alpha)$ to give a polynomial of degree three in α. Again the coefficient of α^j correspond to the elements of the jth component sequence and is given as a sum of products of at most three different phases of the output of LFSR. Again using the results of [1], the linear complexity is upper bounded by:

$$\sum_{i=1}^{3}\binom{4}{i} = 14$$

iii) $w(u)=4$. Similar argument for exponent 15 will give the following upper bound for the linear complexity of component sequences of $s^{(15)}$:

$$\sum_{i=1}^{4}\binom{4}{i} = 15$$

QED

As can be seen the actual linear equivalence of the component sequences depends on $f(x)$. However the above procedure can always be used to obtain the required result. Also Theorem 1 has its counterpart for component sequences.

Theorem 2
Let a LFSR of length m with a primitive characteristic polynomial $f(x)$ generate the state sequence s and let $s^{(u)}$ denote an exponent sequence of s where $w(u)=t$. Then the linear equivalence of the component sequences of $s^{(u)}$ is upper bounded by:

$$\sum_{i=1}^{t}\binom{m}{i}$$

7. Conclusion and Comments

Exponential filtering of the state sequence of LFSR provide an alternative method of generating parallel pseudo-random sequences. For properly chosen exponents, the component sequences have good statistical properties (distribution and inter-dependence) and high linear equivalence. Assessment of a generator and the choice of the set of proper exponent sequences can be facilitated by noting that the exponents that belong to the same cyclotomic cosets have the same complexity. Hence it is sufficient to find the equivalence of one element in each coset which is reduced to finding the product of at most m linear recurrences. If 2^m-1 is a prime number then all the exponent sequences and their associated component sequences have the highest period and good statistical properties. This study suggests the following directions for future research:

i) Exponentiation filters considered in this paper are special cases of the more general class of exponentiation functions over an arbitrary finite field. This can be generalised to the filtering of the state sequence of a LFSR by exponentiation modulo an irreducible polynomial $g(x)$. The properties of the resulting sequences is of interest in this case.

ii) In general a nonlinear filter that can generate n parallel bits from the state sequence of a LFSR of length m corresponds to a mapping from $GF(2^m)$ to $GF(2^n)$.

Characterisation of the mappings which result in component sequences with good statistical properties and high linear equivalence is an open question to be answered.

Acknowledgments

It was a pleasure to have stimulating discussion with Dr. J. Piepryzk on this work. Also I would like to thank Professor J. R. Seberry and other members of the Centre for Computer Security Research (CCSR) for their helpful comments and assistance.

References:

[1] R.A. Rueppel, *Analysis and Design of Stream Ciphers*, Springer Verlag, Berlin, 1986.
[2] Stephen A. Cook, "An Overview of Computational Complexity", *CACM*, vol. 26, no. 6, pp. 401-408, Jan. 1983.
[3] R. Lidl and H. Niederreiter, *Introduction to Finite Fields and Their Applications*, Cambridge University Press, 1986
[4] D. Knuth, *The Art of Computer Programming, Vol. 2*, Addison-Wesley Publishing Company, 1981
[5] S. Golomb, *Shift Register Sequences*, Aegean Park Press,1982
[6] F. J. MacWilliams and N. J. Sloane, *Theory of Error-Correcting Codes*, North-Holland Publishing Company, 1978.
[7] E.S. Selmer, *Linear Recurrence Relations over Finite Fields*, Department of Mathematics, University of Bergen, Norway, 1966.
[8] R. S. Safavi-Naini, *Parallel Genration of Pseudo-Random Sequences*, UNSW, University College, ADFA, TR CS89/24

Large Primes in Stream Cipher Cryptography

Kencheng Zeng[1], C.H. Yang[2], and T.R.N. Rao[2]

[1]Graduate School of USTC
Academia Sinica
P.O. Box 3908
Beijing
People's Republic of China

[2]The Center for Advanced Computer Studies
University of Southwestern Louisiana
P.O. Box 44330
Lafayette, LA 70504-4330
U.S.A.

Abstract. The present research is motivated by the observation that if the period T of a certain binary sequence is a prime, then its linear complexity will be bounded from below by the order of 2 modulo T, i.e., $LC \geqslant Ord_T(2)$. A class of generators with state periods $T(q, n) = q \cdot 2^n - 1$ are constructed for $q = 3, 5, 7, 9$ and arbitrary n on the basis of a pair of m-sequence generators with the same number of stages, each controlling the clock of the other (bilateral stop-and-go clock control). A new test is derived to find the primes among the numbers $T(q, n)$ with the cases $3 \mid q$ and $3 \nmid q$ treated in a unified manner. The orders of 2 modulo some of the primes $T(q, n)$ are given and some additional cryptographic and implementational remarks are made.

I. Introduction

In designing a qualified keystream generator, one of the first concerns of the cryptographer is to guarantee a large enough key-independent lower bound to the linear complexity of the output sequences. More desirably, if such a lower bound can be made to be of an order of magnitude approximately equal to that of the period.

* This research is supported by Board of Regents of Louisiana Grant #86-USL(2)-127-03

One of the traditional ways for attaining this objective [1] is to start with an n-stage LFSR which generates the driving sequence

$$\mathbf{a} = \{ a(0), a(1), \cdots, a(i), \cdots \},$$

and apply to it a nonlinear feedforward transformation of the form

$$b(i) = f_K(a(i), a(i-1), ..., a(i-r+1))$$

to produce the nonlinear feedforward sequence

$$\mathbf{b} = \{ b(0), b(1), \cdots, b(i), \cdots \},$$

where the transfer function $f_K(x_0, x_1, ..., x_{r-1})$ is selected by the key K from a certain family $\{f_K\}$ of algebraic polynomials, which are linear in each one of the r indeterminates separately. It can be shown [2], that the linear complexity LC of the feedforward sequence $\mathbf{b}$ is bounded from above by the inequality

$$LC \leqslant \sum_{k=1}^{d} C_k^d$$

where $d \leqslant r$ is the total degree of the transfer polynomial. This key-dependent upper bound can be attained only for carefully chosen transfer functions f_K and, according to Siegenthaler's theory [3], any attempt aimed at increasing d will lead to a decrease in the degree of correlation-immunity of f_K.

For de Bruijn sequences of period 2^n, we have $LC > 2^{n-1}$, but such sequences either are hard to implement technically or suffer from severe auto-correlation weakness [4].

In a recent work, Gollmann and Chambers [5, 6] proposed another interesting approach by considering a cascade of n clock controlled LFSRs of the same length $p \geqslant 3$, which is chosen to be a prime. The output sequence has been shown to have period p^n and a linear complexity of approximately the same order of magnitude. The drawback is that approximately $p \log T/\log p$ storage elements will be needed in order to produce a sequence of period T, in addition to the fact that several clock cycles will be needed for the generation of one single pseudorandom bit.

There is yet another simple approach which, to our knowledge, has been exploited nowhere for the purpose of achieving a guaranteed lower bound to the linear complexity. Namely, we have the following

Proposition. If a binary sequence $\mathbf{b}$ has an odd prime period T, then its linear complexity will be bounded from below by the order of the number 2 modulo T, i.e.,

$$LC(\mathbf{b}) \geqslant Ord_T(2). \tag{1}$$

Proof. Since T is odd, the minimum polynomial $\mu(x)$ of **b** must have an irreducible factor, say $p(x)$, of degree $d > 1$, and any root α of $p(x)$, as an element in the multiplicative group of its splitting field E, has order T. Thus, if we write $d^* \triangleq Ord_T(2)$, then

$$\alpha, \alpha^2, \cdots, \alpha^{2^{d^*}-1}$$

will be distinct elements in E with $\alpha^{2^{d^*}} = \alpha$. This means

$$p^*(x) \triangleq \prod_{i=0}^{d^*-1}(x - \alpha^{2^i})$$

will be an irreducible polynomial in $F_2[x]$ with a root α in common with $p(x)$. So we must have $p(x) = p^*(x)$ and, in particular, $d = d^*$. consequently,

$$LC(\mathbf{b}) = deg\ (\mu(x)) \geqslant deg\ (p(x)) = d^* = Ord_T(2).$$

A sequence with a prime period has another desirable property that one can subject it to various further cryptographic transformations without influencing the established lower bound to its linear complexity, provided the transformations do not render it into a constant sequence.

However, if $T = 2^n - 1$ happens to be one of the known Mersenne primes, then we shall have $Ord_T(2) = n$. So T should be chosen to be a prime not belonging to the progression $2^n - 1$, $n \geqslant 1$. An idea which naturally arises in this connection is to consider generators with state periods which are primes of the form

$$T(q, n) = q \cdot 2^n - 1, q > 1, q \equiv 1\ mod\ 2.$$

In the sequel, we shall construct generators with state periods $T(q, n)$ for $q = 3, 5, 7, 9$ and arbitrary n. A new algorithm will be designed to find the primes among these numbers, and the orders of 2 modulo the primes thus found will be computed to show that this new approach will be successful.

II. Generators with State Periods $T(q, n)$

We shall construct random-bit generators which will describe cycles of length $T(q, n)$ for $q = 3, 5, 7, 9$ and arbitrary n, in their corresponding spaces of admissible states. But for simplicity of exposition, we shall discuss in details only the case $q = 5$ and shall make an analysis of the state diagram of the generator thus constructed. The cases $q = 3, 7, 9$ can be treated with slight modifications.

In the scheme showed below, we have a pair of m-stage maximal length linear feedback shift registers LFSR-A and LFSR-B. Each of these LFSRs controls the clock pulses to the other in the following way:

(i)   If $(a(t+m-1), a(t+m-2)) = (0,1)$, then the clock pulse to LFSR-B is to be blocked;

(ii)   If $(b(t+m-1), b(t+m-2)) = (0,1)$, but $(a(t+m-1), a(t+m-2)) \neq (0,1)$, then the clock pulse to LFSR-A is to be blocked.

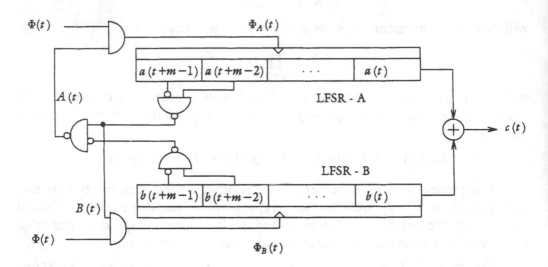

Thus, if we denote the t-th pulse from the clock by $\Phi(t)$, then the $(t+1)$-th pulses to LFSR-A and LFSR-B will be

$$\Phi_A(t+1) = A(t) \cdot \Phi(t+1) , \qquad (2)$$

and

$$\Phi_B(t+1) = B(t) \cdot \Phi(t+1) , \qquad (3)$$

where

$$B(t) = \overline{\overline{a}(t+m-1)a(t+m-2)} \qquad (4)$$

and

$$A(t) = \overline{\overline{a}(t+m-1)a(t+m-2)\,\overline{b}(t+m-1)b(t+m-2)} . \qquad (5)$$

The output signal $c(t)$ of the generator at the moment t is given by

$$c(t) = a(t) + b(t) . \qquad (6)$$

The phase space of the generator consists of vector-pairs of the form $s = (s_A, s_B)$, where

$$s_A \triangleq (a(t+m-1), a(t+m-2), \cdots, a(t))$$

$$s_B \triangleq (b(t+m-1), b(t+m-2), \cdots, b(t))$$

are non-zero vectors of $V_m(F_2)$, so that the number of admissible states of the generator is $(2^m - 1)^2$. The state diagram, however, will compose of branched cycles, and we need the following minimum of working terminology in order to describe its structure.

Definition. We say the admissible state s *precedes* the state $s*$ or $s*$ *succeeds* s, if our generator can go over from s to $s*$ in one step of work. We say a state s is *inaccessible* to the generator if it has no predecessor.

Besides, for any non-zero vector $s \in V_m(F_2)$, we denote by $L_A s$, or $L_B s$, the vector which precedes s when the shift register LFSR-A, or LFSR-B, with undisturbed clock pulses is set to work. Similarly, we denote by $R_A s$, $R_B s$, the successors of s under the work of LFSR-A, LFSR-B, with undisturbed clock pulses.

Theorem 1. (i) There are 2^{2m-4} inaccessible states in the state diagram of the generator, and an admissible state $s = (s_A, s_B)$ is inaccessible to the generator if and only if

$$(a(t+m-1), a(t+m-2)) = (0, 1), \quad (b(t+m-2), b(t+m-3)) = (0, 1) \quad (8)$$

(ii) Every branch to a cycle starts with an inaccessible state and has length 1, and at any cycle state there is at most one entering branch;

(iii) The accessible states in the state diagram fall in

$$T(3, m-2) = 3 \cdot 2^{m-2} - 1 \tag{9}$$

cycles, each of length

$$T(5, m-2) = 5 \cdot 2^{m-2} - 1. \tag{10}$$

Proof. (i) Assume (8) satisfied. If the state s has a predecessor, then it must be one of the states:

$$(L_A s_A, s_B), \quad (s_A, L_B s_B), \quad (L_A s_A, L_B s_B).$$

But the corresponding successors of these states under the work of our generator will be respectively

$$(s_A, R_B s_B), \quad (R_A s_A, L_B s_B), \quad (L_A s_A, s_B),$$

and none of them coincides with s.

Conversely, if at least one of the two conditions in (8) is not satisfied, then similar arguments can be used to find a predecessor for s.

(ii) To see this, we need only observe that no state $s = (s_A, s_B)$ can have more than two predecessors and it will have two if and only if

$$(a(t+m-2), a(t+m-3)) = (0, 1), \quad (b(t+m-2), b(t+m-3)) = (0, 1)(11)$$

where the predecessors are

$$(L_A s_A, s_B), \quad (s_A, L_B s_B),$$

with the first one being inaccessible.

(iii)   To see this, we observe that the clock-controlling sequences $\{A(t)\}$ and $\{B(t)\}$ have the following simple properties:

(a)   If one of the two shift-registers, say LFSR-A, being in the state s_A at a certain moment t, returns to that state for the first time at another mement t^*, then the number of 0's in the segment

$$B(t), B(t+1), ..., B(t^*)$$

is 2^{m-2};

(b)   In both sequences $A(t)$ and $B(t)$ all zero ranges are of length 1.

Now suppose the generator starts at the moment $t = 1$ from the cycle state $s = (s_A, s_B)$ and LFSR-A returns to s_A for the first time at moment t_1, while LFSR-B returns to s_B for the first time at moment t_2, $t_1 \leqslant t_2$. Then

$$t_1 = 2^m + x - 1,$$

where x is the number of 0's in the segment

$$A(1), A(2), \cdots, A(t_1),$$

and LFSR-B will sweep over

$$t_1 - 2^{m-2} = 3 \cdot 2^{m-2} + x - 1$$

distinct m-vectors in the time interval $[1, t_1]$. But this will mean that in the interval $[t_1 + 1, t_2]$ the shift-register LFSR-B has to sweep over the remaining $2^{m-2} - x$ non-zero m-vectors and, at the same time, give rise to the same number of 0's in the segment

$$A(t_1+1), A(t_1+2), ..., A(t_2).$$

So we see from the property (b), that we must have $x = 2^{m-2}$ and hence

$$t_1 = t_2 = 2^m + 2^{m-2} - 1.$$

The number of distinct cycles in the state diagram follows immediately from the identity

$$(2^m - 1)^2 - 2^{2m-4} = (3 \cdot 2^{m-2} - 1)(5 \cdot 2^{m-2} - 1).$$

III. Testing the primality of $T(q, n)$

Necessary and sufficient conditions for numbers of the form $T(q, n)$ with arbitrary q to be prime have been discussed in [7, 8, 9, 10], the computations needed are not expensive, but the cases $3 \mid q$ and $3 \nmid q$ have to be treated separately and the deduction of the conditions is not simple enough as to be put down in a single page. Since in our case q has only one prime factor, we would like to use a new test which, though computationally a little more expensive, treats the two cases in a unified way and can be derived immediately from the elegant elementary proposition of Lenstra [11] quoted below.

Proposition (Lenstra 1982). Let N and s be positive integers, and let A be a commutative ring with 1 containing $Z/(N)$ as a subring (with the same 1). Suppose that there exists $\alpha \in A$ satisfying the following conditions:

(1) $\alpha^s = 1$,

(2) $\alpha^{s/p} - 1 \in A^*$ (the group of units of A) for every prime p dividing s,

(3) the polynomial

$$f(x) = \prod_{i=0}^{t-1}(x - \alpha^{N^i})$$

has coefficients in $Z/(N)$ for some positive integer t.
Then every divisor r of N is congruent to a power of N modulo s.

Theorem 2. Assume $n \geqslant 2$, $q = \pi^a$ is a prime power, and let b be any integer. Define three sequences of integers by the recursive relations:

$$W_{i+1} = bW_i + W_{i-1}, \quad W_0 = 2, \qquad W_1 = b, \quad 1 \leqslant i \leqslant q - 1; \qquad (12)$$

$$U_{i+1} = U_i^2 - 2, \qquad U_1 = W_q^2 + 2, \qquad\qquad 1 \leqslant i \leqslant n - 2; \qquad (13)$$

$$V_{i+1} = V_i^2 - 2, \qquad V_1 = W_{\pi-1}^2 + 2, \qquad\qquad 1 \leqslant i \leqslant n. \qquad (14)$$

(i) If $T(q, n)$ is a prime, of which $b^2 + 4$ is a quadratic nonresidue (QNR), then

$$U_{n-1} \equiv 0 \bmod T(q, n); \qquad (15)$$

(ii) If $U_{n-1} \equiv 0 \bmod T(q, n)$ and

$$(V_{n+1} - 2, T(q, n)) = 1, \qquad (16)$$

then $T(q, n)$ is a prime.

Proof. Write

$$f(x) = x^2 - bx - 1$$

and let α and $\beta = b - \alpha$ be its two roots in the ring $A = Z/(f(x))$.

Evidently, we have

$$W_i = \alpha^i + \beta^i,$$

$$U_i = \alpha^{q\,2^i} + \beta^{q\,2^i},$$

$$V_i = \alpha^{\pi^{s-1}2^i} + \beta^{\pi^{s-1}2^i}$$

(i) In this case, we can replace Z by the prime field $F = Z/(T(q,n))$ and A by the extension of F obtained by adjoining to it the root α of the irreducible quadratic $f(x)$. As is well-known, we must have $\beta = \alpha^{T(q,n)}$ which, together with $\alpha\beta = -1$, means $\alpha^{q\,2^n} = -1$, and hence

$$U_{n-1} = \alpha^{q\,2^{n-1}} + \beta^{q\,2^{n-1}} = \beta^{q\,2^{n-1}}(\alpha^{q\,2^n} + 1) = 0$$

(ii) From $U_{n-1} = 0$ we get $\alpha^{q\,2^n} = -1$, and hence

$$\beta = -\alpha^{-1} = \alpha^{q\,2^n - 1} = \alpha^{T(q,n)}.$$

Thus if we write $N \triangleq T(q,n)$ and $s \triangleq q\,2^{n+1}$, then we shall have

(1) $\alpha^s = \alpha^{q\,2^{n+1}} = 1$;

(2) For the two prime factors 2 and π of s, we have

(2.1). $\alpha^{s/2} - 1 = \alpha^{q\,2^n} - 1 = -2 \in A^*$,

(2.2). $\alpha^{s/\pi} - 1 = \alpha^{\pi^{s-1}2^{n+1}} - 1 = \in A^*$, for we have

$$(\alpha^{\pi^{s-1}2^{n+1}} - 1)(\beta^{\pi^{s-1}2^{n+1}} - 1) = 2 - (\alpha^{\pi^{s-1}2^{n+1}} + \beta^{\pi^{s-1}2^{n+1}}) = 2 - V_{n+1} \in A^*.$$

(3) Moreover, we see

$$f(x) = (x - \alpha)(x - \beta) = (x - \alpha)(x - \alpha^N)$$

has coefficients belonging to $Z/(N)$.

Therefore, we see from the proposition of Lenstra, every divisor r of N will be congruent to 1 or N modulo s. But

$$s = q\,2^{n+1} > q\,2^n - 1 = N,$$

so r must be either 1 or N, which proves the primality of N.

We use (15) to discard the composites and use (16) to confirm the primality of those remain. But (ii) is only a sufficient condition, so in some cases several different b's have to be examined.

Algorithm. Consider a suitably long sequence of integers b_i, $i \geqslant 1$, such that $p_i = b_i^2 + 4$ are primes. The first ones of them are

$$5, 13, 29, 53, 173, 229, 293, 733 \cdots$$

Given the number $T(q,n)$, the primality test proceeds in the following way, with all computations carried out modulo $T(q,n)$:

Step 1. $i \leftarrow 0$.

Step 2. $i \leftarrow i+1$, check whether

$$\left\lfloor \frac{T(q,n) \bmod p_i}{p_i} \right\rfloor = -1.\qquad(17)$$

If yes, go to Step 3. Otherwise repeat Step 2.

Step 3. Check whether (15) is true. If yes, go to Step 4. Otherwise T is composite.

Step 4. If $V_{n+1} - 2 \equiv 0 \bmod T$, then return to Step 2.

Step 5. Check whether (16) is true. If yes, then $T(q,n)$ is prime. Otherwise T is composite.

Since every p_i is a prime of the form $4k+1$, by the Gaussian law of reciprocality, (17) means $p_i \in QNR \bmod T(q,n)$, as required in (i). The functions

$$\left\lfloor \frac{x}{p_i} \right\rfloor, \quad 0 \leqslant x \leqslant p_i - 1,$$

are stored to facilitate the work of Step 2.

The following is a table of all primes of the form $T(q,n)$ with $q = 3, 5, 7, 9$ and $n \leqslant 1000$.

q	n										
3	1	2	3	4	6	7	11	18	34	38	43
	55	64	76	94	103	143	206	216	306	324	391
	458	470	827								
5	2	4	8	10	12	14	18	32	48	54	72
	148	184	248	270	274	420					
7	1	5	9	17	21	29	45	177			
9	1	3	7	13	15	21	43	63	99	109	159
	211	309	343	415	469	781	871	939			

For the cryptographically most interesting case where $q = 5$ the numbers $T(q,n)$ were found to be prime for the following values of $n \leqslant 4000$

$$1340, 1522, 1638, 2014, 2170, 2548, 2622, 2652, 2704.$$

IV. Computing $Ord_{T(q,n)}(2)$

We are interested only in the case where $T(q,n)$ is a prime. Moreover, when $n \geqslant 3$, the number 2 is a quadratic residue modulo $T(q,n)$, so in computing the order of 2 modulo $T(q, n)$ we need only consider the complete factorization of

$$T(q, n-1) = \frac{T(q, n) - 1}{2}.$$

The computations are carried out only for $q = 5$ with $n \leqslant 150$, the case which seems to be cryptographically more interesting than the cases $q = 3, 7, 9$. Instead of $Ord_{T(q, n)}(2)$, the index of the subgroup generated by 2 in the multiplicative group of integers modulo the prime $T(q, n)$ is given in the following table, together with the smallest primitive root modulo that prime.

n	number of digits	index	smallest primitive root
4	2	2	2
8	4	2	3
10	4	2	3
12	5	2	3
14	5	2	3
18	7	6	3
32	11	2	3
48	16	2	3
54	17	2	6
72	23	2	6
148	46	78	7

The computation results meet the expectation stated in Section I satisfactorily.

V. Some Cryptographic Remarks

The point, however, is not in the linear complexity alone. It is a common practice to produce strong keystreams by combining up a certain number of comparatively simple source sequences, and the problem is that the complexities carried by the individual components are in many cases not well alloyed, so that an attack applied to one component may reveal the secrecy in the remaining parts. Take, for example, the so-called multiplexing scheme proposed in [12]. The whole system can be cracked either by the linear consistency test of [13], applied to the LFSR which provides the address numbers, or, as suggested in [14], by the correlational attack applied to the LFSR which provides the signals to be selected.

The idea of bilateral clock-control as describe in the present paper provides a good approach to intensifying the multiplexing scheme. If in our scheme, instead of (6) we produce the output signal $c(t)$ according to the multiplexing scheme with LFSR-A providing the address numbers and

LFSR-B providing the signals to be selected, then the attacks given in [13, 14] will be bound to fail for the two shift-registers are now made inseparable from each other, so that in attacking to one of them the cryptanalyst must take into consideration the other which controls the clock signals to it. The occurrence of a constant output sequence may be avoided by some threshold monitoring device, and so the lower bound to the linear complexity established before will hold true for the new scheme.

References

1. R.A. Rueppel, *Analysis and Design of Stream Ciphers*, Springer-Verlag, 1986.

2. Z.D. Dai et al., "Nonlinear Feedforward Sequences of m-sequences," *Proceedings of Beijing International Workshop of Information Theory*, 1988.

3. T. Siegenthaler, "Correlation-Immunity of Nonlinear Combining Functions for Cryptographic Applications," *IEEE Trans. on Info. Theory*, Vol. **IT-31**, Sep. 1984, pp. 776-780.

4. Z.D. Dai, "On the Construction and Cryptographic Applications of de Bruijn Sequences," submitted to *Journal of Cryptology*, 1989.

5. D. Gollmann and W. G. Chambers, "Stepping Clock Controlled Shift Registers," *EUROCRYPT 89*.

6. D. Gollmann and W.G. Chambers, "Clock-Controlled Shift Registers: A Review," *IEEE J. on Selected Areas in Comm.*, Vol. 7, 1989, pp. 525-533.

7. D.H. Lehmer, "An Extended Theory of Lucas' Functions," *Annals of Math.*, Vol. **31**, 1930, pp. 419-448.

8. H. Riesel, "A note on prime Numbers of the forms $N = (6a+1)2^{2n-1} - 1$ and $M = (6a+1)2^{2n} - 1$," *Ark. för Mat.*, 1955, pp. 245-253.

9. H. Riesel, "Lucasian Criteria For the Primality of $N = h \cdot 2^n - 1$," *Math. Comp.*, Vol. **23**, 1969, pp. 869-875.

10. H. Riesel, *Prime Numbers and Computer Methods for Factorization*, Birkhäuser Boston, Inc., 1985.

11. H.W. Lenstra, Jr., "Primality Test," in: H.W. Lenstra, Jr., R. Tijdeman (eds), *Computational Methods in Number Theory*, Math. Centre Trace **154/155**, Mathematisch Centrum, Amsterdam 1982, pp. 55-77.

12. S.M. Jennings, "A Special Class of Binary Sequences," University of London, 1980, Ph.D. Thesis.

13. Kencheng Zeng, C.H. Yang, and T.R.N. Rao, "On the Linear Consistency Test (LCT) in Cryptanalysis with Applications," presented to the Ninth Annual Crypto Conference, Santa Barbara, California, August 20-27, 1989. To appear in: Advances in Cryptology, Proc. of Crypto'89 (Lecture Notes in Computer Science), Springer-Verlag.

14. S. Mund, D. Gollmann, and T. Beth, "Some Remarks on the Cross Correlation Analysis of Pseudorandom Generators," *EUROCRYPT 87*, 1987, pp. 25-35.

SECTION 6

BLOCK CIPHERS

COMPARISON OF BLOCK CIPHERS

Helen Gustafson* Ed Dawson* Bill Caelli**

* School of Mathematics
Queensland University of Technology
GPO Box 2434
Brisbane Q 4001

** Information Security Research Centre
Queensland University of Technology
GPO Box 2434
Brisbane Q 4001

ABSTRACT

Several DES replacement block ciphers have been published. In this paper a report will be given on the development of a package for analysis and comparison of block ciphers. Experimental results are presented on applying this package to DES, FEAL-N, and Madryga ciphers.

1. INTRODUCTION

There are several properties that a block cipher should satisfy in order to be secure from cryptanalytic attack including plaintext - ciphertext independence, completeness, nonaffine system and the strict avalanche effect. In this paper a report is given on the development of a package for analysis and comparison of block ciphers in terms of the above properties. This package is being designed in such a manner that block ciphers of the same block length can be compared.

Several DES replacement block ciphers with block length 64 have been published including FEAL-N (N = 4, 8, 16, 32) (see [1] and [2]), and Madryga (see [3]). FEAL-N is a Feistel type cipher with N rounds of an F-function. The cipher is faster in encryption and decryption speed than DES in both software and hardware implementation. In [4] Den Boer has a chosen plaintext cryptanalysis of FEAL 4. The Madryga algorithm was designed as a DES replacement block cipher by the Canadian Imperial Bank of Commerce. It is not a Feistel type cipher. In Sections 3 and 4 each of the above block ciphers will be used in a comparison with DES.

In Section 2 several measures of randomness for finite sequences will be discussed consisting of statistical tests, sequence complexity and the binary derivative. In Section 3 it will be shown how these measures of randomness can be used to examine plaintext - ciphertext independence.

In Section 4 it will be shown how to examine the strict avalanche effect by applying the Kolmogrov-Smirnov test to the entries of a dependence matrix defined by avalanche vectors. In addition the dependence matrix defined by plaintext - ciphertext avalanche vectors allows one to determine whether or not the block is complete and nonaffine (for a particular choice of key).

2. MEASURES OF RANDOMNESS

2.1 Statistical Tests

There are several statistical tests described in [5] to measure the randomness of a finite sequence. The tests which were used in the experiments in Section 3 are described below.

Suppose that sequences have length n (in the case used n is 64). Let n_0 and n_1 be the number of zeros and ones respectively in a sequence.

(i) Frequency Test.

Determines whether n_0 and n_1 are approximately equal. The χ^2 statistic is used to examine the hypothesis that $n_0 = n_1$ where

$$\chi^2 = \frac{(n_0 - n_1)^2}{n} .$$

In the experiments described in Section 3 as shown in Tables 2 and 3 values of $n_1 > 40$ or < 24 were used for a level of significance of χ^2 of approximately 5% and values of $n_1 > 42$ and < 22 were used for a level of significance of approximately 1%.

(ii) Serial Test.
Used to ensure that the probability for consecutive entries being equal or different is about the same. Let

n_{00} be the number of 00 entries,
n_{01} be the number of 01 entries,
n_{10} be the number of 10 entries,
n_{11} be the number of 11 entries.

For random sequences

$$n_{00} = n_{01} = n_{10} = n_{11} \approx \frac{n-1}{4} \ (\approx 16 \text{ when } n = 64).$$

The χ^2 statistic is used to test the above hypothesis where

$$\chi^2 = \frac{4}{n-1} \sum_{i=0}^{1} \sum_{j=0}^{1} (n_{ij})^2 - \frac{2}{n} \sum_{i=0}^{1} (n_i)^2 + 1 .$$

In the experiments described in Section 3 as shown in Tables 4 and 5 with two degrees of freedom χ^2 values at the 5% and 1% levels of significance of 5.991 and 9.210 respectively were used.

(iii) Runs Test.
The binary sequence is divided into blocks (runs of ones) and gaps (runs of zeros). The runs test examines the number of runs for random data. This test is only applied if the sequence has already passed the serial test in which case it is known that the number of blocks and gaps are in acceptable limits.
The number of runs is normally distributed with

$$\text{Mean} = 1 + \frac{2 n_0 n_1}{n} ,$$

$$\text{Variance} = \frac{(\text{Mean} - 1)(\text{Mean} - 2)}{n - 1} ,$$

$$z = \frac{\text{Runs - Mean}}{\sqrt{\text{Variance}}} \, .$$

In the experiments described in Section 3 confidence intervals at the 95% and 99% levels were used giving z scores ± 1.96 and ± 2.58 respectively as shown in Tables 6 and 7. Since the number of runs is a discrete variable a continuity correction of ± 0.5 was included in the numerator of the standard normal variable z. Absolute values of z, ABS(z), were obtained where

$$\text{ABS}(z) = \frac{\text{ABS (ABS (Runs - Mean) - 0.5)}}{\sqrt{\text{Variance}}} \, .$$

2.2 Sequence Complexity

A measure for the complexity of a finite sequence s is given in terms of the number of new patterns which appear as we move along the sequence. This number is c(s) called the complexity of s (see references [6] and [7]).

Example
s = 100111101100001110
s = 1 / 0 / 01 / 1110 / 1100 / 001110
c(s) = 6

In [6] it is shown that almost all binary sequences of length n have complexity exceeding

$$\frac{n}{\log_2 n} \, .$$

This value will be used as a threshold of complexity for random sequences. In the case where n = 64 this threshold value is $10\frac{2}{3}$. In the experiments described in Section 3, Table 8 lists the percentage of blocks in each cipher in each experiment having a complexity ≤ 10.

Since the sequence complexity test counts new patterns the poker test and autocorrelation tests from [5] were not included in the statistical tests as described in Section 2.1. The poker test counts the number of similar patterns of a chosen length in the block and the autocorrelation test checks for periodicity in the block.

2.3 Binary Derivative

Given a string of binary digits the first derivative is taken by considering each overlapping pair of digits and recording a zero if they are the same and a one if they are different (see references [8] and [9]). Every successive binary derivative drops one digit. A sequence of length 16 and its first four binary derivatives are given below:

```
1 0 0 0 1 0 0 0 1 1 0 1 0 1 0 1
1 0 0 1 1 0 0 1 0 1 1 1 1 1 1
1 0 1 0 1 0 1 1 1 0 0 0 0 0
1 1 1 1 1 1 0 0 1 0 0 0 0
0 0 0 0 0 1 0 1 1 0 0 0
```

Let $p(i)$ denote the fraction of ones in the ith derivative where $p(0)$ denotes the fraction of ones in original sequence. Suppose that k derivatives are evaluated for a string of length n. Let r denote the range of the $p(i)$ and p denote the average of the $p(i)$ where

$$p_{max} = \text{Max } p(i)$$
$$p_{min} = \text{Min } p(i)$$
$$r = p_{max} - p_{min}$$
$$p = \sum_{i=0}^{k} \frac{p(i)}{k+1}$$

For example, for the above sequence of length 16,
$p(0) = .44, p(1) = .67, p(2) = .43, p(3) = .54, p(4) = .75$
$r = .32$
$p = .57$

In [8] it is stated that one can use $p(0)$, p, r to differentiate between patterned and random strings as shown in Table 1.

Attribute	Patterned Strings	Random Strings
$p(0)$	variable	close to 0.5
p	low	close to 0.5
r	high	low

Table 1

In order to determine how many derivatives to evaluate to differentiate between patterned and random sequences of length n it is suggested in [9] that one use experimental results. To this end the complexity test from Section 3.2 was used on pseudorandom binary data generated by a stream cipher to give 1000 low complexity binary strings of length 64 (Rand 1) and 1000 high complexity binary strings of length 64 (Rand 2) as shown in Figure 1. Binary derivatives were calculated for each of these set of sequences as shown in Figure 2 where the horizontal axis represents the number of derivatives calculated say k and the vertical axis represents the average of the ranges for each set of sequences after k derivatives have been taken. From the graph it appears that between the 5th and 8th derivatives there is the greatest difference between the average ranges for the low and high complexity sequences. It was decided to use seven derivatives to measure randomness.

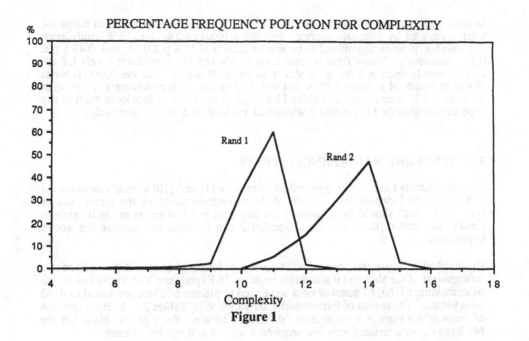

PERCENTAGE FREQUENCY POLYGON FOR COMPLEXITY

Complexity

Figure 1

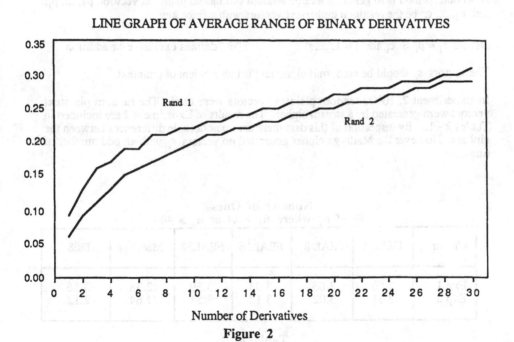

LINE GRAPH OF AVERAGE RANGE OF BINARY DERIVATIVES

Number of Derivatives

Figure 2

In order to measure significance levels for p and r 10000 random blocks of length 64 were generated by a stream cipher. For these blocks 95% and 99% confidence intervals for p were determined to be approximately $0.45 \leq p \leq 0.54$ and $0.43 \leq p \leq 0.55$ respectively. These figures were used as 5% and 1% significant levels for the experiments in Section 3 for p as shown in Tables 9 and 10. For the 10000 random blocks of length 64 generated, 95% and 99% had values of approximately r less than 0.28 and 0.33 respectively. Tables 11 and 12 show the % of blocks in each of the experiments described in Section 3 which had $r > 0.28$ and 0.33 respectively.

3. APPLYING RANDOMNESS TESTS

Several methods have been suggested in references [7] and [10] to examine plaintext and ciphertext independence. Each of these methods involves the generation of sequences which should be random if the plaintext and ciphertext are independent. Hence the random measures from Section 2 can be used to examine the above hypothesis.

One method is to generate nonrandom sequences as binary plaintext. If the ciphertext is independent of the plaintext it should be random. In Experiment 1 all vectors of length 64 containing 64, 63, 62 zeros or ones were used as plaintext. There are a total of 4162 such vectors. The results of Experiment 1 are included in Tables 2 - 12. By inspection of these tables there is no noticeable difference between the ciphers. However the Madryga cipher generated very few sequences with an odd number of ones.

A second method is to generate a large amount of random plaintext vectors $p_1, ..., p_r$. Let $c_1, ..., c_r$ be the resultant ciphertext vectors with a fixed key.

Define $s_i = p_i \oplus c_i$ for $i = 1, ..., r.$ "$\oplus$" defines exclusive-or addition.

The vectors s_i should be random if ciphertext is independent of plaintext.

In Experiment 2, 10000 random plaintext vectors were used. The random plaintext vectors were generated by a stream cipher. The results of Experiment 2 are included in Tables 2 - 12. By inspection of this data there are no noticeable differences between the ciphers. However the Madryga cipher generated no vectors s_i with an odd number of ones.

Number of Ones
% of n_1 where $n_1 < 24$ or $n_1 > 40$

Cipher	FEAL 4	FEAL 8	FEAL16	FEAL32	Madryga	DES
Exp 1	2.90	3.63	3.03	3.39	2.38	3.56
Exp 2	3.31	3.02	3.18	3.51	1.64	2.12

Table 2

Number of Ones
% of n_1 where $n_1 < 22$ or $n_1 > 42$

Cipher	FEAL 4	FEAL 8	FEAL16	FEAL32	Madryga	DES
Exp 1	0.72	0.72	0.60	1.01	0.70	1.01
Exp 2	0.85	0.79	0.84	0.97	0.59	0.82

Table 3

Serial Test
% of $\chi^2 > 5.991$

Cipher	FEAL 4	FEAL 8	FEAL16	FEAL32	Madryga	DES
Exp 1	4.52	4.48	4.73	4.85	5.33	4.88
Exp 2	5.06	4.62	4.81	4.92	4.74	4.82

Table 4

Serial Test
% of $\chi^2 > 9.210$

Cipher	FEAL 4	FEAL 8	FEAL16	FEAL32	Madryga	DES
Exp 1	0.84	0.84	0.74	1.18	1.13	1.20
Exp 2	1.13	1.01	0.99	0.97	1.03	0.93

Table 5

Runs Test
% of $|z| > 1.96$

Cipher	FEAL 4	FEAL 8	FEAL16	FEAL32	Madryga	DES
Exp 1	3.41	3.48	3.63	3.51	3.89	3.44
Exp 2	3.37	3.77	3.52	3.59	3.09	3.39

Table 6

Runs Test
% of | z | > 2.5758

Cipher	FEAL 4	FEAL 8	FEAL16	FEAL32	Madryga	DES
Exp 1	0.67	0.67	0.79	0.67	0.60	0.60
Exp 2	0.76	0.71	0.67	0.61	0.59	0.61

Table 7

Complexity Data
% ≤ 10

Cipher	FEAL 4	FEAL 8	FEAL16	FEAL32	Madryga	DES
Exp 1	3.33	4.08	4.25	3.70	4.08	4.11
Exp 2	3.81	3.99	3.68	4.21	4.15	4.37

Table 8

Average of Binary Derivative
% of p where p < 0.45 or p > 0.54

Cipher	FEAL 4	FEAL 8	FEAL16	FEAL32	Madryga	DES
Exp 1	5.05	4.20	4.71	4.85	4.97	4.32
Exp 2	4.59	5.19	4.52	4.68	4.67	4.93

Table 9

Average of Binary Derivative
% of p where p < 0.43 or p > 0.55

Cipher	FEAL 4	FEAL 8	FEAL16	FEAL32	Madryga	DES
Exp 1	1.15	0.72	1.03	1.15	1.03	0.89
Exp 2	0.91	1.09	0.85	0.96	0.92	1.02

Table 10

Range of Binary Derivative
% of r > 0.28

Cipher	FEAL 4	FEAL 8	FEAL16	FEAL32	Madryga	DES
Exp 1	4.88	5.07	5.17	4.47	4.81	5.21
Exp 2	4.93	4.90	4.78	5.11	5.15	4.96

Table 11

Range of Binary Derivative
% of r > 0.33

Cipher	FEAL 4	FEAL 8	FEAL16	FEAL32	Madryga	DES
Exp 1	1.01	1.25	1.25	0.94	1.18	1.15
Exp 2	1.33	1.19	1.07	1.11	1.06	1.00

Table 12

4. AVALANCHE EFFECT

There are two types of avalanche effects which can be examined:

(i)   the plaintext - ciphertext avalanche effect;
(ii) the key-ciphertext avalanche effect.

4.1 Plaintext - Ciphertext Avalanche Effect

A block cipher satisfies the plaintext - ciphertext strict avalanche effect if each ciphertext bit changes with a probability of one half whenever a single plaintext bit is changed.

From [11] to measure the plaintext - ciphertext strict avalanche effect for a block cipher of length n generate a large number of random plaintext vectors, p_r, for $r = 1, \ldots, m$. Let p_r^j for $j = 1, \ldots, n$ be plaintext vectors that differ from p_r in the j th coordinate. For a fixed key let c_r and c_r^j be ciphertext vectors that result from p_r and p_r^j respectively. Define avalanche vectors $u_r^j = c_r \oplus c_r^j$ for $r = 1, \ldots, m$ and $j = 1, \ldots, n$ where '$\oplus$' defines exlcusive-or addition.

Define an $n \times n$ dependence matrix, A, as follows:
For $r = 1, \ldots, m$ add the n entries of u_r^j to each corresponding entry in column j of A, where the initial values of A are all zero. These entries of A, a_{ij}, $i, j = 1, \ldots, n$, will give the total number of ones for each ciphertext bit corresponding to each of the avalanche vectors for all plaintext strings. The entries refer to the total number of changes in the ciphertext position i when each bit j is changed in the plaintext string, for all m plaintext strings.

The dependence matrix A for the plaintext - ciphertext avalanche effect can be used to decide whether a block cipher is complete and is nonaffine in relation to the key used to define the dependence matrix. A block cipher is said to be complete if each ciphertext bit depends on all of the plaintext bits [12]. Clearly a non zero entry a_{ij} in A indicates that ciphertext bit i depends on plaintext bit j. As shown in [13] if a block cipher is complete then the cipher is nonaffine.

4.2 Key - Ciphertext Avalanche Effect

A block cipher satisfies the key - ciphertext strict avalanche effect if each ciphertext bit changes with a probability of one half whenever a single key bit is changed.
To measure the key - ciphertext strict avalanche effect for a block cipher of length n generate a large number of random key vectors k_r, for r = 1, . . ., m. Let k_r^j for j = 1, . . ., ℓ be key vectors (where the key length is ℓ) that differ from k_r in the j th coordinate. Encrypt a fixed plaintext string p and let c_r and c_r^j be ciphertext vectors that result from k_r and k_r^j respectively. Define avalanche vectors $u_r^j = c_r \oplus c_r^j$ for r = 1, . . ., m and j = 1, . . ., ℓ where '$\oplus$' defines exclusive-or addition. Obtain the dependence matrix A as defined in Section 4.1 where in this case A is an n x ℓ matrix. The entries of A refer to the total number of changes in the ciphertext position i when each bit j is changed in the key string, for all m key strings.

In general a block cipher should satisfy a key - ciphertext complete property in that every ciphertext bit should depend on every key bit. A non zero entry a_{ij} in the dependence matrix for the key - ciphertext effect indicates that ciphertext bit i depends on key bit j.

4.3 Analysis of Results

The chi-squared statistic is used to test the hypothesis that the total number of ciphertext changes (ones) and non-changes (zeros) is the same (= m/2) for each entry in the avalanche matrix.The test statistic used is

$$\chi^2 = (m/2 - a_{ij})^2 / (m/2) .$$

The Kolmogrov-Smirnov Test (See [14]) is used to test the goodness-of-fit of these chi-squared values to a chi-squared distribution with 1 degree of freedom. This test involves comparing the hypothesised cumulative distribution function with the observed relative cumulative frequencies. The hypothesised distribution function is defined as F(x) = P (X ≤ x). Values of F(x) are available in chi-squared tables and the corresponding observed cumulative relative frequencies, CRF(x), are found using the distribution of values of the chi-squared statistic for the entries in the avalanche matrix.
The test statistic for the Kolmogrov-Smirnov test is defined as D and equals the maximum absolute difference between corresponding values of F(x) and CRF(x):

$$D = \max | F(x) - CRF(x) |$$

A large value of D would indicate that the chi-squared values obtained from the avalanche matrix would not fit a chi-squared distribution with 1 degree of freedom. Hence the cipher does not satisfy the strict avalanche effect. The value of D is compared to critical values for selected significance levels.

In order to obtain empirical results on the dependence matrices described in Sections 4.1

and 4.2 a stream cipher was used to generate $m = 10000$ random strings for the ciphers where m is as defined in Section 4.1 and 4.2.

Table 13 contains the level of significance for the Kolmogrov-Smirnov test of 5% and 1% levels. Table 14 contains the experimental results for values of D. As shown by this data the Madryga cipher fails the Kolmogrov-Smirnov test for the goodness of fit of the chi-squared values of the avalanche matrix to a chi-squared distribution with one degree of freedom for the plaintext - ciphertext avalanche effect with the selected key, which was 0123456789ABCDEF in hexadecimal, and the selected random plaintext strings. In Figure 3 Cumulative Relative Frequency graphs are drawn for the plaintext - ciphertext χ^2 values for each cipher and they are all compared to the cumulative χ^2 distribution with 1 degree of freedom, for values chosen from χ^2 tables. The graphs for FEAL-N and DES ciphers all appear close to the actual cumulative distribution curve. It is clearly seen that the Madryga cipher does not fit the χ^2 distribution with 1 degree of freedom, from this graph.

Level of Significance	5%	1%
Critical D value	0.0213	0.0255
* DES Key - Ciphertext Critical D value	0.0227	0.0272

Table 13

	$D = \max \lvert F(x) - CRF(x) \rvert$	
Cipher	Plaintext - Ciphertext	Key - Ciphertext
FEAL-4	0.0173	0.0083
FEAL-8	0.0116	0.0156
FEAL-16	0.0132	0.0066
FEAL-32	0.0063	0.0134
Madryga	0.6453	0.0184
DES	0.0103	0.0111 *

Table 14

CUMULATIVE RELATIVE FREQUENCY GRAPHS AND χ^2 DISTRIBUTION

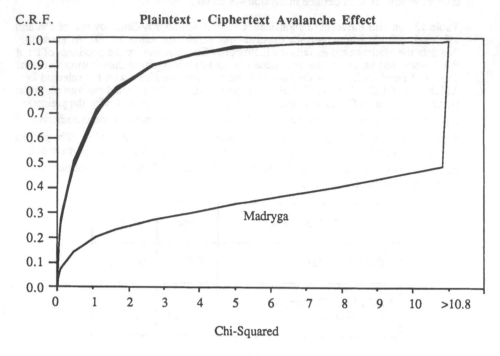

Figure 3

5. CONCLUSION

Based on the experimental results in Sections 3 and 4 it is concluded that the two Feistel ciphers, FEAL-N (N = 4, 8, 16, 32) and DES, satisfy plaintext - ciphertext independence and both the strict plaintext - ciphertext avalanche effect and the strict key - ciphertext avalanche effect. The Madryga cipher based on the results of the strict plaintext - ciphertext avalanche effect and the peculiarity in the number of ones in the plaintext - ciphertext experiments needs further investigation.

The aim of this project is the formation of a package to investigate and compare block ciphers of the same length. A cryptographer could use such a package to help identify a weakness in a newly designed block cipher. However, it is not suggested that the identification of such a weakness would give a systematic method for a cryptanalyst to break the cipher.

REFERENCES

1. A. Shimizu and S. Miyaguchi, 'Fast Data Encipherment Algorithm FEAL', Advances in Cryptology: Proc. Eurocrypt '87, 1988, pp. 267 - 278.

2. S. Miyaguchi, A. Shiraishi, A. Shimizu, 'Fast Data Encipherment Algorithm FEAL-8', Review of the Electrical Communications Laboratories, Vol. 36, No. 4, 1988, pp. 433 - 437.

3. W.E. Madryga, 'A High Performance Encryption Algorithm', Computer Security: A Global Challenge, Elsevier Science Publishers B.V., 1984, pp. 557 - 570.

4. B. Den Boer, 'Cryptanalysis of F.E.A.L.', Advances in Cryptology: Proc. Eurocrypt '88, pp. 293 - 299.

5. H. Becker and F. Piper, 'Cipher Systems: The Protection of Communications', John Wiley and Sons, 1982.

6. A. Lempel and J. Ziv, 'On the Complexity of Finite Sequences', IEEE Trans. on Information Theory Vol. IT - 22, Jan. 1976, pp. 75 - 81.

7. A.K. Leung and S.E. Tavares, 'Sequence Complexity as a Test for Cryptographic Systems', Advances in Cryptology '84, pp. 468 - 474.

8. J.M. Carroll and L.E. Robbins, 'Using Binary Derivatives to Test an Enhancement of DES', Cryptologia, Vol. XII Number 4, Oct. 1988, pp. 193 - 208.

9. J.M. Carroll, 'The binary derivative test for the appearance of randomness and its use as a noise filter', Technical Report No. 221, Dept. of Computer Science, The University of Western Ontario, November 1988.

10. A.K. Leung and S.E. Tavares, 'Sequence Complexity as a Test for Cryptographic Systems', Advances in Cryptology, Crypto '84, pp. 468 - 474.

11. A.F. Webster and S.E. Tavares, 'On the Design of S-Boxes', Advances in Cryptology: Crypto '85, 1986, pp. 523 - 530.

12. A.G. Konheim, 'Cryptography: A Primer', John Wiley and Sons, 1981.

13. J.B. Kam and G.I. Davida, "Structured Design of Substitution - Permutation Encryption Networks", IEEE Trans. on Computer, Vol. C-28, No. 10, Oct. 1979, pp. 747 - 753.

14. D.E. Knuth, "The Art of Computer Programming, Vol. 2: Seminumerical Algorithms", Addison-Wesley, 1973.

KEY SCHEDULING IN DES TYPE CRYPTOSYSTEMS

Lawrence Brown   *Jennifer Seberry*

Department of Computer Science
University College, UNSW, Australian Defence Force Academy
Canberra ACT 2600. Australia

Abstract

This paper reviews some possible design criteria for the key schedule in a DES style cryptosystem. The key schedule involves a Key Rotation component, and the permutation $PC2$. Together these provide for a diffusion of dependency of ciphertext bits on key bits. Some empirical rules which seem to account for the derivation of the key schedule used in the DES are first presented. A number of trials were run with various key schedules, and some further design rules were derived. An alternative form of key schedule was then tested. This used either a null $PC2$, or one in which permutations only occurred within the inputs to a given S-box, and a much larger rotation schedule than used in the DES. This was found to be as effective as the key schedule used in the current DES, and is proposed for use in new cryptosystems.

1. Introduction

The Data Encryption Standard (DES) [NBS77] is currently the only certified encryption standard. It has achieved wide utilization, particularly in the banking and electronic funds transfer areas, and is an Australian standard [ASA85] among others. With the current significant use of DES (especially in banking), there is interest in designing and building a DES-type cryptosystem with an extended key length of either 64 (rather than 56) or 128 bits. This is one of a continuing series of papers [Brow88], [BrSe89], [PiSe89], [Piep89], analysing aspects of the current DES, and indicating criteria to be used in the design of future schemes.

This paper will concentrate on the design of the key schedule, which involves a key rotation component, and the permutation $PC2$. Together these provide for a diffusion of dependency of ciphertext bits in key bits. As a measure of effectiveness, Meyer's analysis of output bit dependence on key bits will be used [MeMa82]. Some empirical rules for the key schedule, derived previously [Brow88], will be presented. A discussion of some alternatives to the current schedule will be presented, followed by the results obtained from testing a number of alternate schedules. A presentation of the implications from these in the design of any extended DES type schemes will conclude the paper.

2. The Key Schedule in DES

The central component of the DES cryptosystem is the function g, which is a composition of expansion function E, eight substitution boxes (S-boxes) S, and a permutation P[1]. Function g has as inputs the plaintext $[L(i-1), R(i-1)]$ from the previous round, and a selection of key bits $K(i)$ (see Fig 1.). This may be written as:

$$g : R(i) = L(i-1) \oplus P(S(E(R(i-1)) \oplus K(i))), L(i){=}R(i{-}1).$$

---

[1] A more detailed description of these functions may be found in [NBS77], [ASA85] or [SePi88].

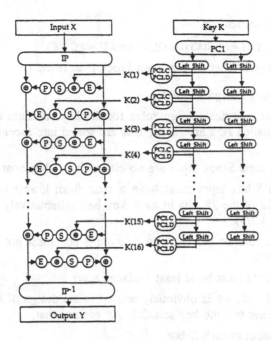

Fig 1. DES as a Mixing Function

The *key schedule* in a DES algorithm is responsible for forming the sixteen 48-bit sub-keys $K(i)$ used in the rounds of the encryption procedure. This function is important since if the same key is used on successive rounds, it can weaken the resulting algorithm (see [GrTu78], [MeMa82], [MoSi87], [MoSi86], and [ASA85]). In detail, the 64-bit key is permuted by $PC1$. This permutation performs two functions: first it strips the eight parity bits out, and then distributes the remaining 56 bits over two 28-bit halves $C(0)$ and $D(0)$. The cryptographic significance of this permutation is questionable [DDFG83]. Subsequently for each round, each 28-bit register is rotated left either one or two places according to the following schedule (subsequently denoted KS):

Table 1 - Key Schedule for DES

Round	1	2	3	4	5	6	7	8	9	10	11	12	13	14	15	16
Shift	1	1	2	2	2	2	2	2	1	2	2	2	2	2	2	1
Total	1	2	4	6	8	10	12	14	15	17	19	21	23	25	27	28

After the shift, the resultant 28-bit vectors are permuted by $PC2$ (which in fact consists of two 28-bit permutations, each of which selects 24 bits) to form the sub-key for that round. This permutation may be written in terms of which S-box each bit is directed to, as shown in Table 2 (nb: an * indicates an autoclave S-box input rather than a message input; an X specifies exclusion of that bit).

Table 2 - Current DES Permutation PC2

C: 1 4* 2* 3 1* 2 4 3* X 2* 1 3 4 1* 2 4* 1 X 3 4 2 X 3* 1 X 3 4 2
D: 8 6* 5 8* 6 7 X 8 5 X 7 6 5* 8 X 7* 6 8* 5 6* 7 8 6 5 7* X 5* 7

The sub-keys $K(i)$ may be written as;

$$K(i) = PC2(KS(U, i)), \quad \text{where } U = PC1(K)$$

and $KS(U,i)$ is the key rotation schedule for input block U at round i.

3. Empirical Key Schedule Design Criteria

In Brown [Brow88], some empirical design rules for the key schedule are presented. The rules presented for permutation $PC2$ are (if the bits are sorted into ascending order of their input bits):

- 1 bits permuted to the same S-box input are no closer than 3 bits apart
- 2 bits permuted to an S-box input must have a span from lowest to highest input bit number of at least 22 of the 28 bits in each key half (alternatively, the average spacing must be at least 3 2/3)
- 3 bits permuted to the selector bits a, f on a given S-box must not be adjacent in the sorted list of input bits
- 4 bits not selected by $PC2$ must be at least 3 places apart

The design of the key schedule KS is obviously related to the design of $PC2$ by rules 1 and 4 given above. Brown notes that the key schedule KS ensures that:

- 1 each bit is used as input to each S-box
- 2 no bit is used as input to the same S-box on successive rounds
- 3 the total number of bits rotated is 56 (which implies that $K(0) = K(16)$, enabling the decryption operation to use right shifts in reverse order).

4. Ciphertext Dependence on Key Bits

This analysis is complex, and is dependent on the choices of permutations P and $PC2$ as well as the KS the key schedule[2]. To quantify this dependency, a 64*56 array F_r is formed, in which element $F_r[i,j]$ specifies a dependency of output bit $X(j)$ on key bit $U(i)$. The vector U is that formed after PC1 is applied, ie $U = PC1(K)$. The number of marked elements in G_r will be examined to provide a profile of the degree of dependence achieved by round r. Details of the derivation of this matrix, and the means by which entries are propagated, may be found in [MeMa82]. This analysis technique will be used as a measure of effectiveness for possible key schedules. In particular, two criteria are used:

- rate of growth of output bit dependence on key bits by any S-box inputs
- rate of growth of output bit dependence on key bits by BOTH message and autoclave S-box inputs

5. Alternatives for the Key Schedule

The purpose of the above rules in designing a key schedule may be summarized as follows:

> to present each key bit to a message input, and to an autoclave input, of each S-box as quickly as possible.

This is achieved by a combination of the key rotation schedule KS, the key permutation $PC2$, and the function $g = S.P.E$. Trials have been performed in which each of these is

---

[2] but not on permutations IP, FP and $PC1$, which only serve to renumber the plaintext, ciphertext, and key bits respectively. The analysis done in this paper ignores these permutations for this reason.

varied in turn, to analyse the effect of each.

In addition to that underlying design purpose, there is a pragmatic decision on the size of the key registers. In the current scheme, the key is divided into two halves. An alternate form could be to have a single large key schedule register. We also wish to extend the size of the key, in order to ensure it is large enough to withstand any foreseeable exhaustive search style attack. One way of providing a measure of this, whilst still maintaining compatibility with existing protocols, would be to remove the notion of parity bits in the key, and use all 64-bits. Combining these two ideas we have the following possibilities for PC2:

28 -> 24 bit
56 -> 48 bit
32 -> 24 bit
64 -> 48 bit

Initially, a key schedule with the same form as the current DES was examined, in order that comparisons with the effectiveness of the current DES scheme could be made. Having obtained some guidelines from these trials, key schedules involving some of the alternatives were then tried.

6. Some Trials on New Key Schedules

In Brown [Brow88], some empirical design criteria for permutation PC2 and the Key Rotation Schedule were presented. The authors have subsequently used these rules to generate a set of permutations PC2. Since all possible 28->24 bit permutations could not be tried, permutations with the form shown in Table 3 were tried (that is all arrangements of the 4 excluded bits, subject to the rules set, were found). This form was chosen in order to distribute key bits to each of the 4 S-boxes being fed by each half of the key schedule as quickly as possible. A total of 7315 permutations were found.

Table 3 - form of generated Permutations PC2
C: 1 2 3 4 1 2 3 4 1 2 3 4 1 2 3 4 1 2 3 4 1 2 3 4 + X X X X
D: 5 6 7 8 5 6 7 8 5 6 7 8 5 6 7 8 5 6 7 8 5 6 7 8 + X X X X

Ciphertext-Key Dependences (CKdep) tests on these permutations produced results shown in Table 4 (with comparisons to the current and worst case PC2 supplied for comparison).

Table 4 - Dependency of Ciphertext bits on Key bits Using Current DES Permutation P and Key Schedule					
Round	Std PC2	Worst PC2	Generated PC2	Regular X 2	Regular X 1, X 3, X 4
1	5.36	5.36	5.36	5.36	5.36
2	39.17	42.19	38.50-39.06	39.06	38.62
3	82.25	81.47	80.25-82.37	82.37	81.47
4	98.44	91.29	96.65-98.66	98.66	98.21
5	100.00	96.21	99.55-100.00	100.00	100.00
6	100.00	99.55	100.00	100.00	100.00
7	100.00	100.00	100.00	100.00	100.00
8	100.00	100.00	100.00	100.00	100.00

Some of these permutations performed better than the PC2 used in the current DES. The best of these were selected, 15 being found. These 15 permutations were all found to have a special form, namely that the excluded bits always fell between bits permuted to S-box 1 and S-box 2 (or 5 and 6 in the D-side). There are thus exactly 15 since $15 = {}^6C_4$. In order to investigate these permutations with a regular placing of the excluded bits, all 60 such permutations were generated. A CKdep analysis of these permutations resulted in only two results, one for permutations with the excluded bit before a bit permuted to S-box 2 (Regular X 2), and one for the others (Regular X 1, X 3, X 4). These results are also shown in Table 4.

So far, we have used the first of the two criteria presented earlier, namely the growth of overall bit dependence of output bits on key bits. If we now consider the alternate measure, namely growth in dependence of output bits on key bits by both message and autoclave S-box inputs, then the results become less clear. As shown in Table 5, whilst growth of overall dependence is greater with the regular $PC2$'s, growth of both is worse.

Table 5 - Dependency of Ciphertext bits on Key bits Using Current DES Permutation P and Key Schedule by Both Message and Autoclave S-box Inputs								
Round	1	2	3	4	5	6	7	8
CKdep	Both,Either	Both,Either	Both,Either	Both,Either	Both,Either	Both,Either	Both,Either	Both,Either
PC2.std	0.0,5.36	2.01,39.17	36.50,82.25	81.03,98.44	95.87,100.0	99.33,100.0	100.0,100.0	100.0,100.0
PC2.worst	0.0,5.36	0.0,42.19	33.71,81.47	73.88,91.29	84.38,96.21	92.86,99.55	98.66,100.0	100.0,100.0
PC2 X 1	0.0,5.36	0.22,38.62	29.91,81.47	65.07,98.21	73.66,100.0	80.13,100.0	87.28,100.0	93.97,100.0
PC2 X 2	0.0,5.36	0.45,39.06	30.36,82.37	67.08,98.66	77.01,100.0	83.26,100.0	90.18,100.0	95.87,100.0
PC2 X 3	0.0,5.36	0.45,39.06	30.13,81.92	66.74,98.21	76.79,100.0	83.04,100.0	89.96,100.0	95.76,100.0
PC2 X 4	0.0,5.36	0.45,39.06	30.13,81.92	66.52,98.21	75.67,100.0	80.36,100.0	86.38,100.0	92.41,100.0

A closer look at the structure of the regular permutations shows that the autoclave input bits are clustered, due to the method used to assign them to S-box inputs. By altering the order of inputs within each S-box, a more regular arrangement of autoclave inputs was obtained. When these were tested, the growth of dependence on both was much greater, thus emphasizing the importance of this criterion on the design of $PC2$.

To obtain an indication of the relative influences of each of the components in the key schedule, a series of trials were run, in which each of the following three components were varied with the specified alternatives:

P based on the results in [BrSe89], two permutations P were used:

  • the current DES P and

  • a strictly regular permutation generated by a difference function on the S-box number of [+1 -2 +3 +4 +2 -1]. Because of its very regular structure, the propagation of dependencies may be more easily calculated.

$PC2$ from the above work, the 4 best performing regular $PC2$ were extracted. Then these were processed to provide three levels of clustering of the autoclave inputs.

KS the key variant in the key rotation schedule appears to be the distribution of shifts of 1 verses 2 places. A set of key schedules with various numbers of shifts of 1 initially were derived as shown in Table 6.

Table 6 - Trial Key Schedules																
Round	1	2	3	4	5	6	7	8	9	10	11	12	13	14	15	16
KS	2	2	2	2	2	2	2	2	2	2	2	2	1	1	1	1
KS	1	2	2	2	2	2	2	2	2	2	2	2	2	1	1	1
KS	1	1	2	2	2	2	2	2	2	2	2	2	2	2	1	1
KS	1	1	1	2	2	2	2	2	2	2	2	2	2	2	2	1
KS	1	1	1	1	2	2	2	2	2	2	2	2	2	2	2	2

When this test was run, the following conclusions were made:

- the current permutation P performed better, possibly because its less than regular structure assisted the distribution of dependencies between the autoclave and message inputs.

- permutations $PC2$ with the best spread of autoclave inputs, performed best as expected.

- a key schedule with as many shifts of 1 initially performed best. This again would appear to be a function of the best method for spreading bits to as many S-box inputs as soon as possible.

7. Design Criteria for New Key Schedules

From the above results, the design principles for designing key schedules can now be summarized as follows:

The key schedule ensures that:

1 each key bit is used as input to each S-box in turn

2 no bit is used as autoclave inputs on successive rounds

3 no bit is excluded on successive rounds

4 the final key register value is identical to the original key register value (to enable easy reversal of the key schedule for decryption)

8. An Alternative Key Schedule Design

In the design of the DES, small key rotations were used, which required the use of permutation $PC2$ to provide a fan-out of key-bits across the S-box inputs, in order to satisfy the above principles. An alternative design can be envisaged in which a large key rotation interval is used, along with a null $PC2$ (ie: so called worst case $PC2$), or a local $PC2$ which only permutes bits within each block of 6 S-box inputs. The two $PC2$ permutations used are shown in Table 7.

Table 7 - Null and Local Permutations PC2 for Alternative Key Schedule
C: 1* 1 1 1 1 1* 2* 2 2 2 2 2* 3* 3 3 3 3 3* 4* 4 4 4 4 4* X X X X
D: 5* 5 5 5 5 5* 6* 6 6 6 6 6* 7* 7 7 7 7 7* 8* 8 8 8 8 8* X X X X
C: 1* 1 1 1 1 1* 2 2 2 2* 2* 2 3 3* 3* 3 3 3 4* 4 4 4 4 4* X X X X
D: 5* 5 5 5 5 5* 6 6 6 6* 6* 6 7 7* 7* 7 7 7 8* 8 8 8 8 8* X X X X

For this design, a constant key rotation of 7 bits was used, both because it is larger than the number of inputs to an S-box, and because after sixteen rounds, the key register contents are the same as the original value (since $7*16=112=4*28=2*56$), for both split key registers or a single large key register. This schedule is shown in Table 8.

Table 8 - Alternative Constant Key Schedule															
Round	1 2 3 4 5 6 7 8 9 10 11 12 13 14 15 16														
KS	7 7 7 7 7 7 7 7 7 7 7 7 7 7 7 7														

The results obtained for these $PC2$ permutations and this key schedule, using both a split key rotation register, and a single key register, are shown in Table 9.

<table>
<tr><td colspan="9" align="center">Table 9 - Dependency of Ciphertext bits on Key bits
Using Current DES Permutation P and the Alternative Key Schedule
by Both Message and Autoclave S-box Inputs</td></tr>
<tr><td>Round</td><td>1</td><td>2</td><td>3</td><td>4</td><td>5</td><td>6</td><td>7</td><td>8</td></tr>
<tr><td>PC2</td><td>Both,Either</td><td>Both,Either</td><td>Both,Either</td><td>Both,Either</td><td>Both,Either</td><td>Both,Either</td><td>Both,Either</td><td>Both,Either</td></tr>
<tr><td colspan="9" align="center">Split Key Register Used</td></tr>
<tr><td>null</td><td>0.0,5.36</td><td>1.56,39.06</td><td>34.82,82.03</td><td>76.56,98.33</td><td>91.52,100.0</td><td>98.21,100.0</td><td>100.0,100.0</td><td>100.0,100.0</td></tr>
<tr><td>local</td><td>0.0,5.36</td><td>2.57,39.06</td><td>38.17,82.03</td><td>83.82,98.33</td><td>98.21,100.0</td><td>100.0,100.0</td><td>100.0,100.0</td><td>100.0,100.0</td></tr>
<tr><td colspan="9" align="center">Single Key Register Used</td></tr>
<tr><td>null</td><td>0.0,5.36</td><td>1.79,38.73</td><td>35.04,81.70</td><td>76.56,98.33</td><td>91.52,100.0</td><td>98.21,100.0</td><td>100.0,100.0</td><td>100.0,100.0</td></tr>
<tr><td>local</td><td>0.0,5.36</td><td>2.79,38.73</td><td>38.39,81.70</td><td>83.82,98.33</td><td>98.21,100.0</td><td>100.0,100.0</td><td>100.0,100.0</td><td>100.0,100.0</td></tr>
</table>

These results are very similar in performance to the key schedule used in the current DES (see Table 5). The null $PC2$ performs slightly worse, whilst the local $PC2$ performs better. Depending on the efficiency required, a tradeoff between best performance and ease of implementation can be made between these. There is very little difference in performance between the split and single key rotation registers, thus either could be chosen, depending on other constraints.

9. Conclusion

The key schedule in the current DES has been analysed, and some empirical principles which could have been used in its design derived. These were used to test a number of alternative key schedules, which led to the development of a new set of generalized principles to be used in the design of a new algorithm. An alternative key schedule which either eliminates permutation $PC2$, or uses a local $PC2$, was tried and found to be as effective as that used in the current DES. This is thus suggested for use in any new algorithm.

Acknowledgements

To the following members of the Centre for Computer Security Research: Leisa Condie, Thomas Hardjono, Mike Newberry, Cathy Newberry, Josef Pieprzyk, and Jennifer Seberry; and to: Dr. George Gerrity, Dr. Andzej Goscinski, and Dr. Charles Newton; for their comments on, suggestions about, and critiques of this paper.
Thankyou.

References

[ASA85] ASA, *"Electronics Funds Transfer - Requirements for Interfaces, Part 5, Data Encryption Algorithm,"* AS2805.5-1985, Standards Association of Australia, Sydney, Australia, 1985.

[Brow88] L. Brown, "A Proposed Design for an Extended DES," in *Proc. Fifth International Conference and Exhibition on Computer Security*, IFIP, Gold Coast, Queensland, Australia, 19-21 May, 1988.

[BrSe89] L. Brown and J. Seberry, "On the Design of Permutation P in DES Type Cryptosystems," in *Abstracts of Eurocrypt 89*, IACR, Houthalen, Belgium, 10-13 Apr., 1989.

[DDFG83] M. Davio, Y. Desmedt, M. Fosseprez, R. Govaerts, J. Hulsbosch, P. Neutjens, P. Piret, J. Quisquater, J. Vanderwalle and P. Wouters, "Analytical Characteristics of the DES," in *Advances in Cryptology - Proc. of Crypto 83*, D. Chaum, R. L. Rivest and A. T. Sherman (editors), pp. 171-202, Plenum Press, New York, Aug. 22-24, 1983.

[GrTu78] E. K. Grossman and B. Tuckerman, "Analysis of a Weakened Feistel-Like Cipher," in *Proc. 1978 IEEE Conf. On Communications*, pp. 46.3.1-5, IEEE, 1978.

[MeMa82] C. H. Meyer and S. M. Matyas, *Cryptography: A New Dimension in Data Security*, John Wiley & Sons, New York, 1982.

[MoSi86] J. H. Moore and G. J. Simmons, "Cycle Structure of the Weak and Semi-Weak DES Keys," in *Eurocrypt 86 - Abstracts of Papers*, p. 2.1, Linkoping, Sweden, 20-22 May 1986.

[MoSi87] J. H. Moore and G. J. Simmons, *Advances in Cryptology: Proc. of CRYPTO'86*, Lecture Notes in Computer Science, no. 263, pp. 9-32, Springer Verlag, Berlin, 1987.

[NBS77] NBS, *"Data Encryption Standard (DES),"* FIPS PUB 46, US National Bureau of Standards, Washington, DC, Jan. 1977.

[PiSe89] J. Pieprzyk and J. Seberry, *"Remarks on Extension of DES - Which Way to Go?,"* Tech. Rep. CS89/4, Dept. of Computer Science, UC UNSW, Australian Defence Force Academy, Canberra, Australia, Feb. 1989.

[Piep89] J. Pieprzyk, "Non-Linearity of Exponent Permutations," in *Abstracts of Eurocrypt 89*, IACR, Houthalen, Belgium, 10-13 Apr., 1989.

[SePi88] J. Seberry and J. Pieprzyk, *Cryptography: An Introduction to Computer Security*, Prentice Hall, Englewood Cliffs, NJ, 1988.

LOKI - A Cryptographic Primitive for Authentication and Secrecy Applications

Lawrence Brown
Josef Pieprzyk
Jennifer Seberry

Centre for Computer Security Research
Department of Computer Science
University College, UNSW
Australian Defence Force Academy
Canberra ACT 2600. Australia

Abstract

This paper provides an overview of the LOKI[1] encryption primitive which may be used to encrypt and decrypt a 64-bit block of data using a 64-bit key. The LOKI primitive may be used in any mode of operation currently defined for ISO DEA-1, with which it is interface compatible [AAAA83]. Also described are two modes of operation of the LOKI primitive which compute a 64-bit, and 128-bit, Message Authentication Code (or hash value). These modes of operation may be used to provide authentication of a communications session, or of data files.

1. Introduction

This paper provides an overview of the LOKI[1] encryption primitive which may be used to encrypt and decrypt a 64-bit block of data using a 64-bit key. It has been developed as a result of work analysing the existing DEA-1, with the aim of designing a new family of encryption primitives [Brow88], [BrSe89], [BrSe90], [PiFi88], [Piep89], [Piep89], [PiSe89]. Its overall structure has a broad resemblance to DEA-1 (see Fig. 1), however the detailed structure has been designed to remove operations which impede analysis or hinder efficient implementation, but which do not add to the cryptographic security of the algorithm. The overall structure and the key schedule has been developed from the work done in [BrSe89] and [BrSe90], whilst the design of the S-boxes was based on [Piep89].

The LOKI primitive may be used in any mode of operation currently defined for ISO DEA-1, with which it is interface compatible [AAAA83]. Also described are two modes of operation of the LOKI primitive which compute a 64-bit, and 128-bit, Message Authentication Code (or hash value) respectively, from an arbitrary length of message input. The modes of use are modifications of those described in [DaPr84], [Wint83], and [QuGi89]. These modes of operation may be used to provide authentication of a communications session, or of data files.

---

[1] LOKI - God of mischief and trickery in Scandinavian mythology. "He is handsome and well made, but of a very fickle mood and most evil disposition. He is of the giant race, but forced himself into the company of the gods, and seems to take pleasure in bringing them into difficulties, and in extracting them out of the danger by his cunning, wit and skill" [Bulfinch's Mythology, Avenel Books, NY 1978].

The LOKI encryption primitive, and the above modes of use have been submitted to the European RIPE project for evaluation [VCFJ89].

2. The LOKI Cryptographic Primitive

2.1. Overview

The LOKI DEA is a family of ciphers designed to encrypt and decrypt blocks of data consisting of 64 bits, under control of a 64-bit key. This Annex defines a common variant of the algorithm for use when compatibility between implementations is required. The same structure, but with alternate substitution functions may be used to build private variants of this algorithm. The same key is used for both encryption and decryption, but with the schedule of addressing the key bits altered so that the decryption process is the reverse of the encryption process. A block to be encrypted is added modulo 2 to the key, is then processed in 16 rounds of a complex key-dependent computation, and finally is added modulo 2 to the key again. The key-dependent computation can be defined in terms of a confusion-diffusion function f, and a key schedule KS. Descriptions of the encryption operation, the decryption operation, and the definition of the function f, are provided in the following sections. The representation of the keys, key values to be avoided and guidelines for the construction of alternate private ciphers, and full results for the tests conducted to date on LOKI, are described in the Appendices.

2.2. Encryption

The encryption computation is illustrated in Fig 1. The 64 bits of the input block to be encrypted are added modulo 2 to the key, processed in 16 rounds of a complex key-dependent computation, and finally added modulo 2 to the key again.

In detail, the 64-bit input block X is partitioned into two 32-bit blocks XL and XR. Similarly, the 64-bit key is partitioned into two 32-bit blocks KL and KR. Corresponding halves are added together modulo 2, to form the initial left and right halves for the following 16 rounds, thus:

$$L_0 = XL \oplus KL_0 \qquad KL_0 = KL \qquad\qquad [Eq.1]$$

$$R_0 = XR \oplus KR_0 \qquad KR_0 = KR$$

The complex key-dependent computation consists (except for a final interchange of blocks) of 16 rounds (iterations) of a set of operations. Each iteration includes the calculation of the encryption function f. This is a concatenation of a modulo 2 addition and three functions E, S, and P. Function f takes as input the 32-bit right data half R_{i-1} and the 32-bit left key half KL_i produced by the key schedule KS (denoted K_i below), and which produces a 32-bit result which is added modulo 2 to the left data half L_{i-1}. The two data halves are then interchanged (except after the last round). Each round may thus be characterised as:

$$L_i = R_{i-1}$$

$$R_i = L_{i-1} \oplus f(R_{i-1}, KL_i) \qquad\qquad [Eq.2]$$

$$f(R_{i-1}, K_i) = P(S(E(R_{i-1} \oplus K_i)))$$

The component functions E, S, and P are described later.

The key schedule KS is responsible for deriving the sub-keys K_i, and is defined as follows: the 64-bit key K is partitioned into two 32-bit halves KL and KR. In each round i, the sub-key K_i is the current left half of the key KL_{i-1}. This half is then rotated 12 bits to the left, and the key halves are interchanged. This may be defined thus:

$$K_i = KL_{i-1}$$

$$KL_i = KR_{i-1} \qquad \text{[Eq.3]}$$

$$KR_i = ROL(KL_{i-1}, 12)$$

Finally after the 16 rounds, the other key halves are added modulo 2 to the data halves to form two output block halves YL and YR which are then concatenated together to form the output block Y. This is defined as follows (note the swap of data and key halves to undo the final interchange in [Eq.2] and [Eq.3]):

$$YL = R_{16} \oplus KR_{16}$$

$$YR = L_{16} \oplus KL_{16} \qquad \text{[Eq.4]}$$

$$Y = YL \mid YR$$

2.3. Decryption

The decryption computation is identical to that used for encryption, save that the partial keys used as input to the function f in each round are calculated in reverse order, and the initial and final additions of key to data modulo 2 use the opposite halves of the key (interchange KL_0 and KR_0 in [Eq.1] and KL_{16} and KR_{16} in [Eq.3]). The calculation of the partial keys for decryption consists of first exchanging key halves, then rotating the left half 12 bits to the right, and then using the left half as the partial key. This is defined as:

$$KR_i = KL_{i-1}$$

$$KL_i = ROR(KR_{i-1}, 12) \qquad \text{[Eq.5]}$$

$$K_i = KL_i$$

2.4. Function f

The encryption function f is a concatenation of a modulo 2 addition and three functions E, S, and P, which takes as input the 32-bit right data half R_{i-1} and the 32-bit left key half KL_i, and produces a 32-bit result which is added modulo 2 to the left data half L_{i-1}. This is shown in Fig 2, and is defined thus:

$$f(R_{i-1}, K_i) = P(S(E(R_{i-1} \oplus K_i))) \qquad \text{[Eq.6]}$$

The modulo 2 addition of the key and data halves ensures that the output of f will be a complex function of both of these values.

The expansion function E takes a 32-bit input and produces a 48-bit output block, composed of four 12-bit blocks which form the inputs to four S-boxes in function f. Function E selects consecutive blocks of twelve bits as inputs to S-boxes S(4), S(3), S(2), and S(1) respectively, as follows:

$[b_3 \, b_2 \; \cdots \; b_0 \, b_{31} \, b_{30} \; \cdots \; b_{24}]$
$[b_{27} \, b_{26} \; \cdots \; b_{16}]$
$[b_{19} \, b_{18} \; \cdots \; b_8]$
$[b_{11} \, b_{10} \; \cdots \; b_0]$

This is shown in Table 1 in full.

Table 1 - LOKI Expansion Function E											
3	2	1	0	31	30	29	28	27	26	25	24
27	26	25	24	23	22	21	20	19	18	17	16
19	18	17	16	15	14	13	12	11	10	9	8
11	10	9	8	7	6	5	4	3	2	1	0

The substitution function S provides the confusion component in the LOKI cipher. It takes a 48-bit input and produces a 32-bit output. It is composed of four S-boxes, each of which takes a 12-bit input and produces an 8-bit output, which are concatenated together to form the 32-bit output of S. The 8-bit output from S(4) becomes the most significant byte (bits [31...24]), then the outputs from S(3) (bits[23...16]), S(2) (bits[15...8]), and S(1) (bits [7...0]). In this Annex, the four S-boxes are identical. However, Appendix 2 details how private versions of this algorithm may be constructed by choosing alternate definitions for the S-boxes, following the guidelines specified there. The form of each S-box is shown in Fig 3. The 12-bit input is partitioned into two segments: a 4-bit row value r formed from bits $[b_{11} \, b_{10} \, b_1 \, b_0]$, and an 8-bit column value c formed from bits $[b_9 \, b_8 \; \cdots \; b_3 \, b_2]$. The row value r is used to select one of 16 S-functions Sfn_r, which then take as input the column value c 'and produce an 8-bit output value. This is defined as:

$$\text{Sfn}_r = (c \oplus r)^{e_r} \bmod \text{gen}_r, \qquad \text{in GF}(2^8) \qquad [\text{Eq.7}]$$

where gen_r is an irreducible polynomial in $\text{GF}(2^8)$, and e_r is the exponent used in forming the rth S-box. The generators and exponents to be used in the 16 S-functions Sfn_r in the standard LOKI are specified in Table 2.

The permutation function P provides diffusion of the outputs from the four S-boxes across the inputs of all S-boxes in the next round. It takes the 32-bit concatenated outputs from the S-boxes, and distributes them over all the inputs for the next round via a regular wire crossing which takes bits from the outputs of each S-box in turn, as defined in Table 3.

Table 2 - LOKI S-box Irreducible Polynomials and Exponents		
Row	gen$_r$	e$_r$
0	375	31
1	379	31
2	391	31
3	395	31
4	397	31
5	415	31
6	419	31
7	425	31
8	433	31
9	445	31
10	451	31
11	463	31
12	471	31
13	477	31
14	487	31
15	499	31

Table 3 - LOKI Permutation P							
31	23	15	7	30	22	14	6
29	21	13	5	28	20	12	4
27	19	11	3	26	18	10	2
25	17	9	1	24	16	8	0

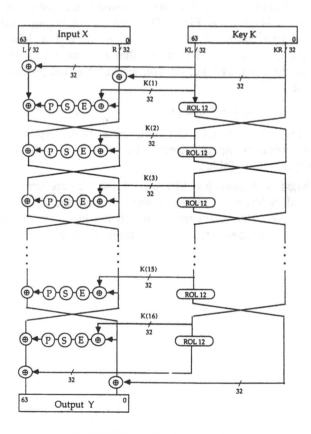

Fig 1. LOKI Encryption Computation

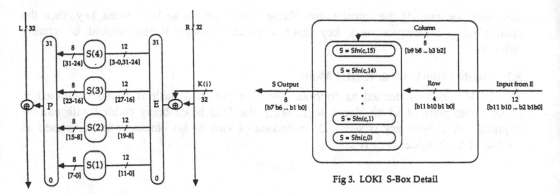

Fig 2. LOKI Function f Detail

Fig 3. LOKI S-Box Detail

3. Additional Modes of Use

The LOKI primitive may also be used in any mode of operation currently defined for ISO DEA-1, with which it is interface compatible [AAAA83]. In addition, two modes of use are defined using the LOKI primitive for the purpose of providing message authentication. The Single Block Hash (SBH) mode computes a 64-bit Message Authentication Code (MAC or hash value), from an arbitrary length of message input. The Double Block Hash (DBH) mode computes a 128-bit MAC from an arbitrary length of message input.

In the following definitions, the LOKI primitive used for encryption is denoted $Y=EL_K(X)$. That is, Y is a 64-bit block formed by encrypting input X using the LOKI primitive with key K.

3.1. Single Block Hash (SBH) Mode

The SBH mode is defined as follows. Data for which a hash is to be computed is divided into 64-bit blocks, the final block being padded with nulls if required. A 64-bit key is supplied, and is used as the initial hash value IV. For each message block M_i: that block is added modulo 2 to the previous hash value to form a key. That key is used to encrypt the previous hash value. The encrypted value is added modulo 2 to the previous hash value to form the new hash value (see Fig 4). The SBH code is the final hash value formed. This process may be summarised as:

$$H_0 = IV$$

$$H_i = EL_{M_i \oplus H_{i-1}}(H_{i-1}) \oplus H_{i-1}$$

$$SBH = H_n$$

The SBH mode is a variant of the Davies and Meyer hash function described in [DaPr84], [Wint83]. The major extension is the addition modulo 2 of the previous hash value to the current message block before using it as key input to the LOKI primitive. This was desired to prevent weak keys being supplied to the primitive when the message data was constant. If the Initialization Value is chosen not to be a weak key, then the chance of generating a weak key from a given message stream should be greatly reduced.

3.2. Double Block Hash (DBH) Mode

The DBH mode is defined as follows. Data for which a hash is to be computed is divided into pairs of 64-bit blocks M_{2i+1}, M_{2i+2}, the final block being padded with nulls if required. A 128-bit key is supplied, composed of two 64-bit blocks, which are used as the initial hash values IV_{-1}, IV_0.

$$H_{-1} = IV_{-1}$$

$$H_0 = IV_0$$

For each pair of message blocks M_{2i+1} M_{2i+2}, the following calculation is performed (see Fig 5):

$$T = EL_{M_{2i+1} \oplus H_{2i-1}}(H_{2i-1} \oplus M_{2i+2}) \oplus M_{2i+2} \oplus H_{2i}$$

$$H_{2i+1} = EL_{M_{2i+2} \oplus H_{2i}}(T \oplus M_{2i+1}) \oplus M_{2i+1} \oplus H_{2i-1} \oplus H_{2i}$$

$$H_{2i+2} = T \oplus H_{2i-1}$$

The DBH block is formed by concatenating the final two hash values as follows:

$$DBH = H_{n-1} \mid H_n$$

The DBH mode is derived from that proposed by Quisquater and Girault [QuGi89]. Again it was extended by the addition modulo 2 of the previous hash value to the current message block before using it as key input to the LOKI primitive.

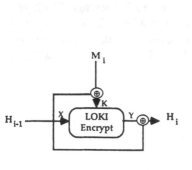

Fig 4. LOKI Single Block Hash (SBH) Mode

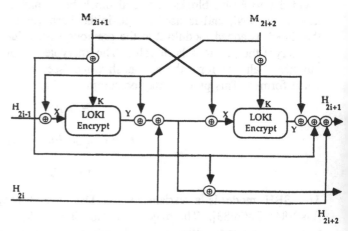

Fig 5. LOKI Double Block Hash (DBH) Mode

4. Conclusion

The LOKI cryptographic primitive, and its associated modes of use for message authentication have been described. This algorithm is currently undergoing evaluation and testing by several parties.

Acknowledgements

To the members of the Centre for Computer Security Research, and the staff of the Department of Computer Science for their help and suggestions. Thankyou.

Bibliography

[AAAA83] *"Information Interchange - Data Encryption Algorithm - Modes of Operation,"* American National Standards Institute X3.106-1983, American National Standards Institute, New York, 1983.

[Brow88] L. Brown, "A Proposed Design for an Extended DES," in *Proc. Fifth International Conference and Exhibition on Computer Security*, IFIP, Gold Coast, Queensland, Australia, 19-21 May, 1988.

[BrSe89] L. Brown and J. Seberry, "On the Design of Permutation P in DES Type Cryptosystems," in *Abstracts of Eurocrypt 89*, IACR, Houthalen, Belgium, 10-13 Apr., 1989.

[BrSe90] L. Brown and J. Seberry, *"Key Scheduling in DES Type Cryptosystems,"* accepted for presentation at Auscrypt90, ADFA, Sydney, Australia, Jan. 1990.

[DaPr84] D. W. Davies and W. L. Price, *Security for Computer Networks*, John Wiley and Sons, New York, 1984.

[PiFi88] J. Pieprzyk and G. Finkelstein, "Permutations that Maximize Non-Linearity and Their Cryptographic Significance," in *Proc. Fifth Int. Conf. on Computer Security - IFIP SEC '88*, IFIP TC-11, Gold Coast, Queensland, Australia, 19-21 May 1988.

[Piep89] J. Pieprzyk, "Non-Linearity of Exponent Permutations," in *Abstracts of Eurocrypt 89*, IACR, Houthalen, Belgium, 10-13 Apr., 1989.

[PiSe89] J. Pieprzyk and J. Seberry, *"Remarks on Extension of DES - Which Way to Go?,"* Tech. Rep. CS89/4, Dept. of Computer Science, UC UNSW, Australian Defence Force Academy, Canberra, Australia, Feb. 1989.

[Piep89] J. Pieprzyk, "Error Propagation Property and Application in Cryptography," *IEE Proceedings-E, Computers and Digital Techniques*, vol. 136, no. 4, pp. 262-270, July 1989.

[QuGi89] J. Quisquater and M. Girault, "2n-Bit Hash Functions Using n-Bit Symmetric Block Cipher Algorithms," in *Abstracts of Eurocrypt 89*, p. 4.5, IACR, Houthalen, Belgium, 10-13 Apr., 1989.

[VCFJ89] J. Vandewalle, D. Chaum, W. Fumy, C. Janssen, P. Landrock and G. Roelofsen, "A European Call for Cryptographic Algorithms: RIPE RACE Integrity Primitives Evaluation," in *Abstracts of Eurocrypt 89*, p. 6.6, IACR, Houthalen, Belgium, 10-13 Apr., 1989.

[Wint83] R. S. Winternitz, "Producing a One-Way Hash Function from DES," in *Advances in Cryptology - Proc. of Crypto 83*, D. Chaum, R. L. Rivest and A. T. Sherman (editors), pp. 203-207, Plenum Press, New York, Aug. 22-24, 1983.

Permutation Generators of Alternating Groups*

Josef Pieprzyk
and
Xian-Mo Zhang

Department of Computer Science
University College
University of New South Wales
Australian Defence Force Academy
Canberra, ACT 2600, AUSTRALIA

Abstract

The problem of generation of permutations from small ones is especially important from a cryptographic point of view. This work explores the case This work addresses the design of cryptographic systems using elementary permutations, also called modules. These modules have a simple structure and are based on internal smaller permutations. Two cases have been considered. In the first, the modules apply internal permutations only. It has been proved that the composition of modules generates the alternating group for the number of binary inputs bigger than 2. In the second, DES-like modules are considered and it has been shown that for a large enough number of binary inputs, they produce the alternating group, as well.

1 Introduction

Coppersmith and Grossman in [2] studied generators for certain alternating groups. They defined k-functions which create corresponding permutations. Each k-function along with its connection topology produces a single permutation which can be used as a generator. The authors proved that these generators produce at least alternating groups using a finite number of their compositions. It means that with generators of relatively simple structure, it is possible to produce at least half of all the permutations using composition.

There is one problem with such generators - they do not have a fixed connection topology. The well-known DES encryption algorithm applies the fixed connection topology. The 64-bit input is divided into halves. The right hand half is used as the input (after the expansion operation) for the eight different S-boxes each of which transforms the 6-bit input into the 4-bit output. The resulting 32-bit string is added modulo 2 to the corresponding bits of the second half. Next the halves are swopped. Even and Goldreich [3]

---
*Support for this project was provided in part by the Australian Research Council under the reference number A48830241 and by TELECOM Australia under the contract number 7027

proved that the DES-like connection topology along with k-functions can also generate alternating groups.

This raises the following question: *Is it possible to generate the alternating groups when k-permutations are used instead of k-functions ?*

2 Background and Notations

Symmetric enciphering algorithms operate on fixed size blocks of binary strings. We assume that the length of the block is N bits, and for a fixed key, algorithms give permutations from the set of 2^N possible elements. The vector space of dimension N over $GF(2)$ contains all binary strings of length N and is denoted as V_N. The following notations will be used throughout the work:

S_X - the group of all permutations on a set X,
S_{V_N} - the group of all permutations on V_N (it consists of $2^N!$ elements),
A_{V_N} - the alternating group of all permutations on V_N (it has $1/2(2^N!)$ elements).

The following definition describes k-permutations and can be seen as a modification of the definition given by Coppersmith and Grossman [2].

Definition 2.1 *Let* $1 \leq k \leq (N/2)$. *By a* k-permutation on V_N, *we mean a permutation* σ *of* V_N *determined by a subset of order* $2k$, $\{i_1, \cdots, i_k, j_1, \cdots, j_k\} \subseteq \{1, \ldots, N\}$ *and a permutation*

$$p : V_N \longrightarrow V_N \qquad (1)$$

as follows:

$$((a_1, \ldots, a_N)\sigma)_m = a_m \qquad \text{for } m \in \{i_1, \ldots, i_k\}$$
$$((a_1, \ldots, a_N)\sigma)_m = a_m \oplus p(a_{i_1}, \ldots, a_{i_k})_m \quad \text{for } m \in \{j_1, \ldots, j_k\}$$

3 Structure without Topology Restrictions

In this section we refer to the results obtained by Coppersmith and Grossman [2]. The k-permutation for a fixed k is the basic module which is used to create our encryption system. Clearly, k-permutations on V_N generate a subgroup of S_{V_N} and the subgroup is denoted as $P_{k,N} \subset S_{V_N}$.

Lemma 3.1

$$P_{1,2} = S_{V_2} \qquad (2)$$

Proof. There are four possible modules and they produce the following permutations:

$$
\begin{aligned}
g_1 &= (0,3,2,1) &= (0)(2)(1,3) \\
g_2 &= (2,1,0,3) &= (1)(3)(0,2) \\
g_3 &= (1,0,3,2) &= (0,1)(2,3) \\
g_4 &= (0,1,3,2) &= (0)(1)(2,3)
\end{aligned}
$$

The above permutations are written in two different ways. The first uses the standard notation which for g_1 is as follows:

$$g_1(0) = 0; g_1(1) = 3; g_1(2) = 2; g_1(3) = 1.$$

The second one applies the reduced cyclic form (see Wielandt [5] pages 1 and 2). It is easy to check that the permutations generate S_{V_2}.

◇

Lemma 3.2 *The 2-permutations generate subgroup* $P_{2,4} \subset S_{V_4}$ *whose coordinates are affine.*

    Proof. Observe that the permutation $p : V_2 \longrightarrow V_2$ must always have linear coordinates.

◇

Lemma 3.3 *For* $k \geq 3$,

$$P_{k,2k} = A_{V_{2k}}. \tag{3}$$

    Proof. First we prove that $P_{3,6} = G_{2,6}$ (note that $G_{2,6}$ is defined as in [2] and means the subgroup of S_{V_N} generated by the 2-functions). It is always possible to generate any $G_{2,6}$ module by composition of a finite number of 3-permutation modules. For example, assume that we would like to obtain a 2-function module over 6 binary variables $(x_1, x_2, x_3, x_4, x_5, x_6)$ which transforms input

$$(x_1, x_2, x_3, x_4, x_5, x_6) \longrightarrow (x_1, x_2, x_3, x_4 \oplus x_1 x_2, x_5, x_6) \tag{4}$$

Applying the composition of two 3-permutation modules whose coordinates are given in the table below, we obtain the required 2-function module.

x_1, x_2, x_3	First Module f_1, f_2, f_3	Second Module g_1, g_2, g_3	Composition $f_1 \oplus g_1, f_2 \oplus g_2, f_3 \oplus g_3$
000	110	110	000
001	111	111	000
011	100	100	000
010	101	101	000
110	011	010	001
111	010	011	001
101	000	000	000
100	001	001	000

The example can easily be extended for arbitrary case. It means that:

$$P_{3,6} \supseteq G_{2,6} = A_{V_6}$$

Using Coppersmith and Grossman theorem [2] (referred to as the C-G theorem), it is obvious that:

$$P_{3,6} \subseteq G_{2,6} = A_{V_6}$$

Combining the two inclusions, we get the final result :

$$P_{3,6} = A_{V_6} \tag{5}$$

In general, the C-G theorem and properties of k-permutations allow us to write the following sequence of inclusions:

$$A_{v_{2k}} \subseteq G_{2,2k} \subseteq P_{3,2k}$$
$$\subseteq P_{k,2k} \subseteq G_{k,2k} \subseteq A_{V_{2k}}$$

where $k \geq 3$ and it proves the lemma.

$\diamondsuit$

The proved lemma and the C-G theorem allow us to formulate the following theorem.

Theorem 3.1 *The group generated by* $P_{k,2k}$ *is:*

- *the group* S_{V_2} *for* $k = 1$,
- *the subgroup of affine transformations for* $k = 2$,
- *the group* $A_{V_{2k}}$ *for* $k \geq 3$.

The theorem can be easily generalized (the proof is omitted).

Theorem 3.2 *Let* $N \geq 2k$. *The group generated by* $P_{k,N}$ *is:*

- *the subgroup of affine transformations for* $k = 2$,
- *the group* A_{V_N} *for* $k \geq 3$.

4 DES Structure

There is an interesting question how the structure of the well-known DES algorithm [4] limits the permutation group generated by DES-like functions on V_{2k}. Even and Goldreich [3] proved that the DES-like functions generate the alternating group for $k > 1$ and the whole permutation group for $k = 1$. In this section we are going to examine a case when the DES is based on permutations that is, the S-boxes realize one-to-one transformations (the existing S-boxes provide inverible mapping of 6-bit input into 4-bit output).

Definition 4.1 *The DES-like permutation* σ *on* V_{2k} *is defined by a composition of two modules:*

- *the first module is determined by permutation* $p : V_k \longrightarrow V_k$ *and transforms the input* $(x_1, \cdots, x_{2k}) \in V_{2k}$ *into:*

$$(x_1 \oplus p_1(x_{k+1}, \cdots, x_{2k}), \cdots, x_k \oplus p_k(x_{k+1}, \cdots, x_{2k}), x_{k+1}, \cdots, x_{2k})$$

where $(p_1, \cdots, p_k)$ *are coordinates of* $p(x_{k+1}, \cdots, x_{2k})$,

- *the second module swops the vector:*

$$(x_1, \cdots, x_k, x_{k+1}, \cdots, x_{2k})$$

with the vector

$$(x_{k+1}, \cdots, x_{2k}, x_1, \cdots, x_k).$$

The group generated by DES-like permutations is denoted by DESP_{2k} *(*$\mathrm{DESP}_{2k} \subset S_{V_{2k}}$*).*

Lemma 4.1

$$\mathrm{DESP}_2 = \mathbf{A}_{V_2} \tag{6}$$

Proof: There are two possible permutations σ_1 and σ_2 generated by $p_1(x_2) = x_2$ (the identity permutation) and $p_2(x_2) = \bar{x}_2$ (the negation permutation), respectively and

$$\begin{aligned}
\sigma_1 &= (0)(1,2,3) \\
\sigma_2 &= (0,1,2)(3)
\end{aligned}$$

It is easy to check that the two permutations generate $\mathbf{A}_{V_2}$.
◇

Lemma 4.2 *The permutation* $\theta_4 \in \mathbf{A}_{V_4}$ *that swops* (x_1, x_2, x_3, x_4) *into* (x_3, x_4, x_1, x_2) *can be expressed by composition of DES-like permutations.*

Proof. First note that

$$\begin{aligned}
\theta_4 &= (0,4,8,12,1,5,9,13,2,6,10,14,3,7,11,15) \\
&= (1,4)(2,8)(3,12)(6,9)(7,13)(11,14)
\end{aligned}$$

We shall show that θ_4 may be obtained using the composition of the following four DES-like permutations:

$$\begin{aligned}
g_1 &= (0,5,10,15,1,4,11,14,2,7,8,13,3,6,9,12) \\
&\quad \text{for } p(x_3, x_4) = I = (0,1,2,3) \\
g_2 &= (0,6,9,15,1,7,8,14,2,4,11,13,3,5,10,12) \\
&\quad \text{for } p(x_3, x_4) = (0,2,1,3) \\
g_3 &= (0,6,11,13,1,7,10,12,2,4,9,15,3,5,8,14) \\
&\quad \text{for } p(x_3, x_4) = (0,2,3,1) \\
g_4 &= (0,7,9,14,1,6,8,15,2,5,11,12,3,4,10,13) \\
&\quad \text{for } p(x_3, x_4) = (0,3,1,2)
\end{aligned}$$

where $p(x_3, x_4)$ are permutations of four binary elements 0,1,2,3 (coded 00,01,10,11). We create 3 intermediate permutations as follows:

$$\begin{aligned}
\alpha &= g_2^{-1} \circ g_1 \\
&= (0,13,14,3,4,9,10,7,8,5,6,11,12,1,2,15) \\
&= (1,13)(2,14)(5,9)(6,10); \\
\beta &= g_3^3 \circ \alpha \circ g_3^{-3} \\
&= (0,4,2,6,1,5,3,7,8,12,10,14,9,13,11,15) \\
&= (1,4)(3,6)(9,12)(11,14); \\
\gamma &= g_4^3 \circ \alpha \circ g_4^{-3} \\
&= (0,1,8,9,4,5,12,13,2,3,10,11,6,7,14,15) \\
&= (2,8)(3,9)(6,12)(7,13);
\end{aligned}$$

and the composition of the last two is:

$$\gamma \circ \beta = (0,4,8,12,1,5,9,13,2,6,10,14,3,7,11,15)$$
$$= (1,4)(2,8)(3,12)(6,9)(7,13)(11,14).$$

◇

Lemma 4.3 *The permutation $\theta_6 \in A_{V_6}$ which swops:*

$$(x_1, x_2, x_3, x_4, x_5, x_6) \rightarrow (x_4, x_5, x_6, x_1, x_2, x_3) \qquad (7)$$

can be expressed by composition of DES-like permutations.

Proof. Any DES-like permutation from S_{V_6} transforms the input sequence

$$(x_1, x_2, x_3, x_4, x_5, x_6) \qquad (8)$$

into

$$(x_4, x_5, x_6, x_1 \oplus p_1(x_4, x_5, x_6), x_2 \oplus p_2(x_4, x_5, x_6), x_3 \oplus p_3(x_4, x_5, x_6)), \qquad (9)$$

where the permutation $p(x_4, x_5, x_6) = (p_1, p_2, p_3)$. We simplify our considerations choosing the permutation $p(x_4, x_5, x_6) = (x_4, p_2(x_5, x_6), p_3(x_5, x_5))$. So, we can independently consider two DES-like permutations. First one σ transforms the sequence (x_1, x_4) into $(x_4, x_1 \oplus x_4)$ and generates the identity permutation after three compositions ($\sigma^3 = I$). The second permutation $\sigma_g : V_4 \rightarrow V_4$ belongs to $DESP_4$. If we select the same sequence of permutations as in the previous Lemma, we can obtain:

$$(x_2, x_3, x_5, x_6) \rightarrow (x_5, x_6, x_2, x_3) \qquad (10)$$

This can be done using 22 compositions (observe that $g_2^{-1} = g_2^5$; $g_3^{-3} = g_3^2$; $g_4^{-3} = g_4^2$). Therefore after 66 compositions it is possible to obtain

$$(x_1, x_2, x_3, x_4, x_5, x_6) \rightarrow (x_1, x_5, x_6, x_4, x_2, x_3) \qquad (11)$$

By repeating the process three times, we get

$$(x_1, \quad x_2, \quad x_3, \quad x_4, \quad x_5, \quad x_6)$$
$$\downarrow$$
$$(x_4, \quad x_2, \quad x_5, \quad x_1, \quad x_3, \quad x_6)$$
$$\downarrow$$
$$(x_3, \quad x_2, \quad x_6, \quad x_1, \quad x_4, \quad x_5)$$
$$\downarrow$$
$$(x_3, \quad x_4, \quad x_5, \quad x_1, \quad x_3, \quad x_2)$$

To obtain the DES swopping operation, we need to exchange bits x_3 and x_2. This can be done using the product of the following permutations:

$$\delta \circ \gamma \circ \beta = (1,2)(5,6)(9,10)(13,14) \qquad (12)$$

where:

$$\beta = g_1^2 \circ \alpha \circ g_1;$$
$$\gamma = g_3^2 \circ \alpha \circ g_3^4;$$
$$\delta = g_2^5 \circ g_4 \circ g_2;$$
$$\alpha = g_6^4 \circ g_5;$$

and:

$$g_1 = (0,5,10,15,1,4,11,14,2,7,8,13,3,6,9,12)$$
$$\text{for } p(x_3,x_4) = I = (0,1,2,3);$$
$$g_2 = (0,5,11,14,1,4,10,15,2,7,9,12,3,6,8,13)$$
$$\text{for } p(x_3,x_4) = (0,1,3,2);$$
$$g_3 = (0,6,9,15,1,7,8,14,2,4,11,13,3,5,10,12)$$
$$\text{for } p(x_3,x_4) = (0,2,1,3);$$
$$g_4 = (1,4,10,15,0,5,11,14,3,6,8,13,2,7,9,12)$$
$$\text{for } p(x_3,x_4) = (1,0,2,3);$$
$$g_5 = (3,4,9,14,2,5,8,15,1,6,11,12,0,7,10,13)$$
$$\text{for } p(x_3,x_4) = (3,0,1,2);$$
$$g_6 = (3,5,8,14,2,4,9,15,1,7,10,12,0,6,11,13)$$
$$\text{for } p(x_3,x_4) = (3,1,0,2).$$

To leave other positions unchanged, it is necessary to apply the above sequence of permutations three times.
◇

Theorem 4.1 *The group* $DESP_{2k}$ *generated by DES-like permutations is:*

(a) *the alternating group* A_{V_2} *for* $k = 1$,

(b) *the group of affine transformations for* $k = 2$,

(c) *the alternating group* $A_{V_{2k}}$ *for* $k \geq 3$.

Proof. The statement (a) has been proved in the lemma 4.1. According to the lemmas 4.2 and 4.3 each swopping module can be expressed as a composition of DES-like permutations for $k \geq 2$. It means that any permutation from $P_{k,2k}$ may be represented by a composition of DES-like permutations, i.e.:

$$DESP_{2k} \supseteq P_{k,2k}. \tag{13}$$

Considering the theorem proved by Even and Goldreich [3] (referred to as the E-G theorem), the following inclusion holds:

$$DESP_{2k} \subseteq DES_{2k} = A_{V_{2k}} \text{ for } k \geq 3 \tag{14}$$

where DES_{2k} is a group generated by DES-like functions given in [3]. Taking 13 and 14, we obtain the statement (c). The statement (b) is obvious.
◇

5 Conclusions

When designing new cryptographic algorithms, we face the problem of selecting the algorithm structure (or the connection topology). Results by Coppersmith and Grossman [2], Even and Goldreich [3] proved that the DES structure is flexible enough as a composition of DES iterations can generate the suitable alternating group while the number of iterations is not limited (the DES uses 16 ones) and functions in S-boxes are not fixed (i.e. they can be freely selected for each iteration).

In this work we have answered the problem of what happens if S-boxes realize one-to-one mapping (the current S-boxes in the DES are one-to-many). Astonishingly, the structure with one-to-one S-box transformations does not restict the number of possible permutations obtained using the composition if only the number of inputs/outputs is equal to or larger than 6 (or $k \geq 3$).

Each iteration may be considered as a generator of the alternating group. We have simply proved that having $(2^{N/2})!$ generators we can produce $(2^N)!$ different permutations. From a practical point of view we would like to have a smaller set of generators. Bovey and Williamson reported in [1] that a ordered pair of generators can produce either A_{V_N} or S_{V_N} with the probability greater than $1 - exp(-log^{1/2}2^N)$. So if we select the pair at random, there is a high probability that it generates at least A_{V_N}. However, we would not like to rely on the probability theory. Instead, we would like to know for certain that the set of generators is complete, i.e. that it generates either A_{V_N} or S_{V_N}.

There remain the following open problem:

- Are the DES generators complete (considering the current S-box structure) ?

ACKNOWLEDGMENT

We would like to thank Cathy Newberry and Lawrie Brown for their assistance during preparation of this work.

References

[1] J. Bovey and A. Williamson. The probability of generating the symmetric group. *Bull. London Math. Soc.*, 10:91–96, 1978.

[2] D. Coppersmith and E. Grossman. Generators for certain alternating groups with applications to cryptography. *SIAM Journal Appl. Math.*, 29(4):624–627, December 1975.

[3] S. Even and O. Goldreich. DES-like functions can generate the alternating group. *IEEE Transaction on Inf. Theory*, IT-29(6):863–865, November 1983.

[4] Data Encryption Standard. National bureau of standards. Federal Information Processing Standard Publication, January 1977. No.46.

[5] H. Wielandt. *Finite permutation groups*. Academic Press, New York, 1964.

SECTION 7

ZERO-KNOWLEDGE PROTOCOLS

Showing Credentials without Identification
Transferring Signatures between Unconditionally Unlinkable Pseudonyms

David Chaum

Introduction

Individuals today have many relationships with organizations. Such relationships are made up of interactions, like simple cash purchases, tax filings, and medical checkups. Many of these interactions generate considerable data about individuals, and automation may greatly extend this data's capture and influence. Different approaches to automating, however, yield quite different results.

One extreme approach requires individuals always to show universal identifiers, such as passport numbers or thumbprints. Perfecting universal identification through automation would make information about individuals fully transferable and authenticatable. This would provide organizations with information they may legitimately need, but it would also deny individuals the ability to monitor, let alone control, transfers of information about themselves.

The opposite extreme approach to automating relationships is complete anonymity. This might be as simple as with cash purchases today, or individuals could choose random account numbers for continuity within some relationships. Although complete anonymity would optimally protect individuals' privacy, it would rule out the often beneficial and even necessary transfers of information between relationships.

In short, these extreme alternatives would make verifiability by organizations or control by individuals mutually exclusive. The solution presented here, however, offers the best of both worlds. Information transfer between relationships is controlled by individuals—yet the transfers are incontestable.

Two assumptions are made initially and relaxed later. The first, in the context of fixed sets of individuals and organizations, allows each individual at most one relationship per organization. Every such relationship is carried out under a unique "account number" called a *digital pseudonym*. The items of information that an individual transfers between these pseudonymous relationships are called *credentials*. The second assumption is that only a particular organization, called Z, has the power to create pseudonyms or to issue credentials; nevertheless, Z learns no more about a person than do other organizations.

In the first section, the basic concept is introduced by analogy and then the actual system using digital signatures is explained. How pseudonyms are authorized is detailed in section 2, which additionally relaxes the second assumption mentioned above. Section 3 relaxes the first assumption and illustrates the result with several applications. With the protocols of Section 4, individuals can transform one credential into another and obtain new credentials by combining and making choices among those they already have. Finally, building on the techniques of the third and fourth sections, the fifth achieves the quite general result of allowing people exclusive control over the database of all their own relationships.

1. BASIC SYSTEM

The essential concept can be illustrated by means of an analogy to window envelopes lined with carbon paper. If a slip of plain paper is placed in such an envelope, and a mark is later made on the outside, the carbon-paper lining transfers the mark's image onto the slip.

Suppose you want to obtain a credential from an organization using this method. First you choose a standard-sized slip and seal it in a carbon-lined envelope whose window exposes a particular zone on your slip. The random pattern of paper fibers in this zone constitutes your pseudonym with the organization. (Note that the fiber pattern, though random, is tied to the otherwise independent patterns on the slip's other zones, in that it is part of the same piece of paper; but each zone's pattern is almost certainly unique.) When the organization recieves the envelope, it makes a special signature mark across the outside, which the carbon lining transfers onto the slip. This signature is the credential, and the type of signature represents the type of credential. Which credential type the organization decides to issue you depends on the history of your relationship.

Upon getting the sealed envelope back from the credential-issuing organization, you remove the slip and verify the credential signature on it. To show the credential to some other organization, you seal it in a different envelope, whose window exposes the zone you use with that organization, together with a part of the signature's carbon image. The receiving organization can verify, by looking through the window, both the pseudonym and the credential. But it cannot link this information to any of your other pseudonyms, since the randomly-chosen envelopes expose only a single disjoint zone whose random paper-fiber pattern is independent of that in any other zone.

An "is-a-person" organization uses physical identification to ensure that it issues no more than one credential to each person. Every organization would then require this credential before accepting a pseudonym as valid. Thus, each person can use at most one slip, while the physical identification remains unlikable to anything else.

1.1 Digital Signatures

The actual credential mechanism is based on digital signatures. Only the particular organization Z, not needed with paper, will form signatures (at least until §2.4); all such signatures are made with a single RSA modulus m, the factorization of which is known only to Z. But since Z will form signatures for the other organizations as they require, the fact that Z actually makes the signatures can be ignored for simplicity.

The original RSA system [Rivest, Shamir, & Adleman] used a public exponent e, and a corresponding secret exponent d. Here, this is extended to h public exponents $e_1,\dots,e_h$, each having a respective secret exponent $d_1,\dots,d_h$. The signature of type j, $1 \le j \le h$, on message a is thus denoted a^{d_j}, with reduction modulo m implicit throughout. This notion of "signature type" will be essential for distinguishing between credentials (and will later also between pseudonyms). For convenience, the product of all the e_l is denoted as C. (Once C becomes significantly larger than m, techniques that could be substituted here allow any h, without increasing the amount of computation <Chaum 87>.)

A few technical requirements must be satisfied: each exponent pair behaves as with the original RSA system, so that $e_j d_j \equiv 1 \pmod{\lambda(m)}$ and $(e_j, \lambda(m)) = 1$, $1 \le j \le h$, where $\lambda(m)$ is the order of the multiplicative group modulo m; and the e_i may be taken as the first h odd primes, though it is sufficient for each e_j to be odd and to have a unique prime divisor.

1.2 System with Numbers

The numeric credential mechanism corresponds in several aspects with the envelope analogy. Your pseudonym p_X with organization X is of the form $u r_X{}^C$ (corresponding, in the analogy, to the slip-in-envelope ensemble), where u is your "universal identifier," and r_X is a random "blinding factor" you create. The role of u

is to allow your credentials to be transferred between your pseudonyms while preventing them from being transferred to the pseudonyms of others; the role of r_X is to provide *blinding*, which is a special kind of hiding, for u. Creating and establishing pseudonyms with organizations (in the analogy, the function of the zones) is accomplished through a special protocol described in §2.

A pseudonym can be considered as an entry in a matrix. Still assuming that each individual has a single relationship with each organization, each row in the matrix is labeled with an individual and each column with an organization. Thus, each row consists of all an individual's pseudonyms, and each column is made up of all the pseudonyms established with an organization.

Individuals of course know the pseudonym entries in their own rows, and which column each entry is in—that is, which organization each of their pseudonyms is used with (just as they know their own zones). Organizations do know all the entries in their own columns. What no organization knows, however, is which row any entry is in and thus which pseudonym is used by which person; even colluding organizations cannot link the pseudonyms of an individual because the blinding r's are chosen independently and uniformly at random over the residues modulo m, and, like a so called one-time pad, they hide the u's perfectly. (Though u roughly corresponds to the slip, nothing about it is revealed by the pseudonym.)

To give you a credential, X issues you a digital signature formed on your pseudonym with it. The type of signature (the type of signature mark in the analogy) corresponds to the type of credential X wishes to issue on your pseudonym. Suppose X is giving you a "good credit risk" credential. The public exponent corresponding to this credential type would be one of the e_i, say e_1—the same exponent used for every person receiving this credential. The signature X issues you would then be of the form $p_X{}^{d1}$.

Knowing both u and r_X, you can express this signature as $(ur_X{}^C)^{d1} = u^{d1}r_X{}^{C/e1}$. And because C/e_1 is an integer, you can form $r_X{}^{C/e1}$ and divide it out of the signature to obtain u^{d1}. (These computations correspond to removing the signed slip from the envelope.)

Now suppose you want to show the credential to organization Y (i.e., to place the signed slip in the envelope you use with Y), which means showing the e_1th root of your pseudonym with Y, $p_Y{}^{d1} = (ur_Y{}^C)^{d1}$. You can form this as the product of u^{d1}, which you have already isolated, and $r_Y{}^{C/e1}$, which you can easily construct. When Y receives the credential from you, Y merely raises it to the e_1 power and checks that the result is p_Y. This process of transforming a credential signature on a pseudonym p_X into the corresponding signature on another pseudonym p_Y will be called *re-blinding*.

1.3 Formal Summary: Part I

To summarize more formally, consider s individuals $I_1,\ldots,I_s$ and t organizations $O_1,\ldots,O_t$. Each individual has a universal identifier u_i, $1 \le i \le s$, and has independently and uniformly chosen blinding factors $r_{i,j}$, $1 \le j \le t$. Thus the pseudonym matrix can be written as $P = (p_{i,j}) = u_i r_{i,j} C$. The ith individual, I_i, knows u_i and the row vector $(r_{i,1},\ldots,r_{i,t})$. The jth organization, O_j, knows the unordered set $\{p_{i,j}\}$, $1 \le i \le s$, but gains no information about the ordering of the pseudonyms from their values.

Organization X, $X \in \{1,\ldots,t\}$, issues credential k to individual i as the message
$$O_X \to I_i: (p_{i,X})^{dk} = u^{dk} r_{i,X} C^{lek}.$$
Individual i may then show credential k to organization Y, $Y \in \{1,\ldots,t\}$, after multiplying with $r_{i,X}^{-C/ek} r_{i,Y} C^{lek}$, by sending
$$I_i \to O_{j,Y} (p_{i,Y})^{dk} = u^{dk} r_{i,Y} C^{lek}.$$
Finally, organization Y raises this message to the e_k power and checks that the result is the pseudonym $p_{i,Y}$ it knows, thereby verifying the credential.

2. VALIDATOR ISSUING & SHOWING

Persons establish pseudonyms with organizations by means of signatures called *validators*. Validators can be thought of as messages sent by Z to other organizations through a channel provided by individuals. The organizations cannot link the messages sent with those received: Z knows only how many validators it issues for each combination of validator type (column designation) and u (row designation); other organizations know only that each message arrives hiding its row designation but authenticating its column designation and its pseudonym.

Two ways to create and show validators are described below. The first is a direct application of some quite general results. It is easily stated and also shows how modest the requirements for such systems are in theory—but the construction is not at all practical. The second is a quite practical protocol. All other sections, however, are quite independent of these particular protocols. This section's final subsection relaxes the remaining assumptions about Z.

2.1 Theoretical Approach

One way to create a validator is for Z simply to make a signature on the row identifier u involved. This signature itself cannot of course be shown directly to the organization with whom the pseudonym is to be established, since the signature's representation is obviously recognizable by Z. But there are techniques, described in <Brassard, Chaum & Crépeau 88>, that allow a person to "prove" possession of a suitable signature, without revealing any information about the signature's actual representation. To establish the signature's suitability with these protocols, a person must show that the signature satisfies an agreed "computational test."

If the validator signature were only issued on u, nothing would prevent it from being shown in creating multiple pseudonyms. This problem is solved by making the signature determine the blinding factor r^C, without revealing this secret to Z. A direct way to include a hidden form of the blinding factor is by using a "commitment," the primitives on which the signature-possession proofs are based. A value is transformed into a suitable commitment by a kind of encryption that cannot be decrypted, even by infinite computing power. The number that Z signs, then, is the concatenation of the row identifier u that Z has chosen and a commitment to r^C that you form.

After receiving such a signature from Z, you can use it to establish a pseudonym with X. First you supply X with your pseudonym, $ur_X{}^C$. Then you use the techniques mentioned above to prove that the signature is valid. The proof's computational test verifies that the pseudonym you gave is the product of two things: the u in the signature and the Cth power of the r_X committed to in the signature.

In principle, you could also use this theoretical approach to show possession of credential signatures. Each organization would then have the freedom to use its own signature scheme to form credentials. The test that would be applied when proving possession of some credential would include showing that all the signatures being checked contain the same u.

2.2 Practical Approach

A quite practical way to create validators was first summarized in <Chaum 85> and then rigorously analyzed in <Chaum & Evertse 86>. It is based on a cut-and-choose (terminology adapted from the well known "I cut and you choose" protocol for eating cake). To get one validator, you must initially send a quantity k of *candidates* to Z. Each candidate is of the form $f(ur_{X_j}{}^C)s_j e^0$, for $1 \le j \le k$. The argument of f is a potential pseudonym, but the blinding factor s you choose perfectly hides the pseudonym from Z.

After receiving your k candidates, Z selects half of them at random and informs you of the selection. You must then provide Z with all the r's and s's used to form the selected candidates. These allow Z to reconstruct all the selected candidates, since Z also knows u and how to compute f. If no discrepancy is found, then Z knows (with exponential certainty) that at least some of the unselected candidates are properly formed. To issue the validator, Z signs the product of all the unselected candidates by raising to d_0 and gives this signature to you.

Upon receiving this, you check its validity with the corresponding public exponent e_0. You can also unblind, by dividing out all the s's in the unselected candidates. In this way, you recover the signed product of images under f, which is the validator.

When showing the validator to establish a pseudonym with organization X, you must supply all the $k/2$ pseudonyms it contains. Then X applies f to each of these, forms the product of all the results, and checks that the validator is a signature on this product. It remains for you to also convince X that if any of the candidates are of the proper form, then they all are. To do this, you select a candidate and show the Cth root of its quotient with every other candidate.

2.3 Formal Summary: Part II

To summarize this validator issuing and showing protocol more formally, let u_i be the identifier by which Z knows the ith individual I_i as before, k be an even integer constant, and e_0 be a public exponent representing the validator type.

Organization Z issues a validator for use with organization X to individual I_i through an exchange of four messages. Initially, I_i sends k candidates c_j to Z,

$$I_i \rightarrow Z: c_j = f(u_i r_{i,x,j}{}^C)(s_{i,x,j})^{e0}, \text{ for } 1 \leq j \leq k.$$

Next, Z selects $k/2$ of these candidates,

$$I_i \leftarrow Z: \{1,...,k\} \supset Q_i, |Q_i| = k/2.$$

The constituents of the selected candidates are supplied by I_i:

$$I_i \rightarrow Z: r_{i,x,j}, s_{i,x,j}, \text{ for } j \in Q_i.$$

If, by reconstructing their values, Z determines that the selected candidates were properly formed, then Z issues the validator

$$I_i \leftarrow Z: \Pi (f(u_i r_{i,x,j}{}^C)s_{i,x,j})^{d0}, \text{ for } j \notin Q_i.$$

After dividing the validator by $s_{i,x,j}$, for $j \notin Q_i$, and re-indexing so that $j \in \{1,...,j/2\}$, I_i can show the validator to organization O_X,

$$I_i \rightarrow O_X: V = \Pi f(u_i r_{i,x,j})^{d0}, 1 \leq j \leq k/2,$$
$$a_j = u_i r_{i,x,j}{}^C, 1 \leq j \leq k/2,$$

$$b_j = (r_{i,x,j}/r_{i,x,1}), \ 2 \le j \le k/2.$$

Finally, O_X checks that V^{e0} is the product of the a_j, and that $a_1 b_j{}^C = a_j$, for $2 \le j \le k/2$.

2.4 Limiting Z's Power

It has been assumed initially that Z would have complete control over the issuing of pseudonyms and credentials. This assumption can be relaxed in several ways. The following discussion is in terms of the practical validator-issuing protocol above, although it could be applied to the theoretical one as well.

As pseudonym holders, individuals should be able to issue signatures or learn the content of confidential messages sent them, while preventing Z from obtaining these abilities. To implement this, f is extended from one argument to two. People choose public keys as the second argument to each f. The public keys of selected candidates in validator issuing are discarded; those remaining are only revealed when the validator is shown, and one of them is designated as the pseudonym's public key. Even if Z attempts to falsely incriminate someone by issuing a validator containing an actual u, everyone would be able to vindicate themselves by showing a special credential issued on all but the pseudonym in question.

While a single Z may be needed to make signatures on the pseudonyms, another organization can be responsible for signing validators. This is because signing and blinding of validators is done on images under f, whereas the pseudonyms are pre-images under f, and the modulus used with the images can differ from that used with the pre-images. In fact, different organizations can issue validators independently or redundantly.

Signing pseudonyms is the exclusive province of Z in the practical protocol. It should be noted that centralization of this power does provide some advantages for organizations, however, in that more effective controls can be placed on the signature issuing of front-line organizations. But this power of Z can be distributed in several ways. One is that some organizations may be allowed to request signatures from Z on blinded messages. This way, Z can be kept from learning what is being signed, and dummies can hide how many and when messages are signed. It is also possible that the pseudonym modulus is established in a way that makes cooperation of multiple organizations necessary to issue credential signatures. For instance, a trusted computation could initially divide the secret exponents into random parts that are distributed as "private outputs" to the signing organizations <Chaum, Damgård & van de Graph 87>. Another way is for each organization to choose its own RSA modulus, with the product taken as the pseudonym modulus.

3. PSEUDONYM TYPES

The type of signature on a validator is the *type* of the corresponding pseudonym. In the "basic model" considered so far, each individual has one pseudonym with each organization and a pseudonym's type corresponds to the organization with which it will be used. Thus, Z gives each person t validators, all containing the person's u (row designation); one of each having a signature type (column designation) corresponding to the name of each organization. Each organization accepts only validators with its own unique type. Issuing and checking validators in this way enforces the basic model.

By changing these validator issuing and checking rules, this section first extends the basic matrix. The credential mechanism is then used quite differently: to build payment systems. A powerful technique based on combining pseudonyms is presented next. Its utility is then illustrated in two example applications that will be extended, using the techniques of the next section, to play an important role in the final section.

3.1 Extending the Basic Matrix

As a first example, consider how the unnecessary data organizations accumulate about individuals can be reduced by periodically adding new validator types. Such a new validator includes not only the organization with whom it is to be used, but also the year of its validity. Thus, a new collection of column vectors, one for each organization, is appended onto the pseudonym matrix each year. This allows a person to establish a whole new set of pseudonyms each year and change over to using them. Since all a person's pseudonyms have the same u, credentials can be passed from those of the previous year to those of the current year. Only information passed forward in this way is associated with the new pseudonyms, thereby stripping away unnecessary detail which might otherwise accumulate.

A second useful extension, returned to in the next section, lets Z give a person more than one pseudonym of the same type. For instance, anyone who has two bank accounts might be entitled to two corresponding pseudonyms with "indistinguishable" validator types, thereby preventing the bank from knowing anything about which accounts belong to the same person. Trying to accomplish this with two different validator types, one of each issued to a person, does not quite work, since the bank would know that pairs of accounts with the same type are not from the same person.

3.2 Payments

The mechanism of validators, pseudonyms, and credentials can, of course, be used in applications quite different from those considered so far. The protocol remains the same; what can be changed is the meaning of the validator and credential types, as well as what combinations of row and column designations Z is willing to issue each person.

Consider using validators and credentials to create a payment system. Two validator types are needed, one for the bank and another one shared by all shops; the resulting pseudonym matrix likewise has two columns, one for the bank and a second shared by the shops. Validators are thus issued in *couples*, one couple per row, such that each couple contains a unique row designation, here denoted v (to distinguish it from its counterpart u of the basic matrix). A single credential type, say "worth one dollar," will be used (though multiple denominations are of course possible). Since a validator here will be of no worth without a credential for it, Z can fulfill all requests for validator couples, without regard for which are obtained by which person (so long as Z ensures that the v of each couple is unique).

To withdraw money from the bank, you supply an unused validator of type "bank." The bank then verifies the validator's signature, decreases your account balance by a dollar, and gives you back a credential signature on the pseudonym. You simply re-blind this credential, as you would in the basic system, so that you can show it on the couple's "shop" pseudonym.

When paying at a shop, you show the validator of type "shop" along with its "worth one dollar" credential. The shop verifies the the credential and validator signatures. To ensure that you cannot use the pseudonym again, the shop broadcasts it to all other shops; the shop also checks the record of all previously broadcast pseudonyms to ensure that you have not used the pseudonym before. (Such broadcast can be avoided by techniques described in <Chaum, Fiat & Naor 88>.) Having successfully completed these tests, the shop hands over your one-dollar purchase.

This simple payment system can be extended to include receipts. A third column, labeled "tax agency," say, is concatenated to the pseudonym matrix; validators are issued not in couples, as before, but in triples. When you pay for something, the shop gives you a receipt credential. This credential might, for instance, be of type "one dollar medical expense receipt." To provide the receipt to the tax agency, you simply re-blind it and show it along with the "tax agency"

validator. Further advantages of this system over already known electronic money
([Chaum 88], e.g.) derive from the extensions of §4.

3.3 Accountability After the Fact

Another use for couples is illustrated in the example of library lending. It can be
applied, however, in many accountability-after-the-fact as opposed to prior-
restraint situations. Initially, Z gives each person a number of validator couples.
The validator type for the first column is "clearinghouse"; that for the second is
"library," which is common to all libraries (just as shops shared a column). To
borrow books, you must supply a validator of type "library," which the lending
library ensures has not been used before (just as shops did with money.) When you
return the books in good condition, the library gives you a "books returned"
resolution credential.

At the end of the year, you could show resolution credentials to the
clearinghouse on all the "clearinghouse" pseudonyms that you used, along with all
the remaining "library" validators that you did not use. The clearinghouse would
verify the signatures on the validators and pseudonyms, and check that the library
validators were unused. As a patron who returned all borrowed books, you would
then be entitled to a new batch of validator couples for the next year.

3.4 Joining Pseudonyms

Couple pseudonyms can be "joined" to your basic pseudonyms. This can ensure,
for example, that money from your account is used to pay only in relationships under
your basic pseudonyms and that the resulting receipts can be shown only for your
tax account.

A credential issued on the *join* of two pseudonyms is a signature on the
product of the two pseudonyms. Thus, when the bank gives you the "one dollar"
credential, they can join it to your basic pseudonym by giving you the signature only
on the product of your bank account pseudonym from the basic matrix and a couple
pseudonym of type "bank." You can then re-blind the product and show it to a shop
as the product of your basic pseudonym with the shop and a couple pseudonym of
type "shop".

More precisely, let $p_X=ur_X{}^C$ and $p_Y=ur_Y{}^C$ be your basic pseudonyms with the
bank and the shop, respectively; and let $p_a=vr_a{}^C$ and $p_b=vr_b{}^C$ be the couple
pseudonym you use to withdraw a particular dollar and to spend that dollar,
respectively. During withdrawal, the bank gives you $(p_Xp_a)^{d1}$. You re-blind this by

multiplying with $(r_X r_a)^{-C/d1}(r_Y r_b)^{C/d1}$, yielding $(p_Y p_b)^{d1}$, which you later show to the shop.

Pseudonyms from the same matrix can also be joined, as illustrated by the marriage of two people. If a signature of type "married to" is issued on the product of their pseudonyms with the marriage license organization, then they can show this credential on the product of their pseudonyms with any other organization.

3.5 Replacing Validator Types

The basic matrix may contain many more columns than are used by any one person, making it impractical to give each person a validator for all columns. This raises privacy problems (that will be solved by building on the techniques of this subsection in §5.1). It also raises the practical problem that an interactive validator issuing protocol will have to be conducted repeatedly by each person.

A solution is based on *generic* pseudonyms. Their matrix has only one column, and each person has only one row. Yet each entry in the matrix contains an ample supply of pseudonyms. Every one of a person's relationships would involve a different pseudonym from that person's entry in the generic pseudonym matrix. Like the multiple bank account example above, in which pseudonyms from the same entry had validators of the same indistinguishable type, all validators for generic pseudonyms would be of the same type.

Restricting the collection of pseudonyms anyone can have, such as by limiting each person to one tax account, is accomplished by requiring credentials on a generic pseudonym joined to an *authorizing* pseudonym. The authorizing pseudonym matrix has the same couple-enforcing structure as the simple money or library borrowing matrices above. When Z determines that you are allowed a given kind of pseudonym, Z issues you the credential of the corresponding type on the join of one of your generic pseudonyms and an unused authorizing one. You re-blind this credential, just as with any other joined-pseudonym, before showing it to the organization. In this way, credentials can be transformed into validated pseudonyms.

4. COMBINING & TRANSFORMING CREDENTIALS

There are a variety of ways people can obtain new credentials based on their credential holdings and also on choices they make. These are naturally divided into those that you can transform all by yourself and those that require some involvement of Z. The latter type can be extended to prevent even rather implausible cheating by Z.

4.1 Transformable Credentials

Many credentials are essentially quantitative, like age, grade-point average, credit rating, and income. A convenient way to represent such a credential, whose values can be encoded with the integers $1,2,\ldots, k$, is with public exponents $e_1{}^1,\ldots,e_1{}^k$. This representation has the double-edged property that giving a person a single credential, say with public exponent $e_1{}^j$, allows the person to compute any credential $e_1{}^i$, $1 \leq i \leq j$, simply by raising the credential signature to the $e_1{}^{j-i}$ power.

Such credentials are quite safe for organizations when they are considered to be *MIN* credentials. These are only accepted as proof that the holder meets or exceeds various minimum requirements. A person could simply show the MIN credential value they hold: doing this can establish that the requirement is met, but the person also reveals the exact value held. If instead a person shows a MIN credential only after reducing it to the minimum acceptable value, nothing more is revealed than the fact that the minimum requirement is met.

For requirements expressed as upper limits, a MIN-credential offers organizations no protection, but a "MAX" credential is quite suitable. The public exponent of a *MAX* credential would be $e_2{}^{k-j+1}$ when it has the same meaning as the MIN credential $e_1{}^j$. Thus both a MIN and a MAX credential could be issued for the same qualification, allowing a person to show that the value held falls within any specified range—without revealing more.

Suppose an organization X were to require that you have each of two credentials, say both that with public exponents e_1 and e_2. You could send X separately $p_X{}^{d1}$ and $p_X{}^{d2}$. It is also possible for you to use the two credentials to form the single credential $p_X{}^{d1d2}$, which will be called their *AND*. Unlike the MIN and MAX, the AND credential does not hide anything from organizations, it merely offers compactness. To create the AND, you: set g to the multiplicative inverse of d_1 modulo d_2; set h to the remainder after dividing gd_1-1 by d_2; and computing $(p_X{}^{d1})g(p_X{}^{d2})$-$h = p_X{}^{d1d2}$.

The MIN, MAX, and AND credentials are quite efficient. They are applicable, though, only if the extent of possible requirements is known in advance and is limited in complexity. The more general situation is of course that organization's requirements are unpredictable and that anticipating all possibilities is infeasible.

4.2 Combined Credentials

In the solutions developed next, people can dynamically create new types of credentials as needed to meet organizations' requirements, still without revealing any unnecessary information.

If X were to require that you have at least one of the two credentials e_1 or e_2, then you could simply show X one of the two. But this is more than you may wish to reveal, since X then obviously learns that you have the one you show. What is needed is a single *OR* credential, say e_3, that you can obtain exactly when you have at least one of e_1 or e_2—but which itself reveals nothing about which of the two you have.

Implementing OR credentials is simple with so called "block ciphers." Such a cipher is typically used to encrypt values with a key so that they can only be decrypted by someone having the same key. More precisely, a block cipher is a class of functions, where the choice of a function from the class is called the "key," and knowing the key is necessary and sufficient to easily evaluate the corresponding function or its inverse.

To issue you an OR credential $p_X{}^{d3}$, X provides you with two encrypted copies of $p_X{}^{d3}$: one with $p_X{}^{d1}$ as key and the other with $p_X{}^{d2}$ as key. If you have $p_X{}^{d1}$ or $p_X{}^{d2}$, only then can you decrypt the corresponding encryption and recover $p_X{}^{d3}$.

Sometimes an organization's rules require a person to make a choice that cannot be changed (at least not unless certain conditions pertain). Examples are the choice of a bank account type, a major area of study for a degree, or an insurance plan. People could of course simply be required to announce such choices to an organization. It is also possible, however, for organizations to ensure that a person has made and will stick with a choice, while the person keeps any organization from learning the choice.

Suppose organization X wishes to allow such a *CHOICE* credential, say between credentials e_1 and e_2. Initially X makes public two corresponding constants, a and b, as well as some public exponent, e_3. To make your choice, you send X message q, which should be either $a r^{e3}$ or $b r^{e3}$, depending on wether you chose e_1 or e_2, respectively. Then X encrypts $p_X{}^{d1}$ with key $(q/a)^{d3}$ and $p_X{}^{d2}$ with key $(q/b)^{d3}$. When you receive these two encryptions from X, you can use r to decrypt the one corresponding to your initial choice, but you cannot find the key to the other because $a^{d3}b^{-d3}$ is never revealed.

A variation is where you are to compute a and b yourself as images under f of designated other credentials; you need the corresponding credential at the time you make the choice—it does not help if you get it later.

4.3 Verifiable Combining

A possible shortcoming of such use of block cipher encryption is that an organization might try to cheat by giving you random garbage in place of some encryptions. Suppose, for instance, that one of the two encryptions issued for an OR credential is garbage. If you lack the credential needed to decrypt that one, you won't even be able to detect the garbage; but if you show the OR credential, you reveal that you have the credential corresponding to the valid encryption. If you have the credential that should decrypt the garbage, and the organization has digitally signed its claim about the encryptions, then of course you can incriminate the organization; but this reveals something about your credential holdings.

A cut-and-choose technique, different from that used above (§§2.2 & 2.3), solves this problem. It can ensure beyond all reasonable doubt that the encryptions are properly formed.

Consider again a single encryption for an OR credential. Instead of the encryption of p_X^{d3} with key p_X^{d1}, you get from X the encryption of $(bp_X)^{d3}$ with key $(ap_X)^{d1}$ as well as the values a and b. You are then allowed to choose between two cases: In the first case, X gives you a^{d1} and b^{d3}, which you check by raising to the corresponding e_i powers. If the encryption was properly formed and you have p_X^{d1}, then you can recover p_X^{d3} by using the product of p_X^{d1} and a^{d1} as decryption key and dividing b^{d3} out of the result. If you choose the second case, X gives you $(ap_X)^{d1}$ and $(bp_X)^{d3}$, which again can check by raising to the corresponding e_i powers. You can also check that the encryption was properly formed, since decrypting it with $(ap_X)^{d1}$ should yield $(bp_X)^{d3}$.

If you are allowed to make k such choices at once (preferably evenly divided between the two cases), then, with probability greater than $1-2^{-k/2}$, either you can obtain p_X^{d3} from p_X^{d1}, or you can show that X cheated without revealing whether you have p_X^{d1}. If X's messages are formed so that both cases can be checked successfully, then you can decrypt the credential from any second case; the more of X's messages that are not properly formed, the more overwhelming the chance of detection. To incriminate X, you merely have to reveal X's signature on an improper message. If you reveal your pseudonym p_X (which is safe because only you retain the corresponding secret key mentioned in §2.4), then all your checks could as well be verified by anyone else. Even infinite computing power cannot help X cheat.

When the same OR credential, for instance, should be available to everyone on their own pseudonyms, X's output for all people could first be published, and some mutually trusted random source could later be used to make all the choices. Thus any observer can check that everyone has access to the properly formed encryptions, and hence can access the credentials to which they are entitled.

5. TYING IT ALL TOGETHER

In order to maintain the database all their own credentials, individuals should be able to show the truthfulness of responses to any query by an organization.

First, by combining techniques presented so far, it is shown how a large class of queries can convincingly be answered. Then, protocols are given that can replace supervisory organizations, which had special access to information in some of the previous techniques. Finally, all reasonable limits are removed from the set of queries that can be answered indisputably.

5.1 Positive Queries

Consider the question of *positive* credentials, those not in a person's interest to conceal. An organization's query about a positive credential merely defines various qualifying subsets of credentials (i.e. it is a monotone predicate). Thus, a query about positive credentials asks you to convince an organization that you do have at least one qualifying subset of credentials (but not that you lack some credential).

You can efficiently respond to any positive query (i.e. a question about a positive credential) by combining the AND and OR credential techniques described above. An AND credential might be one of the alternatives needed to get some OR credential; several OR's might be required to get an AND. More generally, by composing AND and OR in this way, any positive credential can be achieved. Thus, when provided with the appropriate block-cipher encryptions, you can convincingly answer any positive query on your credentials, by showing the corresponding positive credential—without revealing additional information.

5.2 Superfluous Organizations

Ways to use positive queries obviate the organizations—along with their special access to information about individuals—who issue pseudonyms and establish accountability after the fact.

For issuing pseudonyms, the authorizing credential technique allowed credentials to be turned into pseudonyms (§3.4). But these credentials were still issued by an organization who's decisions required knowledge of people's holdings and choices. When organizations instead make authorizing credentials available through positive credentials (and another kind considered later) people can directly compute an authorizing credential for any new pseudonym to which they are entitled. Organizations can still set the rules for who gets which type of authorizing credential, by the encryptions issued, and the rules can still depend on people's other credentials, choices, and pseudonyms.

With after-the-fact accountability, a similar effect is obtained by different means. In the library lending example, each borrower must contact the library clearinghouse at the end of each period in order to get credentials for the next period; furthermore, the clearinghouse periodically learns how many borrowings each person makes. These disadvantages would be removed if people could use positive credentials to directly compute their own "books returned" credentials, but the number of previous library visits is not in this case a positive credential.

The solution, which illustrates a technique generalized later, involves the library clearinghouse giving out block-cipher encryptions. First the clearinghouse imposes a *sequence* order, say, $p_{a1},....,p_{ak}$, on the couple pseudonyms you use with it. Then it gives you k-1 encryptions. The first encryption allows you to change a "returned" credential on the first pseudonym of the sequence into an "all ok" credential on the second; the second transform allows you to change a "returned" credential on both of the first two pseudonyms of the sequence into an "all ok" credential on the third; and so on, until the final transform allows a "returned" credential on all the first k-1 pseudonyms to be changed into an "all ok" credential on the kth.

The jth time you borrow a book, you show the validator for the second column of the couple pseudonym $p_{b,j}$ and the "all ok" credential on it. You can produce this credential simply by using the j-1th encryption, since you have the "returned" credential on all borrowings from the first to the j-1th. (For j=1, the "all ok" credential is assumed available.) It is interesting to note that the effect of OR credentials is achieved here without them, by the indistinguishability of the couple pseudonym rows.

You may naturally wish to have books borrowed from more than one library at a time. To permit this, MIN "due date" credentials can be issued when a book is loaned. When the book is later returned, a "plus infinity due-date" credential is given. These resolution credentials are combined just as were the "returned" credentials. Thus, with only an initial issue of encryptions from the clearinghouse, you will be able to show any library that you have no overdue books.

Sometimes it may be necessary, or more efficient, for the encryptions used in combining credentials to be issued by the organization making a query to a limited number of people. To allow this, an extra column concatenated to the "couple" matrix is used by the organization. If Z initially gives a unique credential on each row for its sequence number, then you can answer any positive query on your library history with out pre-arrangement with Z.

5.3 Controlling Your Own Database

For positive credentials, the techniques shown above are quite adequate. It is only those that a person may wish to hide, called "negative" credentials, for which a satisfactory solution is still needed.

A simple way to implement negative credentials is by government mandate. Everyone might, for example, be expected to have a current "has filed a tax return" credential. No new mechanism is needed to handle this. Moreover, this type of solution becomes unworkable as the number of organizations who can issue negative credentials grows: "everything ok" credentials would continually have to be obtained from all such organizations and nothing prevents an organization from withholding credentials from people without any reason.

A better solution creates the potential for negative credentials by agreement, between a person and an organization. Such an establishing agreement would be consummated during an interaction; its optional cancellation would be a positive credential. The remaining problem is how to let people convince organizations that they are not omitting any such "establishing agreement" credential (or any other subsequently issued negative credential) from their responses to queries.

The general solution requires issuing in each interaction an *undilutable* credential that cannot be hidden or changed into another (less negative) one. If no potentially negative credentials result from an interaction, then the undilutable credential has a special value, say e_1. If, on the other hand, a negative credential results from an interaction, then no e_1 credential is issued. Instead, the undilutable credential takes on a value reserved for the particular negative credential. (An efficient encoding of such credentials is as the set of left- or right-decedent decisions to reach the credential considered as a leaf in a binary search tree.)

264

In each interaction of the solution, as in the library lending example, a validator from a sequenced couple is shown and tested for uniqueness. When the transaction is consummated, in addition to any positive credentials, an undilutable credential is issued on the join of the basic and couple pseudonyms. Thus, any query can be realized by an appropriate combination of encryptions that check the undilutable credential on all earlier couples in the sequence.

A (non-practical) three-pass identification protocol using coding theory

Marc Girault

Service d'Etudes communes des Postes et Télécommunications

42 rue des Coutures

BP 6243, 14066 Caen, France

ABSTRACT

At EUROCRYPT'89, Stern has presented an identification scheme whose security is based on general linear decoding problem, which is NP-hard. The number of passes of the protocol he designed is five. In this paper, we propose an alternative protocol which works with only three passes and is much simpler to describe (that is the good news) but which is not practical (the bad news).

1. INTRODUCTION

The idea of using error-correcting codes to build identification protocols is due to Harari [Ha]. But the first secure and (more or less) practical such protocol has been presented by Stern at EUROCRYPT'89 [St1]. The underlying hard problem is the problem of finding words of small weight which have a given syndrome. It is very closely connected to general linear decoding problem, whose NP-hardness is well-known [BMT].

Stern's protocol is a five-pass one, designed to be zero-knowledge, even if no formal proof of this fact has been provided. We address here the question of building a simpler protocol based on the same type of hard problems, hoping that it is also simpler to analyse (but not necessarily easier to implement!). We propose a solution based on a three-pass protocol which appears to be zero-knowledge, as long as a certain problem, stated later and used to commit the random choices of the prover, is really a hard one.

In contrast with Stern's solution, ours one is only of theoretical interest, since a prohibitive number of bits has to be exchanged. But it may constitute a step towards a future protocol, for which simplicity, security and practicability will be simultaneously achieved.

Our identification protocol can be turned into a signature scheme, following the Fiat-Shamir technique exposed in CRYPTO'86 [FS]. An important consequence is the possibility for an authority to authenticate users' public keys.

In last section, we also show how this protocol can be applied to recent Shamir's identification scheme [Sh], based on the so-called Permuted Kernel Problem.

2. THE IDENTIFICATION PROTOCOL

The starting point is the same as Stern's one : all the users share a common (k-n) matrix G over the two-element field GF(2), which has been randomly built (e.g. by a commonly trusted authority). The secret key of any user is an n-bit word s, randomly chosen by him, whose weight $|s|$ is equal to a prescribed number p, also fixed by the authority. In practice, p must be small compared to n, but not too small (see discussion on the security of the scheme). The public key of the same user is the word i given by :

$$i = G(s)$$

If G is seen as the parity-check matrix of a linear error-correcting code, computing s from G and i comes to finding a word of given small-weight and given syndrome i, a NP-hard problem.

Now, suppose that Alice wants to identify herself to Bob, who knows her public key i (we will consider later the problem of the authenticity of this public key). The goal of Alice is to prove her knowledge of the secret key s without revealing it. More precisely, she has to convince Bob that she knows a word s of weight equal to p such that G(s) = i, without leaking any information about this word.

This can be done in the following manner :

* *Step 1*: Alice secretly chooses a random (n-n) permutation matrix P and a random (k-k) non-singular matrix S. She calculates the matrix G' = SGP, the word i' = S(i), then sends G' and i' to Bob.

* *Step 2*: Bob chooses a random bit c and sends it to Alice.

* *Step 3a*: If c=0, Alice replies by delivering S and P to Bob, who checks that SGP = G' and S(i) = i'.

* *Step 3b*: If c=1, Alice replies by delivering s' = $P^{-1}(s)$ to Bob, who checks that $|s'|$=p and G'(s') = i'.

Now, this protocol is repeated t times in order to reach a level of security equal to $1-2^{-t}$ (see section 3.2).

This protocol is a three-pass one and is very simple to describe. Unfortunately, it is not practical at all since Alice must send to Bob very big matrices, whose size is equal to or greater than 128 Kbits. Concerning exclusively the computation time, an implementation on SUN 3/160 (68020 microprocessor) [DG] showed that four minutes were necessary to reach a level of security equal to $1-2^{-25}$ (30 seconds for Stern's protocol).

3. SECURITY OF THE SCHEME

We now discuss the security of the scheme.

Concerning possible collisions between public keys (two different secret keys may a priori have the same image by G) and the difficulty of retrieving s from G and i, we refer to Stern's paper [St1]. Let us just recall the values of n, k and p recommended by him : n=512, k=256 and p=30 ; or n=1024, k=512 and p=40. With these (small) values of p, collisions are strictly impossible for almost all matrices G.

But, conversely, p should not be too small because Stern has also shown that words s of too small weight (e.g. p=20) may presumably be retrieved from G and i [St2]. In fact, an implementation in our laboratory [DG] of improved Stern's method has shown that p should be approximately chosen between 40 and 55 (in the first case), or between 35 and 100 (in the second case).

3.1 Completeness

It is child's play to show that our protocol is complete, i.e. that Alice will always be "accepted" by Bob if the two partners correctly follow the protocol. In the case c=0, it is obvious. In the case c=1, it is also true because the weight of s' is equal to the weight of s (since s' is obtained by permuting the bits of s) and because : G'(s') = SGP(P^{-1}(s)) = SG(s) = S(i) = i'.

3.2 Soundness

We now show why a person different from Alice (call him Charlie) will be accepted by Bob with probability substantially smaller than one. More precisely, we show that Charlie has, at most, essentially one chance over two to be accepted by Bob (hence one chance over two to be rejected).

Charlie can prepare G' and i' such that his Step 3a-reply be correct. He just has to do in Step 1 exactly as Alice would have done herself. Charlie can also prepare G' and i' such that his Step 3b-reply be correct. For that, he randomly chooses a word s' of weight p and a (k-n) matrix G'. Then he calculates i' = G(s') and sends G' and i' to Bob.

But Charlie cannot prepare G' and i' such that his reply be always correct, whatever the value of c is. Indeed if he could, it would mean that he found S, P and s' such that G' = SGP, i' = S(i), |s'|=p and G'(s') = i'.

But all these equalities imply: $SGP(s') = S(i)$ and $|s'|=p$, equivalent to: $G(P(s')) = i$ and $|P(s')|=p$ (since S is non-singular and P is a permutation). In other words, Charlie would have found a word s" such that $G(s") = i$ and $|s"|=p$, exactly what he is supposed not to be able to calculate ! (Of course, the word s" is very presumably nothing but s).

It results that probability of success of Charlie is at most 1/2. By repeating the protocol a number t of times (with independent random choices), this probability falls down to $(1/2)^t$. In this case, the level of security is said to be $1-(1/2)^t$.

3.3 Zero-knowledge

It remains to see whether the protocol leaks or not any information about Alice's secret key. Clearly, it will be the case if and only if it is computationally infeasible to deduce anything about the permutation P from G' and i'. For any information about P leads, via $s' = P^{-1}(s)$, to an information about s.

Unfortunately, we actually lack some theoretical as well as practical arguments, which could "prove" the difficulty of solving the equation $G' = SGP$ in unknowns S and P. We will only make three remarks.

First, the knowledge of one of the unknown matrices leads to the knowledge of the other one. Indeed if S is known, the equation : $G' = SGP$ can be rewritten : $G' = HP$ (where $H = SG$) and the task of finding a permutation matrix which turns a matrix into another one is generally easy. If P is known, the equation : $G' = SGP$ can be rewritten : $SG = H$ (where $H = G'P^{-1}$) and this is only a (redundant) linear system to solve. This rewriting shows what makes the problem presumably difficult : the enemy has a linear system to solve -an easy task- but he has no idea which one !

Secondly, we remark that the problem of finding S and P such that $G'= SGP$ can be polynomially reduced to the problem of finding the first column of P. Indeed, let AL be an algorithm which, given G and G', provides the first column of P. Let us denote by $G'^{(i)}$ the matrix obtained by applying n left-rotations to the columns of G. The first column of P is obtained by performing AL on inputs G and G'. The second column is obtained by performing AL on G and $G'^{(1)}$. More generally, the (i+1)-th column of P is obtained by performing A on inputs G and $G'^{(i)}$. Now, since we know P, we easily find S (see previous remark).

Finally, let us note that security of Mac Eliece's public-key cryptosystem [Mc] is based on the hardness of solving the same equation, except that G is only known to be the coding matrix of a Goppa code, instead of being completely known.

4. SIGNATURE SCHEME

We can transform our identification scheme into a signature scheme with Fiat-Shamir's method [FS]. In order to sign a message M, Alice does the following :

* *Step 1*: Alice chooses (say) 64 random (n-n) permutation matrices P_j and 64 random non-singular (k-k) matrices S_j. Then she calculates, for each j, $G'_j = S_j GP_j$ and $i'_j = S_j(i)$.

* *Step 2*: Alice feeds a one-way hash-function with inputs M and (G'_j, i'_j) for j=1..64, and picks up the 64 first bits c_j of the hash-result.

* *Step 3*: Alice sends to Bob the message M along with the signature which is composed of all the (G'_j, i'_j) and : (S_j, P_j) when $c_j=0$, or $s'_j = P_j^{-1}(s)$ when $c_j=1$.

Now, Bob verifies the signature by checking that for each j : $S_j GP_j = G'_j$ and $S_j(i) = i'_j$ when $c_j=0$, or $|s'_j|=p$ and $G'_j(s'_j) = i'_j$ when $c_j=1$.

Of course, this leads to an incredibly long signature. But this solves, at least theoretically, the problem of authentication of user public keys : each public key can now be authenticated with a certificate delivered by the trusted authority. This certificate is composed of the user's identity, one's public key, the date of validity of the certificate (may be also other elements) and a signature of the whole package by the authority, whose public key must be universally known.

5. ANOTHER RELATED RESULT

Recently, Shamir presented a new identification scheme [Sh], based on the Permuted Kernel Problem : given a prime p, a (k-n) matrix G and a n-vector i (both with coefficients in GF(p)), find a permutation Q such that s = Q(i) lies in the kernel of G.

This problem is also a NP-hard one, and can be used to design identification schemes (as before, s is user's secret key and i is his public key). We show that our protocol can also be used for this problem, after a very slight adaptation :

* *Step 1*: Alice secretly chooses a random (n-n) permutation matrix P and a random (k-k) non-singular matrix S. She calculates the matrix G' = SGP and sends G' to Bob.

* *Step 2*: Bob chooses a random bit c and sends it to Alice.

* *Step 3a*: If c=0, Alice replies by delivering S and P to Bob, who checks that SGP = G'.

* *Step 3b*: If c=1, Alice replies by delivering $R = P^{-1}Q$ to Bob, who checks that G'(R(i)) = 0.

Despite the great resemblance between this protocol and the one of section 2, an important theoretical point distinguishes them. Explicitly, there is an argument (of NP-complete nature) which gives a lot of confidence in the commitment used in step 1 of the above protocol. This will be developed in a future paper [Gi].

6. BIBLIOGRAPHY

[BMT] E.R. Berlekamp, R.J. McEliece and H.C.A. van Tilborg, "On the inherent intractability of certain coding problems", IEEE Trans. Inform. Theory, vol.IT-24, pp. 384-386, May 1978.

[DG] J.L. Duras and M. Girault, "Etude et implémentation d'algorithmes d'authentification basés sur les codes correcteurs d'erreurs", Technical report, SEPT, 1989.

[FS] A. Fiat and A. Shamir, "How to prove yourself : Practical solutions to identification and signature problems", Proc. of CRYPTO '86.

[Gi] M. Girault, "Other protocols for Shamir's identification scheme", in preparation.

[Ha] S. Harari, "Un algorithme d'authentification sans transfert d'information", Proc. of Trois journées sur le codage, Toulon, France, 1988.

[Mc] R. J. Mac Eliece, "A public-key cryptosystem based on algebraic coding theory", DSN Progress Report, Jet Propulsion Laboratory, CA, Jan. & Feb. 1978, pp. 42-44.

272

[Sh] A. Shamir, "An efficient identification scheme based on permuted kernels", extended abstract presented at CRYPTO'89 rump session, Proc. to appear.

[St1] J. Stern, "An alternative to the Fiat-Shamir protocol", EUROCRYPT'89, Proc. to appear.

[St2] J. Stern, "A method for finding codewords of small weight", Proc. of Trois journées sur le codage, Toulon, France, 1988.

Demonstrating Possession without Revealing Factors

and its Application

Hiroki Shizuya*   Kenji Koyama†   Toshiya Itoh‡

*Education Center for Information Processing,
Tohoku University
Kawauchi, Aoba-ku, Sendai, 980 Japan

† NTT Basic Research Laboratories,
Nippon Telegraph and Telephone Corporation
3-9-11, Midori-cho, Musashino-shi, Tokyo, 180 Japan

‡ Faculty of Engineering,
Tokyo Institute of Technology
2-12-1, O-okayama, Meguro-ku, Tokyo, 152 Japan

Abstract   This paper presents a zero-knowledge interactive protocol that demonstrates that *two* factors, a and b, of a composite number n ($= ab$) are actually known by the prover, without revelation of the factors themselves and where a and b are not necessarily primes. The security of this protocol is based on the difficulty of computing a discrete logarithm modulo a large prime. As an extension of this, a protocol that can demonstrate that *two* or *more* factors are known by the prover is shown and applied to a weighted membership protocol with hierarchical classes within a group.

1. Introduction

Feige, Fiat, and Shamir [FFS] formalized a zero-knowledge interactive proof of *knowledge* by extending the concept of *language membership* introduced by Goldwasser, Micali, and Rackoff [GMR]. To prove knowledge, a prover convinces a verifier that a secret is *known* without revealing any actual information related the secret. Several practical protocols have been proposed for this type of knowledge proof. Some are for number theoretic problems and are used to demonstrate knowledge of square roots modulo a composite [GMR, FFS] and discrete logarithm modulo a prime [CEGP].

In 1987, Tompa and Woll [TW] showed that any random self-reducible problem has a zero-knowledge interactive proof. They proved that the computation of both square roots modulo a composite and discrete logarithms modulo a prime are in the random self-reducible class. They also showed that a practical zero-knowledge interactive proof exists to convince the verifier that the prover knows all the prime factors of composite n. This protocol is based on the fact that computing square roots modulo n is polynomial (in $\log n$) time equivalent to factoring n.

Suppose all prime factors of n are known, then knowledge can actually be demonstrated to the third party by executing the protocol in [TW]. However, if only partial factors of n are known, then that protocol can no longer be applied to demonstrate knowledge. Though it is easily seen that dividing any n into two non-trivial factors is polynomial (in $\log n$) time equivalent to factoring n completely, knowledge of the partial factors does not always result in the ability to compute the remaining factors. In fact, large composites have often been partly factored in the Cunningham Project [BLSTW] managed by American Mathematical Society (AMS), although the best known algorithm and the fastest available computers are used. For example, a 30-digit prime of a 200-digit composite target is discovered, and the remaining 170-digit composite cannot be factored yet. Thus, a more general problem was considered by Koyama [Ko] in 1988 by extending a *complete factoring problem* to a *partial factoring problem*. The former problem requires that all the factors must be primes, the latter problem allows that factors of a composite may or may not be primes. A question is whether or not it is possible to construct a practical zero-knowledge interactive proof that convinces the verifier that the prover knows at least two of the factors for n, where the factors are not necessarily primes. Note that it was theoretically shown in [GMW] that there exist zero-knowledge interactive proofs for the problems in $\mathcal{NP}$. Since this problem is also in $\mathcal{NP}$, a general (but complicated) zero-knowledge interactive proof can be constructed. However, it is not clear whether or not a practical protocol exists for this problem.

There have been a few related studies on zero-knowledge proof for the number of primes for a given integer. Blum, Feldman, and Micali [BFM] formalized a complexity assumption that says it is computationally hard to distinguish the integer product of 2 primes from that of 3 primes. Based on this assumption, they proposed a non-interactive zero-knowledge proof. Brassard and Crépeau [BC] proposed a protocol in which the prover convinces the verifier that a given integer is the product of exactly k distinct primes. This protocol requires primality tests on the Boolean circuit for the encrypted factors.

This paper presents an interactive protocol that demonstrates factors a and b of n ($= ab$) are actually known by the prover, where the factors are not necessarily primes nor distinct, and also shows that it is a zero-knowledge interactive proof. The sequel to this paper presents an extension of this protocol that demonstrates that two or more factors are known by the prover, and shows that the extended protocol is applicable to a weighted membership protocol with hierarchical classes in a group.

2. The Protocol

2.1. Preliminaries and Basic Idea

Hereafter, the prover will be denoted as P and the verifier as V. P and V share composite number n as an open problem. V is assumed to have only probabilistic polynomial (in log n) computation resources, and P to have probabilistic polynomial (in log n) computation resources as well as knowledge of non-trivial factors a and b of n. Other number theoretic notations used in this paper are defined as follows:

$\mathbf{Z}$ denotes integers.

$\mathbf{N}$ denotes positive integers.

$\mathbf{Z}_\ell$ denotes a ring of integers modulo $\ell \in \mathbf{N}$, that is,
$$\mathbf{Z}_\ell = \{x \mid x \in \mathbf{Z}, \ 0 \le x \le \ell - 1\}.$$

$\mathbf{Z}_\ell^*$ denotes the multiplicative group over $\mathbf{Z}_\ell$, that is,
$$\mathbf{Z}_\ell^* = \{x \mid x \in \mathbf{Z}_\ell, \ \gcd(x, \ell) = 1\}.$$

$\langle g \rangle$ denotes a group $\subseteq \mathbf{Z}_\ell^*$ generated by $g \in \mathbf{Z}_\ell^*$, that is
$$\langle g \rangle = \{x \mid x = g^i \bmod \ell, \ g \in \mathbf{Z}_\ell^*, \ i \in \mathbf{Z}\}.$$

PRIME   denotes the set of primes $\in \mathbf{N}$.

poly(x)   denotes any polynomial of x.

Suppose P happens to know factors a and b of n $(= ab)$, and intends to convince V that P knows two factors of n, without revealing the factors themselves. Our basic implementation idea is as follows:

1: P hides the factors in discrete logarithms modulo a prime so that $u = g^a \bmod q$ and $v = g^b \bmod q$, where q is an appropriate prime and g is a generator of $\mathbf{Z}_q^*$, and sends q, g, u and v to V.

2: P randomly picks R from $\mathbf{Z}_{q-1}$, computes $y = (uvg^R)^R \bmod q$, and sends y to V.

3: V randomly picks e from $\{0, 1\}$ and sends e to P.

4: P sends x to V, where $x = R$ if $e = 0$ and $x = a + R \bmod q - 1$ if $e = 1$.

5: V checks that $y \equiv (uvg^x)^x \pmod q$ if $e = 0$ and $y \equiv (u^{-1}vg^x)^x g^{-n} \pmod q$ if $e = 1$. (Note that $(u^{-1}vg^x)^x g^{-n} \equiv g^{(a+R)(b+R)-n} \equiv g^{(a+b+R)R} \equiv y \pmod q$.) If this check fails, V halts the interaction; otherwise P and V go on to 2.

To summarize, in our protocol P forces V to check if $g^{ab} \equiv g^n \pmod q$, without revealing a and b to V.

Using this basic idea, P with no factor of n can cheat V if P computes u and v such that $u = g^w \bmod q$ and $v = g^{nw^{-1}} \bmod q$, where w is in $\mathbf{Z}_{q-1}^*$. We make such cheating ineffective by choosing prime q of such a specific form that $q = kn + 1$, $\gcd(k, n) = 1$. Since $kn \equiv 0 \pmod{q-1}$, V can now detect P's cheating by checking to see if $u^k \not\equiv 1 \pmod q$ and $v^k \not\equiv 1 \pmod q$.

The other defect of this basic idea is the random value R picked by P. While R is neither observable nor controllable for V, P can possibly be to intentionally change the value of R at any step of the protocol. Thus, additional steps are required so that R becomes uncontrollable for both P and V. In the same way, generator g can be determined by a coin-flipping subprotocol so that g becomes uncontrollable for both P and V.

The complete protocol description is shown in the next section and reflects the above discussions. Hereafter, this protocol will be called the basic SKI protocol. This name relates to the authors and implies the later extension.

2.2. Basic SKI Protocol

The following is a zero-knowledge interactive proof system where P can demonstrate to V that P knows non-trivial factors a $(1 < a < n)$ and b $(1 < b < n)$ of n, where n is odd. All prime factors of n are assumed to be greater than poly(log n).

[Basic SKI Protocol]

Input to (P, V) : $n\ (= ab)$

    Step 1.1 P and V share the least prime, denoted by $q\ (\in \text{PRIME})$, such that
$$q = kn + 1, \quad \gcd(k, n) = 1,$$
and P and V factorize k as
$$k = p_1^{m_1} p_2^{m_2} \cdots p_s^{m_s},$$
where $p_i \in \text{PRIME}\ (1 \le i \le s)$.

    Step 1.2 P and V execute a coin-flipping protocol (for example, the protocol in [B]), and share random number $g\ (\in \mathbf{Z}_q^*)$.

    Step 1.3 P and V check to see if g satisfies
$$g^{(q-1)/p_i} \not\equiv 1 \pmod{q} \quad \text{for all prime divisors } p_i\ (1 \le i \le s) \text{ of } k.$$
If this check fails, return to Step 1.2.

    Step 2.1 P randomly picks $\lambda \in \mathbf{Z}_{q-1}^*$, then computes u and v such that
$$u = g^A \bmod q, \quad \text{where } A = \lambda a \bmod q - 1,$$
$$v = g^{B} \bmod q, \quad \text{where } B = \bar{\lambda} b \bmod q - 1, \bar{\lambda} = \lambda^{-1} \bmod q - 1,$$
and sends u and v to V.

    Step 2.2 V checks to see that u and v satisfy
$$u^k \not\equiv 1 \pmod{q} \text{ and } v^k \not\equiv 1 \pmod{q}.$$
If this check fails, V detects cheating and halts.

Repeat Steps 3.1 to 3.6 for t rounds.

    Step 3.1 P randomly picks $\sigma \in \mathbf{Z}_{q-1}$, computes $h = g^\sigma \bmod q$, and sends h to V.

    Step 3.2 V randomly picks $r \in \mathbf{Z}_{q-1}$, and sends r to P.

    Step 3.3 P computes
$$y = (uvg^R)^R \bmod q,$$
where $R = \sigma + r \bmod q - 1$, and sends y to V.

    Step 3.4 V randomly picks $e \in \{0, 1\}$ and sends e to P.

    Step 3.5 P sends x to V if $e \in \{0, 1\}$, where
$$x = \begin{cases} R, & \text{if } e = 0; \\ R + A \bmod q - 1, & \text{if } e = 1. \end{cases}$$

Otherwise, P halts.

Step 3.6 V checks

$$
\begin{cases}
g^{x-r} \equiv h \pmod{q}, \quad y \equiv (uvg^x)^x \pmod{q}, & \text{if } e = 0; \\
g^{x-r} \equiv hu \pmod{q}, \quad y \equiv (u^{-1}vg^x)^x g^{-n} \pmod{q}, & \text{if } e = 1.
\end{cases}
$$

If these check fail, then V detects cheating and halts.

end-Repeat

Notes :

(1) In Step 1.1 of the protocol, P and V agree on q and k. Note that k can be factored in a polynomial time of $\log n$. The reason for this is described in Section 3.1.

(2) In Step 1.2, P and V choose as a candidate of a generator of Z_q^*. Once g passes Step 1.3, g is a generator with overwhelming probability, as shown in Lemma 1 in Section 3.2. The checks in Step 2.2 by V exclude a cheating prover who does not know a and b such that $n = ab$, but does know $\tilde{a}$ and $\tilde{b}$ such that $n \equiv \tilde{a}\tilde{b} \pmod{q-1}$. The reason for this is described in Lemmas 2 and 3 in Section 4.2.

(3) In Step 2.1, values a and b are randomized by λ and $\bar{\lambda}$ such that
$$
A = \lambda a \bmod q - 1, \quad B = \bar{\lambda} b \bmod q - 1.
$$

(4) Throughout Steps 3.1 to 3.3, P computes random variable $R = \sigma + r \bmod q - 1$, where σ and r are generated by P and V, respectively. Furthermore, the R value is uncontrollable for P and V, and is unobservable for V. This property is essential to the soundness and zero-knowledge of the protocol.

(5) In Step 3, P convinces V that P knows a and b such that $n = ab$. During the checks in Step 3.6 by V, $e = 1$ and $x \equiv R + A \pmod{q-1}$ imply that
$$
g^{x-r} \equiv g^{R+A-r} \equiv g^{\sigma+r+A-r} \equiv g^{\sigma+A} \equiv hu \pmod{q},
$$
$$
y \equiv (u^{-1}vg^x)^x g^{-n} \equiv g^{(R+B)(R+A)-n} \equiv g^{(A+B+R)R} \equiv (uvg^R)^R \pmod{q}.
$$

(6) Security parameter $t = O(\log n)$, which is proportional to required time and space, is set before the protocol is started. Successive t round executions in Step 3 forces V to accept the exponentially small chance that P is cheating, i.e. that P is pretending to know non-trivial a and b but in fact does not. More precisely, since the probability that P will succeed in cheating for each round in Step 3 is $1/2$, the overall probability that P will succeed in cheating during successive t rounds in Step 3 is 2^{-t}.

3. Remarks

3.1. Modulus

In the protocol, modulus q is restricted to a prime in the form of $kn + 1$. However, it is not difficult to find such a prime since it is assured by Dirichlet's theorem and related research.

[Dirichlet's theorem] : *There are infinitely many primes in any arithmetic progression*

$$c, \; c + d, \; c + 2d, \; c + 3d, \; \cdots$$

if gcd(c, d)=1, and d > 0.

In the present case, $c = 1$ and $d = n$, so there is an infinite number of primes in the form of $kn + 1$ for a given n. It is almost verified that $k = O(\log^2 n)$ for the least prime of $kn + 1$ for a given n [AM, H]. Therefore, least prime q can be found in the probabilistic polynomial time of $\log n$ by applying polynomial time primality testing algorithm [CL] for $kn + 1$ ($k = 2, 4, 6, \ldots$). After determining k, it can also be factored in the polynomial time of $\log n$ because it is as small as $O(\log^2 n)$.

3.2. Generator

If $q - 1$ can be completely factored, it can easily be verified to whether or not integer g is a generator of $\mathbf{Z}_q^*$, i.e., primitive element modulo q [Kn]. The number g is a generator of $\mathbf{Z}_q^*$ (q: odd prime) if and only if $g \not\equiv 0 \pmod{q}$, and $g^{(q-1)/\delta} \not\equiv 1 \pmod{q}$ for any prime divisor δ of $q - 1$. In our protocol, however, $q - 1$ does not need to be completely factored. If g successfully passes the check in Step 1.3, it is a generator of $\mathbf{Z}_q^*$ with overwhelming probability.

Lemma 1. Let g be an element of $\mathbf{Z}_q^*$ that has successfully passed the check in Step 1.3, and let γ denote the probability that g is a generator of $\mathbf{Z}_q^*$. Then,

$$\gamma > 1 - \frac{1}{\text{poly}(\log n)}.$$

Proof: Since the number of generators of $\mathbf{Z}_q^*$ is $\varphi(q - 1)$,

$$\gamma = \frac{\varphi(q - 1)}{\varphi(q - 1) + |\tilde{G}|} = 1 - \frac{1}{\dfrac{\varphi(q - 1)}{|\tilde{G}|} + 1},$$

where φ denotes Euler's totient function and $|\tilde{G}|$ denotes the cardinality of the set $\tilde{G}$ that consists of all g's that are non-generators satisfying the check at Step 1.3.

Now assume that $n = r_1^{\alpha_1} r_2^{\alpha_2} \cdots r_c^{\alpha_c}$ where $r_i \in \text{PRIME}$ and $r_i > \text{poly}(\log n)$ $(i = 1, 2, \ldots, c)$. Then, we have

$$\varphi(q - 1) = \varphi(kn) = \varphi(k)\varphi(n) = kn \prod_{i=1}^{s}(1 - 1/p_i) \prod_{j=1}^{c}(1 - 1/r_j).$$

Since the set $\tilde{G}$ is defined as

$$\tilde{G} = \{g \mid g \in \mathbf{Z}_q^*, \ g^{(q-1)/p_i} \not\equiv 1 \pmod{q} \text{ for all } i \ (i = 1, \ldots, s)\}$$

$$\cap \left\{ \bigcup_{j=1}^{c} \{g \mid g \in \mathbf{Z}_q^*, \ g^{(q-1)/r_j} \equiv 1 \pmod{q}\} \right\},$$

$|\tilde{G}|$ is bounded as

$$|\tilde{G}| \leq \sum_{j=1}^{c} |\{g \mid g \in \mathbf{Z}_q^*, \ g^{(q-1)/r_j} \equiv 1 \pmod{q}\}| = \sum_{j=1}^{c}(q-1)/r_j = kn \sum_{j=1}^{c} 1/r_j.$$

Hence,

$$\frac{\varphi(q-1)}{|\tilde{G}|} \geq \frac{\prod_{i=1}^{s}(1 - 1/p_i) \prod_{j=1}^{c}(1 - 1/r_j)}{\sum_{j=1}^{c} 1/r_j}.$$

Let $r = \min\{r_1, r_2, \ldots, r_c\}$ and let $p = \min\{p_1, p_2, \ldots, p_s\}$. Note that $p = 2$ for any k because $q(= kn + 1)$ and n are odd numbers. Then, we have

$$\frac{\prod_{i=1}^{s}(1 - 1/p_i) \prod_{j=1}^{c}(1 - 1/r_j)}{\sum_{j=1}^{c} 1/r_j} \geq \frac{(1 - 1/p)^s (1 - 1/r)^c}{c/r} = \frac{(1 - 1/r)^c}{2^s c/r}$$

$$\geq \frac{1}{k} \cdot \frac{1 - c/r}{c/r} = \frac{1}{k}\left(\frac{r}{c} - 1\right),$$

where the following two inequalities are used and easily verified :

$$(1 - 1/r)^c \geq 1 - c/r, \quad 2^{-s} \geq 2^{-\log k} = 1/k.$$

Since $r > \text{poly}(\log n)$ (by our assumption. Note that another assumptions were described in [IT] and [Ka].), $c \leq \log n$ and $k = O(\log^2 n)$, the right side of the inequality to be concerned is dominated by $r = 2^{O(\log n)}$. Therefore, we finally have

$$\frac{\varphi(q-1)}{|\tilde{G}|} \geq \frac{1}{k}\left(\frac{r}{c} - 1\right) > \text{poly}(\log n).$$

Hence,

$$\gamma = 1 - \frac{1}{\frac{\varphi(q-1)}{|\tilde{G}|} + 1} > 1 - \frac{1}{\text{poly}(\log n)}. \quad \square$$

4. Discussions

4.1. Security based on Discrete Log Problem

The security of this protocol is based on the difficulty of computing a discrete logarithm modulo a large prime. More precisely, it is based on the difficulty of the following problems.

SKI Problem-1 (Product Factoring Problem)

INSTANCE: $q \in \text{PRIME}$, $n \in \mathbf{N}$, $g \in \mathbf{Z}_q^*$, u, $v \in \langle g \rangle$, where $q = kn + 1$, $\gcd(k,n) = 1$.

SOLUTION: a and b such that $n = ab$, $u = g^{\lambda a} \bmod q$, and $v = g^{\bar{\lambda} b} \bmod q$,

where $\lambda \bar{\lambda} \equiv 1 \pmod{q-1}$.

Comment: a and b are obtained if n is factored or multiple discrete log problem [CEG] is solved. Note that the best known algorithm for factoring $n = ab$ requires a running time of $\exp((1 + o(1))\sqrt{\log n \ \log\log n})$ [L]. Since n is assumed to be unsmooth, $q - 1$ is also so. Note that the best known algorithm for computing discrete logarithms modulo q requires a running time of $\exp((1+o(1))\sqrt{\log q \ \log\log q})$ [COS]. Since $q > n$, the running time for solving the multiple discrete log problem modulo q is greater than that for factoring n. If both a and b are large and the multiple discrete log problem is difficult, there seems no way to compute a and b from just n, $g^{\lambda a} \bmod q$ and $g^{\bar{\lambda} b} \bmod q$. It is conjectured but not proven that there is no way to solve SKI problem-1 without solving the factoring problem or the multiple discrete log problem.

SKI Problem-2 (Product Recognition Problem)

INSTANCE: $q \in \text{PRIME}$, $n \in \mathbf{N}$, $g \in \mathbf{Z}_q^*$, x, $y \in \langle g \rangle$, where $q = kn + 1$, $\gcd(k,n) = 1$.

SOLUTION: $z = 1$, if $\alpha\beta = n$; $z = 0$, otherwise;

where $x = g^{\lambda \alpha} \bmod q$, $y = g^{\bar{\lambda} \beta} \bmod q$, and $\lambda \bar{\lambda} \equiv 1 \pmod{q-1}$.

Comment: SKI problem-2 is a recognition problem that asks whether or not x and y are generated from two factors of n. This is closely related to SKI problem-1. If the solution for SKI problem-2 is obtained as $z = 1$, then the condition for solving SKI problem-1 is satisfied. It is conjectured but not proven that no probabilistic polynomial time algorithm exists to solve SKI problem-2 without solving either the integer factoring problem or the multiple discrete log problem.

4.2. Detection of cheating

This section describes the detection mechanism for cheating in Step 2.

Consider a direct product set where,

$$D = \{(A, B)\} \in \mathbf{Z}_{q-1} \times \mathbf{Z}_{q-1}.$$

Then, D can be broken down into two exclusive subsets S and S^c such that

$$S = \{ (A, B) \mid AB \equiv n \ (\text{mod } q - 1), \ (A, B) \in D \},$$
$$S^c = \{ (A, B) \mid AB \not\equiv n \ (\text{mod } q - 1), \ (A, B) \in D \}.$$

Furthermore, S can be broken down into two exclusive subsets T and T^c such that

$$T = \{ (A, B) \mid (A, B) \in S, \ A \notin \mathbf{Z}_{q-1}^*, \ B \notin \mathbf{Z}_{q-1}^*, \ A \not\equiv 0 \ (\text{mod } n), \ B \not\equiv 0 \ (\text{mod } n)\},$$

and

$$T^c = S - T.$$

If $(A, B) \in T^c$, then either A or B is in $\mathbf{Z}_{q-1}^*$ or a multiple of n. Note that, if $(A, B) \in T$, both A and B are expressed as

$$A = \lambda a \bmod q - 1, \quad B = \bar{\lambda} b \bmod q - 1,$$

where $\lambda \bar{\lambda} \equiv 1 \ (\text{mod } q - 1)$, and a and b are non-trivial factors of $n \ (= ab)$.

Since $D = S + S^c = T + T^c + S^c$, any (A, B) should belong to one of the three exclusive subsets T, T^c, or S^c. Then, u or v generated from $(A, B) \in T^c$ is always rejected by V in Step 2.2 (see Lemma 2). Conversely, u and v generated from $(A, B) \in T$ passes check in Step 2.2 with overwhelming probability (see Lemma 3). The other u and v generated from $(A, B) \in S^c$ are rejected by V in Step 3.6 with overwhelming probability (see Section 4.3), even if they pass V's checks at Step 2.2.

Lemma 2. If $g \in \mathbf{Z}_q^*$ satisfies the following condition (i), then either u or v generated from g and $(A, B) \in T^c$ will never pass checks (ii) or (iii).

    (i) $g^{(q-1)/p_i} \not\equiv 1 \ (\text{mod } q)$ for all prime divisors $p_i \ (1 \leq i \leq s)$ of k,

    (ii) $u^k \not\equiv 1 \ (\text{mod } q)$, where $u = g^A \bmod q$,

    (iii) $v^k \not\equiv 1 \ (\text{mod } q)$, where $v = g^B \bmod q$.

Proof : The following cases, (1) and (2), are analyzed.

(1) When either A or B is in $\mathbf{Z}_{q-1}^*$, it can be assumed that $A \in \mathbf{Z}_{q-1}^*$, without loss of generality. Then, $B = nA^{-1} \bmod q - 1$ and, since $q - 1 = kn$, we have

$$v^k \equiv (g^B)^k \equiv (g^{nA^{-1}})^k \equiv (g^{q-1})^{A^{-1}} \equiv 1 \ (\text{mod } q).$$

Therefore v will never pass check (iii).

(2) When neither A nor B is in $\mathbf{Z}^*_{q-1}$ but $A \equiv 0 \pmod{n}$ or $B \equiv 0 \pmod{n}$, it is obvious that $u^k \equiv 1 \pmod{q}$ or $v^k \equiv 1 \pmod{q}$. Therefore, u and v will never pass checks (ii) or (iii).

Thus, Lemma 2 is proven. $\square$

Lemma 3. If $g \in \mathbf{Z}^*_q$ satisfies the following condition (i), then both u and v generated from g and $(A, B) \in T$ pass checks (ii) and (iii) with overwhelming probability.

  (i) $g^{(q-1)/p_i} \not\equiv 1 \pmod{q}$ for all prime divisors p_i $(1 \leq i \leq s)$ of k,

  (ii) $u^k \not\equiv 1 \pmod{q}$, where $u = g^A \bmod q$,

  (iii) $v^k \not\equiv 1 \pmod{q}$, where $v = g^B \bmod q$.

Proof : Recall Lemma 1 in which g is a generator of $\mathbf{Z}^*_q$ with overwhelming probability if g satisfies condition (i). If g is a generator of $\mathbf{Z}^*_q$, the orders of u and v generated from $(A, B) \in T$ are kB and kA, respectively, and both are greater than k. Thus, Lemma 3 is proven. $\square$

4.3. Proof of ZKIP for the Protocol

This section shows that the proposed protocol constructs a ZKIP (zero-knowledge interactive proof) system. Three protocol conditions, completeness, soundness, and zero-knowledge, are proven as follows.

Completeness :

Assume that P has factors a and b of n, and that P and V follow the specified protocol. Here we show that (P, V) accepts input n $(= ab)$ with overwhelming probalibity.

The proof of validity for Step 1.1 is trivial (see the related description in Section 3.1). Once g passes the check in Step 1.3, the check in Step 2.2 is passed with overwhelming probability (see Lemma 3), because (A, B) is in T. The validity of Step 3 can be verified as follows. When $e = 0$, the checks at Step 3.6 are successfully passed as

$$g^{z-r} \equiv g^{R-r} \equiv g^\sigma \equiv h \pmod{q},$$

$$y \equiv (uvg^z)^z \equiv (uvg^R)^R \pmod{q}.$$

When $e = 1$, the checks in Step 3.6 are successfully passed as described in Note (5) in Section 2. Therefore, if P actually knows that a and b are such that $ab = n$, and P and V follow the protocol, P's proof is accepted by V with overwhelming probability. $\square$

Soundness :

To prove protocol soundness, the definition by Feige, Fiat, and Shamir [FFS] is adopted.

[Definition of Soundness]

$\exists M \, \forall P^* \, \forall RP^* \, \exists c \, \forall \log n > c$

$\Pr[(P^*, V) \text{ accepts } n] > 1/\text{poly} (\log n) \implies$

$\Pr[\text{output of } (P^*, M) \text{ on } n \text{ is a non-trivial factor of } n] > 1 - 1/\text{poly} (\log n),$

where P^* represents a polynomial time cheater who does not necessarily have factors of n, RP^* is the random tape of P^*, and M is a probabilistic polynomial time Turing machine with complete control over P^*.

In other words, it is proven that if (P^*, V) accepts n with non-negligible probability, probabilistic polynomial time Turing machine M can output a non-trivial factor of n.

Assume that (P^*, V) accepts n with non-negligible probability in every round of execution in Step 3. Then, there exist x and y that can pass the checks in Step 3.6 without respect to the value of $e \in \{0, 1\}$. That is, P^* can generate x and y to satisfy

$$\begin{cases} g^{x-r} \equiv h \pmod{q}, \quad y \equiv (uvg^x)^x \pmod{q}, & \text{if } e = 0; \\ g^{x-r} \equiv hu \pmod{q}, \quad y \equiv (u^{-1}vg^x)^x g^{-n} \pmod{q}, & \text{if } e = 1. \end{cases}$$

For exponents to base g, we have

$$\begin{cases} g^{x-r} \equiv h \pmod{q}, & \text{if } e = 0; \\ g^{x-r} \equiv hu \pmod{q}, & \text{if } e = 1, \end{cases}$$

$$\implies \begin{cases} x \equiv R^* \pmod{q-1}, & \text{if } e = 0; \\ x \equiv R^* + A^* \pmod{q-1}, & \text{if } e = 1, \end{cases}$$

where g is a generator of $\mathbf{Z}_q^*$ with overwhelming probability according to Lemma 1. As a result, y satisfies

$$y \equiv (uvg^{R^*})^{R^*} \equiv (u^{-1}vg^{R^*+A^*})^{R^*+A^*} g^{-n} \pmod{q}.$$

and this yields

$$(A^* + B^* + R^*)R^* \equiv (-A^* + B^* + R^* + A^*)(R^* + A^*) - n \pmod{q-1}.$$

Hence,

$$A^* B^* \equiv n \pmod{q-1}.$$

This implies that if (P^*, V) accepts n with non-negligible probability, (A^*, B^*) is in S. However, if (A^*, B^*) is in $T^c \ (= S - T)$, P^* is always rejected by V (see Lemma 2). Thus, (A^*, B^*) is in T, that is, (A^*, B^*) satisfies $A^* B^* \equiv n \pmod{q-1}$, $A^* \notin \mathbf{Z}_{q-1}^*$, $B^* \notin \mathbf{Z}_{q-1}^*$, $A^* \not\equiv 0 \pmod{n}$, and $B^* \not\equiv 0 \pmod{n}$. If (A^*, B^*) is in T, both A^* and B^* are multiples of non-trivial factors of n. Therefore, Turing machine M can output a non-trivial factor of n by computing $\gcd(A^*, n)$ or $\gcd(B^*, n)$. This gcd can then be computed in polynomial time of $\log n$. □

Zero-Knowledge :

Let V^* be any probabilistic polynomial time algorithm for the verifier, and let ρ be a random string such that $\rho \in \{0, 1\}^{\text{poly}(\log n)}$. Then, there is a probabilistic polynomial time Turing machine M_{V^*}, on input (n, ρ), which simulates communication between (P, V^*). The procedure executed by M_{V^*} is listed in the appendix, where M_{V^*} uses V_ρ^* as a subroutine. M_{V^*} runs in a probabilistic polynomial time of $\log n$.

It is clear that anyone with probabilistic polynomial time computation power can not distinguish the simulated (u, v) from the real (u, v) in (P, V^*) if SKI problem-2 is difficult. Furthermore, it is also impossible to distinguish the simulated x from the real x between (P, V^*) because both are subject to the same probabilistic distribution. Consequently, the zero-knowledge is proven. □

5. Extension of the Basic SKI Protocol

5.1. Basic Idea

The basic SKI protocol was designed to demonstrate that *two* non-trivial factors ($\neq 1$, n) of a composite number n are actually known by the prover. By extension of the basic SKI protocol, a protocol is proposed that demonstrates that the non-trivial d (≥ 2) factors of a composite n ($= a_1 a_2 \cdots a_d$) are actually known by the prover, without revealing the factors themselves. Furthermore, the factors a_i ($1 \leq i \leq d$) of n are not necessarily primes nor distinct.

The strategy of the extended protocol is to repeatedly call on the basic SKI protocol for each pair of factors, and to convince the verifier that the following two propositions hold simultaneously.

Proposition 1 : a_j and b_j ($= n/a_j$) ($1 \leq j \leq d$) are known to the prover and are non-trivial factors of n. That is, $n = a_1 b_1$, $n = a_2 b_2$, $\ldots$, $n = a_d b_d$.

Proposition 2 : a_j and c_j ($= c_{j-1}/a_j$) ($1 \leq j \leq d-1$, $c_0 = n$) are known to the prover and are non-trivial factors of c_{j-1}. That is, the tree-like structure of factors shown below holds.

$$(n =)c_0 \to c_1 \to c_2 \to \cdots \to c_{d-2} \to c_{d-1}(= a_d)$$

$$\searrow \quad \searrow \quad \searrow \quad \cdots \quad \searrow$$

$$a_1 \quad a_2 \quad\quad\quad a_{d-1}$$

The former is implemented as Block 2 in the extended protocol, and as Block 3 for the latter. Such cross-checking of factors is essential to the soundness of the protocol. The reason for this will be shown in Note (1) following the description of the protocol.

5.2. The Extended SKI protocol

P and V share a composite number n and a number of factors d. All prime factors of n are assumed to be greater than poly$(\log n)$. P and V are assumed to have only probabilistic polynomial (in $\log n$) computation resources. The following is a zero-knowledge interactive proof system where P demonstrates to V that P knows non-trivial d factors $a_1, a_2, \ldots, a_d$, whose product is n.

[Extended SKI protocol]

Input to (P, V) : $n \ (= a_1 a_2 \cdots a_d)$, d

Block 1:

P and V execute Step 1 of the basic SKI protocol.

end of Block 1

Block 2:

P and V execute Steps 2 and 3 of the basic SKI protocol for each pair of
$$(a_j, \ b_j(= n/a_j)) \ (1 \le j \le d).$$

Instead of u and v, P and V use u_j and v_j such that
$$u_j = g^{A_j} \bmod q, \text{ where } A_j = \lambda_j a_j \bmod q - 1,$$
$$v_j = g^{B_j} \bmod q, \text{ where } B_j = \overline{\lambda}_j b_j \bmod q - 1.$$

For $j = d(> 2)$, P takes λ_d as $\lambda_d = \prod_{\ell=1}^{d-1} \overline{\lambda}_\ell \bmod q - 1$.

end of Block 2

if $d = 2$ then stop otherwise continue.

Block 3:

P and V execute Steps 2 and 3 of the basic SKI protocol for each pair of
$$(a_j, \ c_j(= c_{j-1}/a_j)) \ (2 \le j \le d-1, c_1 = b_1).$$

Instead of u and v, P and V use u_j and w_j such that

u_j is the same as used in Block 2,

$w_j = g^{C_j} \bmod q$, where $C_j = \bar{\Lambda}_j c_j \bmod q - 1$.

Here $\bar{\Lambda}_j = \prod_{\ell=1}^{j} \lambda_\ell \bmod q - 1$. Each λ_j is the same as used in Block 2, hence P does not pick λ_j in Block 3.

V's check in Step 2.2 is omitted, and Step 3.6 is slightly modified to

$$g^{z-r} \equiv h u_j \pmod{q}, \quad y \equiv (u_j^{-1} w_j g^z)^z w_{j-1}^{-1} \pmod{q}, \quad \text{if } e = 1.$$

Here $w_1 = v_1$, which appeared in Block 2.

end of Block 3

Notes :

(1) Block 2 demonstrates that a_j $(1 \le j \le d)$ is a factor of total product n $(= a_1 a_2 \cdots a_d)$. Block 3 demonstrates that a_j $(2 \le j \le d-1)$ is a factor of partial product $a_j a_{j+1} \cdots a_d$. Variable u_j $(= g^{\lambda_j a_j} \bmod q)$ links Blocks 2 and 3. The values for u_j and λ_j $(1 \le j \le d)$ generated in Step 2 of Block 2 are used again in Block 3.

If only Block 2 is carried out, a cheating prover who knows only factors a_1 and a_2, such that $n = a_1 a_2$, can demonstrate that the prover knows d (≥ 2) factors a_j, such that $n = a_1 a_2 \cdots a_d$. The cheating prover can then generate any successful pair $(\tilde{A}_j, \tilde{B}_j)$ $(j \ge 2)$ by computing

$$\tilde{A}_j = \theta_j a_1 \bmod q - 1, \quad \text{where } \theta_j \in \mathbf{Z}_{q-1}^*,$$

$$\tilde{B}_j = \bar{\theta}_j a_2 \bmod q - 1, \quad \text{where } \theta_j \bar{\theta}_j \equiv 1 \pmod{q-1}.$$

Since $\tilde{A}_j \tilde{B}_j \equiv n \pmod{q-1}$ and

$$u_j^k \equiv g^{\tilde{A}_j k} \equiv g^{\theta_j a_1 k} \not\equiv 1 \pmod{q}, \quad v_j^k \equiv g^{\tilde{B}_j k} \equiv g^{\bar{\theta}_j a_2 k} \not\equiv 1 \pmod{q},$$

the checks in Step 2.2 of Block 2 are successfully passed for pair

$$(\tilde{u}_j, \tilde{v}_j) = (g^{\tilde{A}_j} \bmod q, \ g^{\tilde{B}_j} \bmod q).$$

If only Block 3 for $j = 2, ..., d-1$ and Block 1 for $j = 1$ are carried out, a cheating prover who knows only factors a_1 and a_2, such that $n = a_1 a_2$, can demonstrate that the prover knows d (≥ 2) factors a_j, such that $n = a_1 a_2 ... a_d$. The cheating prover can then generate any successful pair $(\tilde{A}_j, \tilde{C}_j)$ $(j \ge 2)$ by computing

$$\tilde{A}_j = \theta_j, \quad \text{where } \theta_j \in \mathbf{Z}^*_{q-1},$$

$$\tilde{C}_j = \bar{\theta}_j \tilde{C}_{j-1} \bmod q - 1, \quad \text{where } \tilde{C}_1 = a_2, \ \theta_j \bar{\theta}_j \equiv 1 \pmod{q-1}.$$

Since $\tilde{A}_j \tilde{C}_j \equiv \tilde{C}_{j-1} \pmod{q-1}$, the checks in Block 3 are successfully passed for pair $(\tilde{u}_j, \tilde{w}_j) = (g^{\tilde{A}_j} \bmod q, g^{\tilde{C}_j} \bmod q)$.

The executions in both Blocks 2 and 3 can detect the above cheating, i.e. if P pretends to know non-trivial pairs (a_j, b_j) and (a_j, c_j).

(2) If $d > 2$, Block 3 for $j = 1$ is unnecessary since the same check is performed in Block 2. The value of w_{d-1} in Block 3 is the same as that of u_d which is sent to V in Block 2. It is clear that Block 3 for $j = d$ is unnecessary.

In Step 3.6 of Block 3, $e = 1$ and $x \equiv R + A_j \pmod{q-1}$ imply that

$$(u_j^{-1} w_j g^x)^x w_{j-1}^{-1} \equiv g^{(R+A_j)(R+C_j) - A_j C_j} \equiv (u_j w_j g^R)^R \pmod{q}.$$

(3) For $d > 2$, the basic SKI protocol is called on in Blocks 2 and 3 for d and $d - 2$ times, respectively, hence the total is $2d - 2$ times. If $d = 2$, it is so trivial that the basic SKI protocol is called on only once in Block 2.

5.3. Example

A summary of the flow of the extended SKI protocol is shown for when $d = 4$ and $n = a_1 a_2 a_3 a_4$. For simplicity, the mod q terms are omitted.

In Block 1, P and V share q, k and g.

In Block 2 for when $j = 1$, P and V apply repetition of Step 3 for

$$(u_1, v_1, g^n) = (g^{\lambda_1 a_1}, g^{\bar{\lambda}_1 a_2 a_3 a_4}, g^{a_1 a_2 a_3 a_4}), \ u_1^k \neq 1, v_1^k \neq 1.$$

In Block 2 for when $j = 2$, P and V apply repetition of Step 3 for

$$(u_2, v_2, g^n) = (g^{\lambda_2 a_2}, g^{\bar{\lambda}_2 a_1 a_3 a_4}, g^{a_1 a_2 a_3 a_4}), \ u_2^k \neq 1, v_2^k \neq 1.$$

In Block 2 for when $j = 3$, P and V apply repetition of Step 3 for

$$(u_3, v_3, g^n) = (g^{\lambda_3 a_3}, g^{\bar{\lambda}_3 a_1 a_2 a_4}, g^{a_1 a_2 a_3 a_4}), \ u_3^k \neq 1, v_3^k \neq 1.$$

In Block 2 for when $j = 4$, P and V apply repetition of Step 3 for

$$(u_4,\ v_4,\ g^n) = (g^{\overline{\lambda}_1\overline{\lambda}_2\overline{\lambda}_3a_4},\ g^{\lambda_1\lambda_2\lambda_3a_1a_2a_3},\ g^{a_1a_2a_3a_4}),\ u_4^k \neq 1, v_4^k \neq 1.$$

In Block 3 for when $j = 2$, P and V apply repetition of Step 3 for

$$(u_2,\ w_2,\ w_1(= v_1)) = (g^{\lambda_2a_2},\ g^{\overline{\lambda}_2a_3a_4},\ g^{\overline{\lambda}_1a_2a_3a_4}).$$

In Block 3 for when $j = 3$, P and V apply repetition of Step 3 for

$$(u_3,\ w_3,\ w_2) = (g^{\lambda_3a_3},\ g^{\overline{\lambda}_1\overline{\lambda}_2\overline{\lambda}_3a_4},\ g^{\overline{\lambda}_1\lambda_2a_3a_4}).$$

As a result, the basic SKI protocol is called on 6 $(= 2 \cdot 4 - 2)$ times in the extended SKI protocol.

6. Application of Extended SKI Protocol

Consider a membership protocol which requires the following:
- a member (prover) can convince a verifier that the prover is a member of the group
- a member does not reveal his identity
- a non-member cannot pretend to be a member after execution of the protocol

One scheme for membership protocols has been proposed by Kurosaki, Zheng, Matsumoto and Imai [KZMI]. Their scheme is based on the RSA cryptosystem; however, it has never been proven to be a zero-knowledge protocol. In this section, a zero-knowledge membership protocol is proposed, whose the scheme is based on the SKI protocol.

6.1. Group Membership Protocol

A membership protocol can be realized by using the basic SKI protocol as follows. This membership protocol is implemented in two phases: the first phase is carried out at a trusted center, and the second phase at each member's location. In phase 1, a center randomly picks *two* large primes a_1 and a_2, and computes the product n. Note that the size of each a_i $(i = 1,\ 2)$ must be determined so that factoring n is difficult. It is recommended that each a_i be a number greater than 256 bits. Because the center can generate primitive element g definitely, the basic SKI protocol can be simplified by deleting the generation of g. The center registers a public group key $(n,\ q,\ k,\ g)$ in a public file. The center then sends membership key a_1 secretly to each member. Each member of the group knows factors a_1 and n/a_1 of n. The member can now demonstrate that he knows the non-trivial factors a_1 and a_2 of n in the scheme of basic SKI zero-knowledge interactive protocol. That is, the member (prover) can convince a verifier that the prover is a member of the group in a zero-knowledge way.

6.2. Membership Protocol with Hierarchical Classes within a Group

Members within a group often belong to different classes. For simplicity, membership is set up in two classes, higher and lower. How can we distinguish members of the higher class from members of the lower class within the same group? One simple approach is to use the extended SKI protocol, where members of the higher class hold more information than members of the lower class. A summary of the membership protocol with classes within a group is as follows. In phase 1, a center randomly picks *three* large primes a_1, a_2, and a_3, and computes product n. The center registers a public group key (n, q, k, g) in a public file. The center sends a_1 secretly to each member of the lower class. The center sends a_1 and a_2 secretly to each member of the higher class. Each member of the lower class knows *two* factors, a_1 and n/a_1 of n. Each member of the higher class knows *three* factors, a_1, a_2, and $n/a_1 a_2$ of n. Note that, even if all members of the lower class conspire, they are able to know only *two* factors. In phase 2, Blocks 2 and 3 of the extended SKI protocol on input $(n, d = 3)$ is applied between the member and the verifier. Finally, a member of the higher class can convince a verifier that the member belongs to the higher class of the group by demonstrating the possession of three factors in a zero-knowledge way. Note that this approach can easily be generalized into a structure having three or more hierarchical classes.

7. Conclusion

A zero-knowledge interactive protocol (Basic SKI Protocol) has been proposed to demonstrate that a prover really knows factors a and b of a composite number n $(= ab)$, without revealing the factors themselves. Since this protocol demonstrates the possession of factors that do not necessarily have to be primes, it is more widely applicable in some sense than Tompa and Woll's zero knowledge interactive protocol for demonstrating the possession of all primes of a composite.

Furthermore, the Extended SKI Protocol that has been proposed demonstrates that two or more factors of a given composite are known. This extended SKI protocol is applicable to a weighted membership protocol with hierarchical classes within a group.

Acknowledgments

The authors are grateful to Kazuo Ohta, Tatsuaki Okamoto, Kaoru Kurosawa, Atsushi Fujioka, Mitsunori Ogiwara, Hideo Suzuki, Youichi Takashima and Hiroyuki Masumoto for their helpful comments on the earlier version of this paper.

References

[AM] Adleman, L. M. and McCurley, K. S., "Open problems in number theoretic complexity," *Proc. Japan-U.S. Joint Seminar on Discrete Algorithms and Complexity, Academic Press*, pp.237-262 (1987).

[B] Blum, M., "Coin flipping by telephone," *Proc. IEEE COMPCON*, pp.133-137 (1982).

[BC] Brassard, G. and Crépeau C., "Non-Transitive Transfer of Confidence: A Perfect Zero-Knowledge Interactive Protocol for SAT and Beyond," *Proc. of FOCS*, pp.188-195 (1986).

[BFM] Blum, M., Feldman, P., and Micali, S., "Non-Interactive Zero-Knowledge and its Applications," *Proc. of STOC*, pp.103-112 (1988).

[BLSTW] Brillhart, J., Lehmer, D. H., Selfridge, J. L., Tuckerman, B., and Wagstaff, Jr., S. S., "Factorizations of $b^n \pm 1$, $b = 2, 3, 5, 6, 7, 10, 11, 12$ up to high powers," *Contemporary Mathematics*, Vol. 22, (Second Edition) American Mathematical Society (1988).

[CEG] Chaum, D., Evertse, J., and van de Graaf, J., "An improved protocol for demonstrating possession of discrete logarithms and some generalizations," *Proc. EUROCRYPT'87*, pp.127-142 (1987).

[CEGP] Chaum, D., Evertse, J., van de Graaf, J., and Peralta, R., "Demonstrating possession of a discrete logarithm without revealing it," *Proc. CRYPTO'86*, pp.200-212 (1986).

[CL] Cohen, H. and Lenstra, Jr., H. W., "Primality testing and Jacobi sums," *Math. Comp.*, Vol.42, No.165, pp.297-330 (1984).

[COS] Coppersmith, D., Odlyzko, A. M., and Schroeppel, R. "Discrete logarithms in GF(p)," *Algorithmica*, pp.1-15 (1986).

[FFS] Feige, U., Fiat, A. and Shamir, A., "Zero knowledge proofs of identity," *Proc. 19th STOC*, pp.210-217 (1987).

[GMR] Goldwasser, S., Micali, S., and Rackoff, C., "The zero-knowledge complexity of interactive proof-systems," *Proc. 17th STOC*, pp.291-304 (1985).

[GMW] Goldreich, O., Micali, S., and Wigderson, A., "Proofs that yield nothing but their validity and a methodology of cryptographic protocol design," *Proc. 27th FOCS*, pp.174-187 (1986).

[H] Heath-Brown, D.R., "Almost-primes in arithmetic progressions and short intervals," *Math. Proc. Cambridge Philos. Soc.*, 83, pp.357-375 (1978).

[IT] Itoh, T. and Tsujii, S., "How to generate a primitive root modulo a prime," *SIG Notes of IPS of Japan*, SIGAL9-2 (1989).

[Ka] Kaliski, Jr., B.S., "A pseudo-random bit generator based on elliptic logarithms," *Proc. CRYPTO'86*, pp.84-103 (1986).

[Kn] Knuth, D.E., "The art of computer programming", Vol. 2 Addison-Wesley (1981).

[Ko] Koyama, K., "Speeding the elliptic curve method and its examination of factoring" (in Japanese), *IEICE Technical Report*, ISEC 88-19 (1988).

[KZMI] Kurosaki, M., Zheng, Y. Matsumoto, T. and Imai, H. "Simple Protocol for Showing Membership of Several Groups" (in Japanese), *Proc. 11th Symposium on Information Theory and Its applications (SITA'88)*, pp.585-590 (1988).

[L] Lenstra, Jr. H. W., "Elliptic curve factorization and primality testing," *Proc. of Computational Number Theory Conference* , (1985).

[TW] Tompa, M. and Woll, H., "Random self-reducibility and zero knowledge interactive proofs for possession of information," *Proc. 28th FOCS*, pp. 472-482 (1987).

Appendix

The procedure executed by M_V.

Stage 1 :

$k \leftarrow 0;\ \ell_0 \leftarrow 0;\ \ell_1 \leftarrow 0$

do until $\ell_0 = 1$

    $k \leftarrow k + 2$

    $q \leftarrow kn + 1$

    if $q \in$ PRIME then $\ell_0 \leftarrow 1$

end do

Factorize k as $k = p_1^{m_1} p_2^{m_2} \cdots p_s^{m_s}$.

do until $\ell_1 \neq 0$

    Pick $g \in \mathbf{Z}_q^*$.

    $\ell_1 \leftarrow \prod_{i=1}^{s}(g^{(q-1)/p_i} - 1) \bmod q$

end do

end-Stage 1

Stage 2 :

$\ell_2 \leftarrow 0;\ \epsilon \leftarrow n \bmod k$

do until $\ell_2 \neq 0$

    Pick λ randomly from $\mathbf{Z}_{q-1}^*$.

    $A \leftarrow \lambda \epsilon \bmod q - 1$

    $B \leftarrow \lambda^{-1} \bmod q - 1$

    $u \leftarrow g^A \bmod q;\ v \leftarrow g^B \bmod q$

    $\ell_2 \leftarrow (u^k - 1)(v^k - 1) \bmod q$

end do

end-Stage 2

Stage 3 :

repeat for t rounds

    Pick σ randomly from $\mathbf{Z}_{q-1}$.

    $h \leftarrow g^\sigma \bmod q$

    Evaluate $r = V_\rho^*(n, H, h)$.

    (H is a history just before Step 3.2.)

    $R \leftarrow \sigma + r \bmod q - 1$

    $\ell_3 \leftarrow 0$

    do until $\ell_3 = 1$

        Pick e' randomly from $\{0, 1\}$.

        if $e' = 0$ then

            $x \leftarrow R$

            $y \leftarrow (uvg^z)^z \bmod q$

        else

            $x \leftarrow R + A \bmod q - 1$

            $y \leftarrow (u^{-1}vg^z)^z g^{-n} \bmod q$

        end if

        Evaluate $e = V_\rho^*(n, H, y)$.

        if $e \notin \{0, 1\}$ then

            Append (H, y) to its record.

            Halt.

        else

            if $e' = e$ then

                Append (H, y, e', x) to its record.

                $\ell_3 \leftarrow 1$

            end if

        end if

    end do

end repeat

end-Stage 3

Output its record and halt.

Anonymous One-Time Signatures
and
Flexible Untraceable Electronic Cash

Barry Hayes*
Xerox Palo Alto Research Center

Drop'd at the Exchequer, or thereabouts, or left in a Coach, on the 30th of *July* 1703, an Order on the Third Quarterly Poll N. 826. payable to *Francis Ruffell* Efq; Affign'd to *John Landfell,* and from him to *John Peters*, with other Papers of Concern, being of no Value to any Perfon except the Owner.

The Daily Courant, London, August 4, 1703

1 Introduction

Traditional systems of exchange have many weaknesses. Cash is anonymous, but is prone to loss and theft. Signed notes, such as checks and letters of credit, are only valuable to the payee, but carry an audit trail which tells of past transactions, giving up the privacy of those transactions. Also, security rests on making currency hard to duplicate, and making signatures hard to forge.

This paper introduces a transferable currency based on digital signatures giving theft-resistance and anonymous private transactions. The protocol resembles endorsements of cashier's checks, with the additional feature that, like letters of credit, some portion of the value of the check can be spent and the remainder retained. The currency is easy to duplicate, being just a bit string, but the special structure of the digital signatures will publicly reveal the identity of anyone who assigns the same note more than once.

*This work was funded by Xerox, and prepared under a fellowship from the Northern California Chapter of ARCS Foundation, Inc., and DARPA contract N00014-87-K-0828. The author's current address is Stanford University, Department of Computer Science, Stanford, California 94309, USA.

1.1 Overview

In Section 2 is a short list of previous relevant work. Section 3 describes how conditionally anonymous one-time revokable signatures can be used to build a protocol for payments with most of the advantages of cash, and theft-resistance as well. These signatures, if abused, will reveal information otherwise computationally hidden, and that information can be used to revoke the signatures. A quick description of the digital coins developed by Chaum, Fiat, and Naor is in Section 4. When these coins are used as signatures, they have the properties required. Section 5 presents an improvement which allows all of a user's signature keys to be revoked if that user cheats the system. The protocols for creating signature keys, using them to sign messages, and revoking them are in Section 6.

2   Previous Work

The seminal paper by Diffie and Hellman [DH76] first consolidated the issues involved in digital signatures. To make minimal-knowledge proofs, Rabin [Rab78] introduced *cut and choose,* which we use here to prove that a signature is correctly formed. He also noted that signing a hashed version of a message is as good as signing the message but is more efficient for large messages. Lamport [Lam79] first showed how to use a one-way function to build digital signatures. Chaum, Fiat, and Naor [CFN88] introduced *electronic coins,* concentrating on unconditional anonymity [Cha85], and much of their mechanism is recycled here as signatures. Okamoto and Ohta [OO89] introduced similar coins with the added advantage of *transferablity.* Nontransferable coins, once spent, can only be returned to the bank. Transferable coins can be passed from transaction to transaction. The improvements to revokable keys presented in Section 5 rely on breaking a secret key into pieces. Boyd [Boy88] has explored other protocols involving such keys.

3   Credit and Signatures

Simple and flexible systems of credit have long been built on a system of signed endorsements [AR04]. Some of the participants in the system, the *banks,* are presumed to be solvent and their notes of credit, *bank notes,* will be accepted without question by anyone.

The other users of the system are not presumed solvent and must show bank notes to establish their credit. For example, if Alice has a bank note for $100, she can issue a letter of credit to Bob, and attach her bank note. Bob now has some credit he can prove by showing the bank note and Alice's letter to him. He can even use this credit as backing for a letter of his own. When the chain of notes gets too long the current owner can exchange it for a fresh bank note at the bank.

Theft of these notes can be made unrewarding by including the name of the payee on the note. Any note of credit is payable to only one person, and that person must sign to use the credit or to assign it to another person. The system described so far resembles paper cashier's checks with endorsements.

3.1 Security Without Copy Protection

Cashier's checks are protected by making them difficult to copy, since a $100 note can not validly back up two $100 transactions. When duplication of notes is easy, the bank can not prevent users from using a single bank note repeatedly, but the bank can still discover frauds if the bank notes have a limited lifetime. When a bank note expires and all the notes backed by it have been returned to the bank, a simple audit will tell if anyone has used a note to back more credit than it was worth. For each instance of cheating the audit will find a chain of notes payable to some user, and two or more different notes signed with that signature, passing the value to several different people. Much as with credit cards, the bank must be able to find and penalize a cheater discovered by the audit.

In addition, having notes which are easy to duplicate makes transactions for non-exact amounts simple. Alice can write notes for, say, $30 and $70, attaching these to two copies of a $100 bank note. When the bank runs an audit, rather than just checking that the note was not duplicated, it must check that the sum of all notes backed by this one is $100 or less.

3.2 Aliases and Privacy

While cashier's checks have a great deal of flexibility, there is no privacy. Anyone who has a chain of notes of credit can examine it and learn about previous transactions. Privacy could be restored by letting each user choose a new alias for each transaction, and sign each alias exactly once. The signatures could neither be linked back to the user, nor could two aliases be shown to belong to the same user.

Requiring aliases to be used only once does not reduce the flexibility of the system, since there is never any need for a user to sign two notes with the same alias. If Alice wants to assign the whole amount of a note to Bob, she has spent her credit, and no longer needs the alias associated with the note. In a transaction for a non-exact amount Alice can pay $70 of $100 to Bob and $30 of $100 to herself, but only identifies herself by another alias. When she pays out the remaining $30 she does it using a different alias than she had as the original owner of the $100 note.

Aliases mix well with other signatures in this system. Those not concerned with privacy can use the same notes but sign with their public signatures.

If Alice is concerned about the validity of the note offered by Bob, she can request a bank note made out to her, or one of her aliases, rather than accepting Bob's note. She could insist that Bob go to the bank and sign his note over to the bank in exchange for a fresh note made out to Alice. To the bank, it is just as if Bob is exchanging a used note for a fresh one. Alice gets a note which depends only on the bank's solvency, but involves another party in the transaction. When to use an alias rather than a public name, and when to trust notes taken in good faith are social questions that can only be answered by knowing who ends up paying for overspending and other social decisions. The social engineering needed for these protocols is beyond the scope of this paper.

3.3   Signature Guarantors

Allowing users to generate their own aliases would make signers impossible to identify, but when the returned money is audited, no useful information would be found. The alias of the cheater could have been generated by any user.

There is a middle ground which allows users to generate their own aliases, and still lets an audit work. Since signing more than one note with the same alias is something that is never required, but is something that is associated with cheating, what's needed is an alias which is anonymous if used once, but reveals the identity of the signer if used more than once, a *conditionally anonymous* alias. These can be generated by a user in cooperation with a *signature guarantor*.

In a paper-based system the guarantor could be trusted to enforce the conditional anonymity of the customers. Customers could submit key numbers, which the guarantor would record and print on stickers which are difficult to forge. In accepting a payment, Alice could have it made out to some sticker number, and in signing it over to Bob, would endorse the payment by affixing that sticker. Anyone can make copies of signed documents, but since there is only one sticker, there would be no cheaters for an audit to find.

If Alice were allowed to have several copies of each sticker, she would never need to use more than one copy of each, but if she used the duplicates to overspend, the audit would catch her. Given the proof of cheating, the guarantor would reveal the name of the cheater.

With stickers, the conditional anonymity is in the hands of the guarantor, and the guarantor must be trusted not to reveal the identity of any signer. The guarantor must also be trusted not to make duplicates of Alice's stickers for its own use, or overspend and then falsely accuse Alice of cheating. With digital signature keys replacing the stickers, the conditional anonymity can be built into the signature keys, and the guarantor can not manufacture false accusations.

While each signature key is used only once, we can create n-time signature keys simply by taking a set of $n + 1$ one-time keys and using one of them to sign a single message stating that the other n signatures are to be considered equivalent. Such n-time signature keys can be used to take n notes, each made out to a different one-time alias, and paste them together into one note for use in a large transaction.

3.4 Revokable Signatures

When a cheater is revealed, the guarantor would like to publish the numbers of all the stickers issued to the cheater, revoking their validity as signatures. Bob must check this *hot list* of invalid signatures when getting a signed note from Alice. Time stamps on the transactions and the hot listed signatures could help establish social conventions for making good on bad transactions. The social problem is no worse than credit card abuse.

Signatures for a system of signed notes useful for cash-like anonymous transactions with hot listing of cheaters must be both conditionally anonymous and revokable. A *conditionally anonymous revokable one-time signature key set* (or just *key set*) is defined as a set of signature keys associated with some secret, called the *hot list secret*, such that two or more different messages signed with the same key reveal the secret, but any set of messages, each signed with a different key from the set, do not. In addition, not knowing the secret, it is impossible to identify two keys as belonging to the same key set, but knowing the secret, all keys from the set can be identified.

4 A Protocol for Conditional Anonymity

Chaum, Fiat, and Naor first constructed key sets with the required properties, and used them as coins rather than signatures. What follows is a summary of their construction. These keys are not revokable, but are conditionally anonymous. Revokability will be built on top of this scheme.

In this protocol, f and g are two-argument, collision-free, one-way functions. The function $\oplus$ represents exclusive-or, and $\|$ represents concatenation. The security parameter k is the number of bits signed by each key. A reasonable value would be 200. The guarantor has a public key, the number n_G, and keeps the factorization of n_G a secret.

The m_is are messages which Alice does not, in general, want to reveal. Some will be revealed when the signature key is created, and others will be revealed if Alice cheats. These messages should not be easy to guess, since they are evidence of cheating, but the guarantor must be able to map any message to the customer who created that message.

Signature keys are created with the following protocol:

1. Alice chooses random values a_i, c_i, d_i, and r_i for $i = 1, \ldots, k$.

2. Alice forms and sends to the guarantor k *blinded candidates*

$$B_i = r_i^3 f(g(a_i, c_i), g(a_i \oplus m_i, d_i)).$$

3. The guarantor chooses a random subset R of $k/2$ candidates, and sends this choice to Alice.

4. Alice reveals a_i, c_i, m_i, d_i, and r_i for $i \notin R$.

5. The guarantor checks that the values revealed generate B_i, and that m_i is correctly formed.

6. The guarantor gives Alice

$$\sqrt[3]{\prod_{i \in R} B_i} \quad (\bmod \ n_G),$$

its signature on the chosen product using n_G as a signature key.

7. Alice divides by the r_is to get

$$S_A = \frac{\sqrt[3]{\prod_{i \in R} B_i}}{\prod_{i \in R} r_i} = \prod_{i \in R} \sqrt[3]{f(g(a_i, c_i), g(a_i \oplus m_i, d_i))} \quad (\bmod \ n_G),$$

her signature key.

Alice reorders the $k/2$ factors of S_A so that they are in ascending order. This is the ordering used on the candidates when a message is signed. For simplicity, assume that $R = \{1, \ldots, k/2\}$ and no reordering was required.

To sign a message of length $k/2$ bits,

1. Alice sends S_A to Bob.

2. Bob sends the meassage $\epsilon_1, \ldots, \epsilon_{k/2}$ to Alice.

3. Alice responds, for all $i = 1, \ldots, k/2$:

   a. If $\epsilon_i = 1$, then Alice sends Bob $g(a_i \oplus m_i, d_i)$, a_i, and c_i.

   b. If $\epsilon_i = 0$, then Alice sends Bob $g(a_i, c_i)$, $a_i \oplus m_i$, and d_i.

4. Bob verifies that S_A is of the proper form, and that Alice's responses fit S_A.

Longer messages can be signed by hashing the message to $k/2$ bits, and signing the hashed value. Neither creating signature keys nor signing messages requires any private communication.

If Alice uses the same signature on two different messages then she has revealed, for some i, both a_i and $a_i \oplus m_i$. She also revealed $k/2$ of the ms when the key was created. By showing these $2/k + 1$ messages the guarantor can prove that she has cheated, provided that the ms can not be forged.

If Alice has a digital signature, σ_A, she can choose random integers z_i' and z_i'' and give the guarantor

$$\sigma_A(g(z_1', z_1'') \parallel \cdots \parallel g(z_k', z_k''))$$

along with the blinded candidates. She will be protected from false accusations if she uses

$$m_i = \text{"Alice's account number"} \parallel \text{"transaction number"} \parallel z_i' \parallel z_i''.$$

The guarantor can check that the ms revealed when the key is created are of the correct form, but can not forge them. If Alice cheats, the $k/2+1$ pre-images of g are damning evidence.

5 Revokable Keys

While the basic scheme is very elegant, there is no provision for hot listing a cheater. The original hot listing method [CFN88] creates a batch of signature keys such that abuse of any one of them allows the guarantor to hot list all of them. Unfortunately, the guarantor can not hot list other batches it has made for the cheater.

This section outlines a new hot listing method which allows keys from a key set to be generated independently. The protocol will remain basicly the same, except that the candidates are arranged in an array. The hot list secret is broken up and spread across each row of the array such that revealing all of the ms in any row, will reveal the secret associated with the key set. All but one from each row will be revealed when the signature key is created, and the last will be revealed if the signature is misused. The number of candidates in each row is defined by the security parameter l. A reasonable choice for l would be 2.

The grouping of candidates into an array eliminates the need for reordering the chosen candidates between creating the key and using the key. Instead, the function f is extended with an argument giving, for each candidate, the row that candidate is in. The generated key will have one candidate chosen from each row, and the order is verified by this argument.

5.1 How to Break Up a Secret

As a concrete example, an RSA encryption key [RSA78] can be used as the hot list secret, since the trapdoor information lets the owner break the private key into l pieces which reveal the trapdoor information if and only if they are all revealed.

To do this Alice chooses some $n = pq$, where p and q are large prime numbers suitable for RSA encryption, and $\mathcal{D}$ and $\mathcal{E}$ are such that $\mathcal{D}\mathcal{E} \equiv 1 \pmod{\varphi(n)}$ where φ is Euler's totient function. She gives the guarantor her public key, $\mathcal{E}$ and n, and $\mathcal{D}$ will be her secret.

To break $\mathcal{D}$ into l parts

1. The guarantor chooses $u < n$ and sends this to Alice.

2. Alice finds $\mathcal{D}_1 \times \cdots \times \mathcal{D}_l \equiv \mathcal{D}$ (mod $\varphi(n)$). She can do this by choosing all but one randomly and dividing mod $\varphi(n)$ to get the last. For $j = 1, \ldots, l$, she sends $V_j \equiv u^{\mathcal{D}_j}$ (mod n) to the guarantor.

3. The guarantor chooses one j' from $1, \ldots l$.

4. Alice reveals all the $\mathcal{D}_j$ except $\mathcal{D}_{j'}$.

5. For each $j \neq j'$, the guarantor checks that $V_j \equiv u^{\mathcal{D}_j}$ (mod n). To check $V_{j'}$ it forms $\mathcal{D}_{j'}^{-1} = \mathcal{E} \times \prod_{j \neq j'} \mathcal{D}_j$, and verifies that $V_{j'}^{\mathcal{D}_{j'}^{-1}} \equiv u$ (mod n).

This protocol can be repeated, generating a set of derived secrets like $\mathcal{D}_j$, any one of which will reveal $\mathcal{D}$. For signature keys, one derived secret will be generated for each row of each key, and a single row signs a bit.

A similar construction applies to any trapdoor permutations for which a small number of applications can verify the structure of the permutation. In addition, each $\mathcal{D}_j$ must be hard to compute, even given its inverse $\mathcal{D}_j^{-1}$ and its value at the points chosen to verify the structure $\mathcal{D}_j(u)$.

5.2   The Hot List

The secret revealed when a signature is misused is also used in making the hot list. Redundant information based on the secret is built into each key as yet another argument of f. When the secret is not known the information appears random, but when the secret is revealed by a cheater, the information can be calculated and published.

Again, assume the hot list secret is an RSA key $\mathcal{D}$. Given a one-way function h, and values t_i, then $h(\mathcal{D}(t_i))$ would appear random to someone who knew t_i and h, but not $\mathcal{D}$. However, if $\mathcal{D}$ is revealed, then anyone can calculate $h(\mathcal{D}(t_i))$. The function h and the permutation $\mathcal{D}$ must be chosen to have no strange interactions. For example, h may be $\mathcal{D}^{-1}$. In this case, $h(\mathcal{D}(t_i))$ is far from random.

Since the t_is are made public, there is no need for them to be random. They could be generated by some one-way function on i chosen for each transaction. If this is done, the t_is need not be recorded, just the function. If the function also takes some transaction identifier as input, it can be used for any transaction. Once $\mathcal{D}$ is revealed, only the transaction identifier is needed to regenerate the t_is and generate the redundant information.

6 Putting It All Together

A guarantor chooses security parameters k, the number of bits a signature key signs, and l, the number of parts the hot list secret will be broken into. Reasonable values would be $k = 200$ and $l = 2$. It also chooses collision-free one-way functions f, g, and h. Intuitively, these could be "good" one-way hash functions. To open an account with a guarantor, Alice, who has digital signature σ_A, chooses a trapdoor permutation $\mathcal{E}$ over some set U. She can calculate $\mathcal{D} = \mathcal{E}^{-1}$, which is her hot list secret. She sends the guarantor $\sigma_A(\mathcal{E} \parallel U)$.

Signature keys are created with the following protocol:

1. The guarantor chooses a unique transaction identifier $u \in U$ and sends this to Alice.

2. Alice finds k ways to break up $\mathcal{D}$ into l parts such that for all $i = 1, \ldots, k$

$$\mathcal{D}_{i,1} \circ \cdots \circ \mathcal{D}_{i,l} = \mathcal{D},$$

where each $\mathcal{D}_{i,j}$ must be hard to discover given its inverse and its value for u.

3. Alice forms and sends to the guarantor a k-by-l verifier array, V, such that

$$V_{i,j} = \mathcal{D}_{i,j}(u),$$

4. Alice chooses random values $a_{i,j}$, $c_{i,j}$, $d_{i,j}$, $t_{i,j}$, $z_{i,j}$, and $r_{i,j}$ for all $i = 1, \ldots, k$ and $j = 1, \ldots, l$.

5. Alice forms and sends to the guarantor a k-by-l candidate array, C, where each candidate has the form

$$B_{i,j} = r_{i,j}{}^3 f(g(a_{i,j}, c_{i,j}), g(a_{i,j} \oplus m_{i,j}, d_{i,j}), h(\mathcal{D}(t_{i,j})), i)$$

and

$$m_{i,j} = u \parallel \mathcal{D}_{i,j} \parallel \sigma_A(u \parallel z_{i,j}).$$

6. Alice reveals all the $t_{i,j}$s.

7. The guarantor chooses a function J which picks one element from each row.

8. Alice reveals $a_{i,j}$, $c_{i,j}$, $m_{i,j}$, $d_{i,j}$, $z_{i,j}$, and $r_{i,j}$ for all i and j such that $J(i) \neq j$.

9. The guarantor checks that the values generate $B_{i,j}$, and that $m_{i,j}$ is correctly formed.

10. The guarantor checks that the verifier array is correct. It can directly check $\mathcal{D}_{i,j}$ for $j \neq J(i)$, and check the rest by checking that $\mathcal{D}_{i,J(i)}{}^{-1}(V_{i,J(i)}) = u$.

11. The guarantor gives Alice

$$\sqrt[3]{\prod_{1 \le i \le k} B_{i,J(i)}} \pmod{n_G}.$$

This is the guarantor's signature on the chosen product.

12. Alice divides out her $r_{i,j}$s to get

$$S_A = \sqrt[3]{\prod_{1 \le i \le k} f(g(a_{i,J(i)}, c_{i,J(i)}), g(a_{i,J(i)} \oplus m_{,i,J(i)} d_{i,J(i)}), h(\mathcal{D}(t_{i,J(i)})), i),}$$

her signature key. The vector of the $h(\mathcal{D}(t_{i,J(i)}))$ values is called the *hot list vector* of S_A.

The guarantor need not keep track of anything but the $m_{i,j}$s of the candidates which have been revealed, and the $t_{i,j}$s of the candidates which have not been revealed. It will have to use these to put Alice's signatures on the hot list if she ever cheats.

6.1 Signing a Message

Alice signs a k-bit message using S_A in nearly the same way she signed in Section 4. To sign a message of length k bits,

1. Alice sends S_A to Bob, along with its hot list vector.

2. Bob sends the message $\epsilon_1, \ldots, \epsilon_k$ to Alice.

3. Alice responds, for all $i = 1, \ldots, k$:

   a. If $\epsilon_i = 1$, then Alice sends Bob $g(a_{i,J(i)} \oplus m_{i,J(i)}, d_{i,J(i)})$ $a_{i,J(i)}$, and $c_{i,J(i)}$.

   b. If $\epsilon_i = 0$, then Alice sends Bob $g(a_{i,J(i)}, c_{i,J(i)})$, $a_{i,J(i)} \oplus m_{i,J(i)}$, and $d_{i,J(i)}$.

4. Bob verifies that S_A is of the proper form, that Alice's responses fit S_A, and that S_A has not been put on the hot list.

The probability of creating a false signature is under the control of the guarantor, who fixes the values of l and k. If a signature has two or more ill-formed candidates in the same row, Alice will certainly be caught, since at least one will be revealed when the signature is submitted for endorsement. If there is one ill-formed candidate in each of k' rows, Alice will be caught at creation time with probability $1 - (\frac{1}{l})^{k'}$. If she manages to get an endorsed signature with k' ill-formed factors and signs two random messages with this signature, her identity will be revealed with probability $1 - (\frac{1}{2})^{k-k'}$.

6.2 Hot Listing a Cheater

If Alice signs two random messages with the same signature, then with probability $1 - (\frac{1}{2})^k$ some corresponding bits in the messages will differ. Alice will have then revealed, albeit in two separate transactions, both $a_{i,J(i)}$ and $a_{i,J(i)} \oplus m_{i,J(i)}$. Since the protocols for using these one-time signatures ensure that eventually both these values get back to the bank, at that point the bank can generate

$$m_{i,J(i)} = u \parallel \mathcal{D}_{i,J(i)} \parallel \sigma_A(u \parallel z_{i,J(i)}).$$

The guarantor, given u, can look up the other $\mathcal{D}_{i,j}$s revealed when the signature was created. It can now form Alice's hot list secret, $\mathcal{D} = \mathcal{D}_{i,1} \circ \cdots \circ \mathcal{D}_{i,l}$. To hot list Alice's signatures the guarantor looks up the $t_{i,J(i)}$s for each signature it has issued Alice, calculates the hot list vector for that signature, and thus withdraws its endorsement. The $k + 1$ signed z values prove that the guarantor has not framed Alice.

When Bob accepts a signature, he checks that the hot list vector does not match any of the hot listed vectors. Since Alice may not have generated all of her $t_{i,J(i)}$s as she should have, Bob must also check that no hot listed vector shares a large number of components with the suspect signature. The entire signature verification procedure involves a small fixed number of multiplications, a small fixed number of applications of hash functions, and a search through the hot list, and could easily be done by a small processor.

7 Future Work

Formal proofs for requirements of the permutations used to break up a secret remain to be done, as well as identification of the assumptions needed to use RSA for these permutations. The design of the hash h is tricky, since it interacts with $\mathcal{D}$.

It should be possible to design key sets which do not require the guarantor to interact with its customers for each signature key created without creating all the signature keys in one batch. It would be sufficient if the functions f and g could interact in such a way that $f(g(a, c), (g(a \oplus m, d))$ would reveal just enough about m to prove the validity of the signature, without revealing m itself. The guarantor then has only to issue key sets, and look for cheaters to hot-list. It will still have to keep track of every used signature, to look for multiple uses, but will be out of the loop when a contract is signed.

8 Acknowledgement

It was a pleasure to discuss this work with Ralph Merkle, who asked the right questions at the right time, and made comments on several drafts.

References

[AR04]   Anne, queen of England. An act for giving like remedy upon promissory notes, as is now used upon bills of exchange, and for the better payment of inland bills of exchange. 1704. 3 & 4 Anne, chapter 9.

[Boy88]   Colin Boyd. Some applications of multiple key ciphers. In Christoph G. Gunther, editor, *Advances in Cryptogrophy: EUROCRYPT '88*, pages 455–467, Springer-Verlag, 1988.

[CFN88] David Chaum, Amos Fiat, and Moni Naor. Untracable electronic cash. In *Crypto '88*, 1988.

[Cha85]   David Chaum. Security without identification: transaction systems to make big brother obsolete. *Communications of the ACM*, 1030–1044, October 1985.

[DH76]   Whitfield Diffie and Martin E. Hellman. New directions in cryptography. *IEEE Transactions on Information Theory*, IT-22:644–654, November 1976.

[Lam79] Leslie Lamport. *Constructing digital signatures from a one way function*. Technical Report CSL-98, SRI International, October 1979.

[OO89]   Tatsuaki Okamoto and Kazuo Ohta. Disposable zero-knowledge authentications and their applications to untracable electronic cash. In *Crypto '89*, 1989.

[Rab78]   Michael O. Rabin. Digitalized signatures. In *Foundations of Secure Computing*, pages 155–166, Academic Press, Inc., 1978.

[RSA78] Ronald Rivest, Adi Shamir, and Leonard Adleman. A method for obtaining digital signatures and public-key cryptosystems. *Communications of the ACM*, 120–126, February 1978.

SECTION 8

THEORY

Dyadic Matrices and Their Potential Significance in Cryptography (Abstract)

Yang Yi Xian
P.O. Box 145
Dept of Information Engineering
Beijing Univ. of Posts and Telecomm.
Beijing, P.R.China

The concept of dyadic matrix was initially proposed for the purpose of design orthogonal matrices and multi-valued codes with perfect autocorrelation property [1,2]. By now much work has been done in this area. In the following text we will show some other new results about the reversibility of dyadic metrices and their potential significance in cryptography.

1 New Results On Dyadic Matrices

Definition Let $a(0), a(1), ...$ is a sequence over $GF(p)$ with period 2^n. For any $0 \leq i, j \leq 2^n - 1$, let $A(i,j) = a(i \oplus j)$, then the $2^n \times 2^n$ matrix $A = [A(i,j)]$ is called as the dyadic matrix of the corresponding sequence. Where $i \oplus j$ means the dyadic sum of integers i and j.

Theorem 1 For any dyadic matrix $A = [A(i,j)] = [a(i \oplus j)]$ in $GF(p)$,

$$H \begin{bmatrix} B_0 & & 0 \\ & \cdot & \\ & & \cdot \\ 0 & & B_{2^n-1} \end{bmatrix} = AH$$

Where H is the $2^n \times 2^n$ Walsh-Hadamard matrix, B_i is the i'th Walsh-Hadamard matrix spectrum of the sequence $a(0), a(1), ...,$ i.e. $B_j = \sum_{k=0}^{2^n-1} a(k)(-1)^{k \odot j}$.

Theorem 2 Dyadic matrix $A = [A(i,j)] = [a(i \oplus j)]$ in $GF(p)$, ($p = q^m$, q is an odd prime) is invertable iff none of the W-H spectrum coefficients of a sequence $a(0), a(1), ...$ equals zero.

Note: Theorem 2 is not true in the finite field GF(2).

Theorem 3 If $p = q^m$, q is odd prime and A is a $2^n \times 2^n$ matrix in $GF(p)$ then A is dyadic matrix iff there exist some $b_0, ..., b_{2^n-1}$ in $GF(p)$ such that:

$$A = \frac{1}{2^n} H \cdot \begin{bmatrix} b_0 & & 0 \\ & \cdot & \\ & & \cdot \\ 0 & & b_{2^n-1} \end{bmatrix} \cdot H$$

Note: Theorem 3 is not true in the finite field $GF(2)$.

Theorem 4 *Let A, B are dyadic matrices in $GF(p)$, $(p = q^m$, q is odd prime), then AB and $A + B$ are also dyadic matrices. If A is invertable then A^{-1} is also a dyadic matrix.*

Theorem 5 *Let A is a dyadic matrix in $GF(3)$. If A is invertable then $A^{-1} = A$.*

Theorem 6 *If A is a dyadic matrix in $GF(p)$, $(p = q^m$, q is odd prime), then $A = kA^{-1}$. $(k \in GP(p))$ (i.e. the corresponding sequence is dyadic codes) iff $kB_i^2 = 1$ $(0 \le i \le 2^n - 1)$. Where B_i's are the W-H spectrum coefficients of A.*

All the above results are for the finite field $GF(p)$ $(p = q^m$, q is odd prime). How about the case of $GF(2)$? The following theorems will give the answer.

Theorem 7 *Let $A = [A(i,j)] = [a(i \oplus j)]$ is a dyadic matrix in $GF(2)$, then A is invertable iff $\sum_{i=0}^{2^n-1} a(i) = 1$ (mod 2). i.e. the number of 1's in the first row of A equals odd.*

Theorem 8 *A is a dyadic matrix in $GF(2)$. If A is invertable then $A^{-1} = A$.*

2 Potential Significance in Cryptography

Let $S_m = \{(a(0), ..., a(2^n - 1)) : \sum_k a(k)(-1)^{k \odot m} = 0$ (mod p), $a(i) \in GF(p)\}$ $r_m(0) = (2^n - 1)$ mod p, $r_m(i) = -\text{Wal}(i, m) = -(-1)^{i \odot m}$ mod p, $(1 \le i \le 2^n - 1)$ and let $R = [R(i,j)] = [r_m(i \oplus j)]$ is the dyadic matrix produced from the sequence $r_m(0), r_m(1), ..., r_m(2^n - 1)$. It's easy to show that the dyadic matrix is singular.

Theorem 9 *Let $(a(0), ..., a(2^n - 1)) \in S_m$ and $A = [A(i,j)] = [a(i \oplus j)]$ then dyadic matrix A is singular and $AR = A = RA$.*

From the above theorem 9, it's easy to see that the special dyadic matrix A act as the pseudo-unite matrix.

Theorem 10 *Let $A_m(j) = \sum_i \text{Wal}(j, i) q_i$ mod $p = \sum_i (-1)^{j \odot i} q_i$ mod p, $B_m(j) = \sum_i \text{Wal}(j, i) q_i^{-1}$ mod $p = \sum_i (-1)^{j \odot i} q_i^{-1}$ mod p. $(0 \le j \le 2^n - 1$, $q_i \ne 0)$ and $A = [A(i,j)] = [A_m(i \oplus j)]$, $B = [B(i,j)] = [B_m(i \oplus j)]$ are two dyadic matrices. Then we have*

1. *$AB = R$ where R is the above pseudo-unite matrix.*

2. *A and B are two singular matrices.*

3. *A is unqiuely determined by B, and B is also uniquely determined by A.*

4. *There are $(p-1)^{2^n}$ different pairs of A and B.*

By the above theorems 9 and 10, a new cryptosystem can be constructed as follows:

Let S_m is the message space, A_1, B_1 and A_2, B_2 are two pairs of matrices in theorem 10 such that $A_1 B_1 = A_2 B_2 = R$ and $B_1 B_2 = B_2 B_1$, where R is the pseudo-unite matrix in theorem 9. Take A_1, B_1 as the secret key of user 1 and A_2, B_2 as that of user 2, then by the following four steps the user 1 can securely transfer any message $p = (m(0), m(1), ..., m(2^n - 1)) \in S_m$ to user 2.

Let $M = [M(i,j)] = [m(i \oplus j)]$ is the dyadic martix of the message $p \in S_m$.

Step 1 User 1 calculates $e_1 = A_1 M$ and sends e_1 to user 2 through public channel.

Step 2 User 2 calculates $e_2 = B_2 e_1$ and sends e_2 to user 1 through public channel.

Step 3 User 1 calculates $e_3 = B_1 e_2$ and sends e_3 to user 2 through public channel.

Step 4 User 2 finishes the descryption by calculating $A_2 e_3 = M$.

In fact $A_2 e_3 = A_2 B_1 e_2 = A_2 B_1 B_2 e_1 = A_2 B_1 B_2 A_1 M = (A_2 B_2)(B_1 A_1)M = RRM = RM = M$. (The last equations are due to $B_1 B_2 = B_2 B_1$ and theorem 9 and theorem 10.)

Note: the message space for this cryptosystem is S_m but the 2^n-dimensional linear space.

The security of the above cryptosystem is based on the next facts:

Fact 1 There exists many many possible pairs of matrices A and B (by theorem 10). So it's very difficult for the breaker to determine which pair of matrices are being used.

Fact 2 When matrix A keeps unknown, the breaker can't compute the solutions of the equation $Y = AX$, even if he knows the cipher Y.

Fact 3 For any singular matrix A, there exists many many solutions of the equation $Y = AX$ so the breaker can't find the true original message.

References

[1] Yang Yi Xian & Hu Zheng Ming, 1987 Int. Conference on Commun. Tech. (ICCT'87), Nanjing, 1987.

[2] Hu Zheng Ming, 1988 Int. Sympo. on Inform. Theory, (ISIT'88) Japan, 1988.

A Note on Strong Fibonacci Pseudoprimes

RUDOLF LIDL
Dept. of Mathematics
University of Tasmania
Hobart, Tas. 7001
Australia

WINFRIED B. MÜLLER
Institut für Mathematik
Universität Klagenfurt
A-9022 Klagenfurt
Austria

Abstract: After summarizing some results on pseudoprimes, a characterization is presented of Fibonacci pseudoprimes of the bth kind for all integers b. Subsequently, some generalizations of the concept of strong pseudoprimes are established.

1. Pseudoprimes

Several convential and public-key cryptosystems require the provision of large "random" prime numbers. There also exist a number of probabilistic tests for testing a given large integer for primality. *Fermat's Little Theorem* plays an important role in motivating the definition of pseudoprimes and generalizations. It tells us that for a prime number n and an integer b with $gcd(b, n) = 1$ the following congruence holds

$$b^{n-1} \equiv 1 \bmod n \tag{1.1}$$

An alternative presentation of Fermat's Little Theorem says that if n is a prime then $b^n \equiv b \bmod n$ for any integer b. If n is not prime then it is still possible, but not very likely, that (1.1) holds. Composite integers, i.e. non-prime integers satisfying (1.1) and thus pretending to be prime numbers get a special name.

Definition 1.1. If n is an odd composite number and b is an integer such that $gcd(b,n) = 1$ and (1.1) is satisfied then n is called a *pseudoprime to the base b*.

It can be shown that n does not satisfy congruence (1.1) for at least half of the possible bases b if n fails (1.1) for one $b \in \mathbf{Z}_n^*$. ($\mathbf{Z}_n^*$ denotes the set of all nonzero residue classes modulo the integer n.) We know that n is composite if (1.1) does not hold for any one chosen b. Property (1.1) can be used for a probabilistic primality test as follows: If n is a pseudoprime for k randomly chosen b then the chance that n is still composite (despite satisfying (1.1) for k different bases) is at most $\frac{1}{2^k}$, unless n is a pseudoprime for every $b \in \mathbf{Z}_n^*$. A *Carmichael number* is a composite integer n such that (1.1) holds for every $b \in \mathbf{Z}_n^*$; $n = 561 = 3.11.17$ is the smallest Carmichael number.

Carmichael numbers have the following properties (see KOBLITZ [6]):

1. They are squarefree.
2. A squarefree n is a Carmichael number if and only if $p - 1 \mid n - 1$ for every prime divisor p of n.
3. A Carmichael number must be the product of at least three distinct primes.

In 1988 DI PORTO & FILIPPONI [3] proposed a method for finding large probable primes that generalizes the notion of pseudoprimes. Let b and c be nonzero integers, let α and β be the roots of $x^2 - bx + c$ with nonzero discriminant. The sequence of numbers

$$g_n(b,c) = \alpha^n + \beta^n, \quad n \geq 0$$

is called the *Lucas sequence associated with the pair (b,c)*, see RIBENBOIM [13]. The $g_n(b,c)$ are integers since they can be constructed via the recurrence relation

$$g_{n+2}(b,c) = b\,g_{n+1}(b,c) - c\,g_n(b,c)$$

with initial conditions $g_0(b, c) = c$ and $g_1(b, c) = b$. The special case $g_n(1, -1)$ gives the usual *Lucas numbers*. Lucas numbers and related primality testing algorithms were considered already by BAILLIE & WAGSTAFF. [1].

We can use *Waring's formula* (which represents a relationship between the elementary symmetric functions of the roots of a quadratic polynomial and the sum of the nth powers of these roots, see LIDL & NIEDERREITER [10], p.30 and p.355) to represent the numbers in a Lucas sequence in terms of the polynomial

$$g_n(x, c) = \sum_{i=0}^{[n/2]} \frac{n}{n - i} \binom{n - i}{i} (-c)^i x^{n-2i}$$

evaluated at $x = b$. These polynomials are called *Dickson polynomials (of the first kind) of degree n in x with parameter c*. They have been widely studied in the past three decades, see e.g. [10], pp.355-361. We note that the special parameter $c = 0$ gives the power polynomial $g_n(x, 0) = x^n$. Dickson polynomials have been used in [8] for defining generalizations of pseudoprimes. We follow the definition of DI PORTO & FILIPPONI [3] and [4].

Definition 1.2. An odd composite integer n is called a *Fibonacci pseudoprime* (in short *Fpsp*) *of the bth kind* if

$$g_n(b, -1) \equiv b \bmod n \tag{1.2}$$

Numerical tests show that Fibonacci pseudoprimes are rare.

2. Strong Fibonacci Pseudoprimes

In this section we give the definition and a number theoretic characterization of Fibonacci pseudoprimes of the bth kind for all $b \in \mathbf{Z}$.

Definition 2.1. A Fibonacci pseudoprime n of the bth kind for all b in the range $1 \le b \le M$ (for a given integer M) is called an *M-strong Fibonacci pseudoprime*. For $M = n - 1$ we call it a *strong Fibonacci pseudoprime* (in short *strong Fpsp*).

FILIPPONI [5] calculated that there are no *strong Fpsp* $\le 10^8$. (According to a private communication from Filipponi there do not exist any Carmichael numbers $n \le 10^{13}$ which satisfy the congruence (1.2) for all $b \in \mathbb{Z}_n$. Hence, by the following Corollary 2.3 there are no *strong Fpsp* $n \le 10^{13}$.) DI PORTO & FILIPPONI [4] found a 29-*strong Fpsp* with 98-digits which is not a 30-*strong Fpsp*. The main point of this note is to present a characterization of *strong Fpsp* which should make it possible to exhibit further properties of such numbers. For this task it is convenient to utilize properties of the Dickson polynomials $g_n(x, -1)$. We note

Lemma 2.1. *The following conditions are equivalent:*

(i) $g_n(b, -1) \equiv b \bmod n$ for all $b \in \mathbb{Z}_n$.

(ii)   *The map* $g_n : b \rightarrow g_n(b, -1)$ *induced by* $g_n(x, -1)$ *on* $\mathbb{Z}_n$ *is the identity.*

(iii) $g_n : b \rightarrow g_n(b, -1)$ *is a permutation of* $\mathbb{Z}_n$ *with n fixed points.*

It is known (see NÖBAUER [12]) that g_n is a permutation of $\mathbb{Z}_n$ if and only if

$$gcd(n, \prod p_i^{e_i-1}(p_i - 1)) = 1,$$

where $n = \prod p_i^{e_i}$ is the canonical factorization. An explicit formula for the numbers of fixed points of the permutation g_n is also known, see [12]. Therefore we can show

Theorem 2.2. *An odd integer n is a strong Fibonacci pseudoprime if and only if*

(i) n *is square-free,*

(ii) $(p_i - 1) \mid (n - 1)$ *for every prime p_i dividing n,*

(iii) *either n is the product*

of an arbitrary number of distinct odd primes $p_i \equiv 1$ mod 4 which satisfy

$$(p_i + 1) \mid (n - 1) \quad or \quad (p_i + 1) \mid (n + 1)$$

and of an even number (including 0) of primes $p_i \equiv 3$ mod 4 which satisfy

$$2(p_i + 1) \mid (n - 1)$$

or n is the product of an odd number of primes $p_i \equiv 3$ mod 4 which satisfy

$$2(p_i + 1) \mid (n - p_i) .$$

As simple consequences of this result (see [9] for detailed proofs) we state

Corollary 2.3.

(i) *Any strong Fpsp is a Carmichael number and thus must be a product of at least three distinct primes.*

(ii) *Any strong Fpsp n of the form $n \equiv 3$ mod 4 must be the product of at least five distinct primes.*

(iii) *The highest power of 2 dividing n must also divide $p_i + 1$ for each prime divisor p_i of n.*

3. Possible Generalizations

The approach of section 2 for defining *Fpsp* and *strong Fpsp* can be generalized by using Dickson polynomials in n variables. We briefly indicate the approach but omit further details. Let $b_1, ..., b_{m+1}$ be nonzero integers, and let $\alpha_1, ..., \alpha_{m+1}$ be the roots of the polynomial $x^{m+1} - b_1 x^m + ... + (-1)^{m+1} b_{m+1}$. Then

$$g_n^{(i)}(b_1, ..., b_m, b_{m+1}) = \sigma_i(\alpha_1^n, ..., \alpha_{m+1}^n), \quad i = 1, ..., n + 1, \tag{3.1}$$

defines a set of integers, where σ_i denotes the ith elementary symmetric function of the nth powers $\alpha_1^n, ..., \alpha_{m+1}^n$ of the roots $\alpha_1, ..., \alpha_{m+1}$. Replacing $b_1, ..., b_m$ by $x_1, ..., x_m$, in turn, and using Waring's formula we obtain *Dickson polynomials in m variables*. If the parameter b_{m+1} is not equal to -1 these polynomials are also referred to as *Chebyshev polynomials*, see LIDL & WELLS [11], LIDL [7] for explicit expressions and recurrence relations.

An odd composite n can be regarded as a *generalized Fpsp of the kind* $(b_1, ..., b_m, -1)$ if

$$g_n^{(i)}(b_1, ..., b_m, -1) \equiv b_i \bmod n \qquad (3.2)$$

for $i = 1, 2, ..., m$.

The integer n is a *strong generalized Fpsp* if (3.2) holds for all $1 \leq b_i \leq n - 1$.

Another generalization is based on rational functions with coefficients in **Z**. We refer to [8] for more details. Let t be a nonsquare in **Z**, $\Theta = \sqrt{t}$, then the Redei function $f_n(x)$ is defined as

$$f_n(x) = \Theta \frac{(x + \Theta)^n + (x - \Theta)^n}{(x + \Theta)^n - (x - \Theta)^n}$$

VON ZUR GATHEN [14] gives the following recursive method for finding $f_n(x)$, where $f_n(x) = \frac{r_n(x)}{s_n(x)}$:

$$\begin{pmatrix} r_n(x) \\ s_n(x) \end{pmatrix} = \begin{pmatrix} x & \Theta \\ 1 & x \end{pmatrix}^n \begin{pmatrix} 1 \\ 0 \end{pmatrix}$$

We define a *Redei pseudoprime* (in short *Rpsp*) *of the kind* b, t as a composite n such that

$$f_n(b) \equiv t^{\frac{1-n}{2}} b \bmod n \qquad (3.3)$$

A *strong Rpsp* n satisfies (3.3) for all $b \in \mathbf{Z}_n$.

It is an open problem to give a characterization for *strong Rpsp*, although criteria for permutations f_n and fixed points of f_n are known.

References

[1] BAILLIE, R., WAGSTAFF JR., S.S.: Lucas pseudoprimes. Math.of Comp. **35**, 1391–1417 (1980).

[2] DICKSON, L.: The Analytic Representation of Substitutions on a Power of a Prime Number of Letters with a Discussion of the Linear Group I. Ann.of Math. **11**, 65–120 (1896).

[3] DI PORTO, A., FILIPPONI, P.: A Probabilistic Primality Test Based on the Properties of Certain Generalized Lucas Numbers. In: Advances in Cryptology – Eurocrypt'88, Lecture Notes in Computer Science **330**, Springer–Verlag, New York–Berlin–Heidelberg, pp. 211–223, 1988.

[4] DI PORTO, A., FILIPPONI, P.: Extended Abstract, Eurocrypt'89, Houthalen (Belgium), April 10–13, 1989.

[5] FILIPPONI, P.: Table of Fibonacci Pseudoprimes to 10^8. Note Recensioni Notizie **37**, no. 1–2, 33–38 (1988).

[6] KOBLITZ, N.: A Course in Number Theory and Cryptography. Springer–Verlag, New York–Berlin–Heidelberg, 1987.

[7] LIDL, R.: Tschebyscheffpolynome in mehreren Variablen. J.reine angew.Math. **273**, 178–198 (1975).

[8] LIDL, R., MÜLLER, W.B.: Generalizations of the Fibonacci Pseudoprimes Test. To appear.

[9] LIDL, R., MÜLLER, W.B., OSWALD, A.: Some Remarks on Strong Fibonacci Pseudoprimes. To appear in Applicable Algebra in Engineering, Communication and Computer Science **1** (1990).

[10] LIDL, R., NIEDERREITER, H.: Finite Fields. Addison Wesley, Reading, 1983. (Now published by Cambridge University Press, Cambridge.)

[11] LIDL, R., WELLS, C.: Chebyshev polynomials in several variables. J.reine angew.Math. **255**, 104–111 (1972).

[12] NÖBAUER, R.: Über die Fixpunkte einer Klasse von Dickson–Permutationen. Sb.d.Österr.Akad.d.Wiss., math.-nat.Kl., Abt.II, Bd. **193**, 521–547 (1984).

[13] RIBENBOIM, P.: The Book of Prime Number Records. Springer–Verlag, New York–Berlin–Heidelberg, 1988.

[14] VON ZUR GATHEN, J.:Testing permutation polynomials. Proc. 30 Annual IEEE Symp. Foundations of Computer Science, pages 88–92, Research Triangle Park, NC, 1989

On the Significance of the Directed Acyclic Word Graph in Cryptology

CEES J.A. JANSEN

Philips USFA B.V.

P.O. Box 218, 5600 MD Eindhoven

The Netherlands

DICK E. BOEKEE

Technical University of Delft

P.O. Box 5031, 2600 GA Delft

The Netherlands

Summary

Blumer's algorithm can be used to build a Directed Acyclic Word Graph (DAWG) in linear time and memory from a given sequence of characters. In this paper we introduce the DAWG and show that Blumer's algorithm can be used very effectively to determine the maximum order (or nonlinear) complexity profile of a given sequence. We also show that this algorithm can be used to determine the period of a periodic sequence in linear time and memory. It also appears that the DAWG is an even more efficient means of generating the sequence, given a number of characters, than e.g. the nonlinear feedback shift register equivalent of that sequence, as it always needs the least amount of characters to generate the remainder of the sequence.

1 Introduction

In [Jans 89] a new complexity measure for sequences is proposed, called Maximum Order Complexity, which is equal to the length of the shortest (possibly nonlinear) feedback shift register that can generate a given sequence. Analogous to Rueppel's linear complexity profile [Ruep 84], the maximum order complexity profile is proposed as a measure of "goodness" for sequences, i.e. a measure which shows how well a given sequence resembles a real random sequence. For the purpose of determining the M.O. complexity profile an algorithm is proposed which is linear in the sequence length with respect to total processing time and memory usage. It turns out that this algorithm has many other interesting properties, such as the ability to determine the period of a periodic sequence, and generating the sequence based on the least number of observed characters.

2 The Directed Acyclic Word Graph

In [Blum 83] Blumer et al. describe a linear-time and -memory algorithm to build a *Directed Acyclic Word Graph* (DAWG) from a given string of letters. This DAWG is then used to recognize all substrings (or words) in the string, or for source coding purposes as in [Blum 85].

The DAWG consists of at most $2l$ nodes connected by at most $3l$ edges, where l is the length of the string. The nodes represent equivalence classes of substrings and the edges are labeled with string letters. An edge points from one node to another if and only if the first equivalence class contains a substring, which extended with the edge's letter belongs to the other equivalence class. The suffix pointer is an edge which points from a node to the node representing the equivalence class with the longest common suffix of all strings of the first node's equivalence class. Two substrings are defined to be equivalent if and only if their endpoint sets are equal. An endpoint set of a given substring is defined as the set containing all positions within a string where the given substring ends.

Example 1 The string $w = 110100$ gives rise to the following set of all possible substrings, denoted $SUB(w)$:

$$SUB(w) = \{\lambda, 1, 0, 11, 10, 01, 00, 110, 101, 010, 100,$$
$$1101, 1010, 0100, 11010, 10100, 110100\}.$$

Here λ denotes the empty string. The endpoint positions within a string are denoted as follows:

$$
\begin{array}{ccccccccl}
1 & & 1 & & 0 & & 1 & & 0 & & 0 & & : w, \\
0 & 1 & & 2 & & 3 & & 4 & & 5 & & 6 & & : \text{endpoints.}
\end{array}
$$

For the given string w the endpoint sets for all substrings are:

$E_w(\lambda) = \{0,1,2,3,4,5,6\}$,
$E_w(1) = \{1,2,4\}$, $E_w(0) = \{3,5,6\}$,
$E_w(11) = \{2\}$, $E_w(10) = \{3,5\}$,
$E_w(01) = \{4\}$, $E_w(00) = \{6\}$,
$E_w(110) = \{3\}$, $E_w(101) = \{4\}$, $E_w(010) = \{5\}$, $E_w(100) = \{6\}$,
$E_w(1101) = \{4\}$, $E_w(1010) = \{5\}$, $E_w(0100) = \{6\}$,
$E_w(11010) = \{5\}$, $E_w(10100) = \{6\}$,
$E_w(110100) = \{6\}$.

From these endpoint sets the following substrings are seen to be equivalent:

$$
\begin{array}{ccccccccc}
01 & \equiv_w & 101 & \equiv_w & 1101, & & & & \\
00 & \equiv_w & 100 & \equiv_w & 0100 & \equiv_w & 10100 & \equiv_w & 110100, \\
& & 010 & \equiv_w & 1010 & \equiv_w & 11010. & &
\end{array}
$$

As can be seen the 17 substrings form 9 equivalence classes. It is customary to represent each equivalence class by its shortest substring. Doing so we have the following set of equivalence classes of w, denoted by $EQ(w)$:

$$EQ(w) = \{\lambda, 1, 0, 11, 10, 01, 00, 110, 010\}.$$

Figure 1 shows the corresponding DAWG of w, where the dotted lines with arrows are the suffixpointers.

The edges of a DAWG are divided into primary and secondary edges. An edge is called primary if and only if it belongs to a primary path, which is the longest path from the source to a node. With the length of a path the number of edges in that path is meant. The depth of a node is the length of the primary path from the source to that node. In the foregoing example the depths of the nodes, denoted by $d(\cdot)$ are as follows:

$d(\lambda) = 0$, $d(1) = 1$, $d(0) = 1$,
$d(11) = 2$, $d(10) = 2$, $d(01) = 4$, $d(00) = 6$,
$d(110) = 3$, $d(010) = 5$.

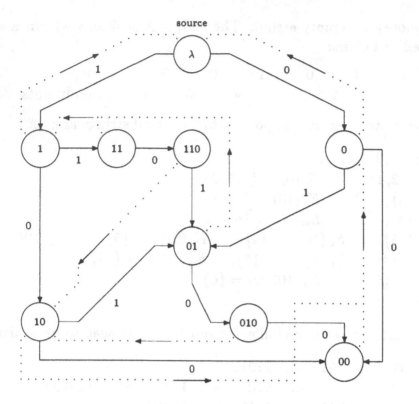

Figure 1: DAWG of 110100

Looking at the nodes with more than one outgoing edge, i.e. λ, 1,0 and 10, it can be seen that 10 is the deepest node with this property.

Let $BN(w)$ be defined as the set of equivalence classes that have more than one outgoing edge. We call this set the set of branchnodes of w. Clearly, $BN(w) \subset EQ(w)$. The maximum depth $\hat{d}(w)$ of a string w is defined as:

$$\hat{d}(w) := \max_{a \in BN(w)} d(a).$$

In Example 1 we have $\hat{d}(110100) = 2$.

Blumer's algorithm, which builds a DAWG in linear time and memory is presented in [Blum 83]. It consists of three procedures that can readily be programmed in a computer programming language like PASCAL or C.

3   The DAWG and the Complexity Profile

From the foregoing section it should already be clear that the DAWG is a useful tool to determine the maximum order complexity of a given sequence. In this section it is shown that Blumer's algorithm can indeed be exploited to determine the maximum order complexity profile of any sequence in linear time and memory.

For the sake of uniformity in notation we will speak of a sequence $\underline{s}$ instead of a string w and of a subsequence instead of a word or substring. All set definitions of the previous subsection are translated in this way. The following proposition relates the complexity of a sequence to its maximum depth in the DAWG.

Proposition 1 *The complexity $c(\underline{s})$ of a sequence $\underline{s}$ with characters from some finite alphabet $\mathcal{A}$ satisfies:*

$$c(\underline{s}) = \begin{cases} 0; & BN(\underline{s}) = \emptyset, \\ \hat{d}(\underline{s}) + 1; & else. \end{cases}$$

Proof. If there are no branchnodes, the sequence $\underline{s}$ consists of only one, possibly repeated character and therefore has zero complexity. If there are branchnodes, the deepest node in fact labels the longest subsequence which occurs at least twice with a different successor character. From the definition of complexity and its properties as given in [Jans 89, pg. 32] it follows immediately that in this case $c(\underline{s}) = \hat{d}(\underline{s}) + 1$.   □

Proposition 2 *Blumer's algorithm can be used to determine the complexity profile of a sequence $\underline{s}$ with characters from some finite alphabet $\mathcal{A}$ in linear time and memory.*

Proof. Blumer's algorithm builds a DAWG in linear time and memory. It therefore suffices to show how the maximum depth can be determined from the DAWG in linear time and memory. By looking at the algorithm in detail, it can be seen that only if a secondary outgoing edge from a node is created (step 5d of procedure *update* in [Blum 83]), a possible change in the maximum depth occurs. Hence, it is only necessary to keep track of the maximum depth by testing the depth of the node being equipped with an additional edge against the existing maximum depth. This testing operation clearly does not change the order of the algorithm.   □

As Blumer et al. did in their paper, it should be noted that the linearity of their algorithm is with regard to the <u>total</u> processing time related to the length of the sequence.

4   The DAWG of a Periodic Sequence

Blumer's algorithm can also be used very well to determine the period of a periodic sequence, as can be seen from the DAWG in the next example.

Example 2 The string $w = 100100100$ is periodic with period 3. Carrying out the partitioning into equivalence classes as demonstrated in Example 1, the result becomes:

$$
\begin{aligned}
EQ(w) &= \{\lambda, 1, 0, 10, 00, 01, 010, 001, 0100, 0010, 00100\}, \\
BN(w) &= \{\lambda, 0\}, \\
\hat{d}(w) &= 1.
\end{aligned}
$$

The import of this example lies in the behaviour of the suffix pointers, as can be seen from Figure 2. The number of edges between e.g. the last node and the node pointed to by the last node's suffix pointer is equal to 3, which is the period of the string.

Proposition 3 *Blumer's algorithm can be used to determine the period of a periodic sequence in linear time and memory. To this end, it is necessary and sufficient to examine $p + \hat{d} + 1$ characters, so at most $2p - 1$ characters, where p denotes the period of the sequence.*

Proof. From Example 2 it can be seen that the number of edges between a node and the node pointed to by the first node's suffix pointer (we will call this the suffix length of a node) can be determined in linear time and memory. Hence, it suffices to show that the suffix length of the last node after processing $p + \hat{d} + 1$ characters is equal to the period.

1. Consider the following sequence:

$$
\underline{s} = (\alpha_0, \ldots, \alpha_{p-1}, \underbrace{\alpha_0, \ldots, \alpha_{j-1}}_{\hat{d}+1}).
$$

As $\hat{d} + 1 = c$, the complexity of $\underline{s}$, it follows that the second subsequence $\alpha_0, \ldots, \alpha_{j-1}$ is unique up to an exact copy at the beginning of $\underline{s}$. This implies that the DAWG's last equivalence class contains $\alpha_{p-1}, \alpha_0, \ldots, \alpha_{j-1}$ as the shortest subsequence, since this subsequence has $p + j$ as its endpoint within $\underline{s}$. This shortest subsequence has

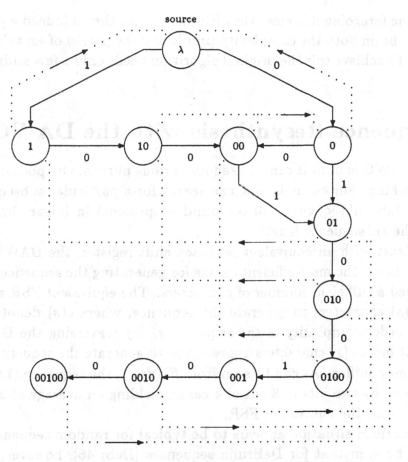

Figure 2: The DAWG of a periodic string

$\alpha_0, \ldots, \alpha_{j-1}$ as its longest suffix, which occurs at first at the beginning of $\underline{s}$, exactly p places back in the equivalence class with endpoint j. Therefore the difference in depth between the endnode and the node pointed to by the endnode's suffix pointer is equal to the period p.

2. Next let $c < j < p$. Because of the uniqueness of subsequences of length c the longest suffix of the shortest subsequence in the endnode's equivalence class occurs only p places back in the sequence, hence the suffix length being again equal to p.

3. Now let $j < c$. In this case the uniqueness of any subsequence of length j is not guaranteed, implying that the suffix length may be shorter than p.

□

From the foregoing it is clear that Blumer's algorithm is indeed a powerful tool to obtain both the complexity profile and the period of an arbitrary sequence. To achieve this the original algorithm needs only a few additional rules.

5 Sequence Resynthesis with the DAWG

Once the DAWG is built it can be used for various purposes by postprocessing or searching. For example, one can search for a particular subsequence and again this subsequence will be found (if present) in linear time, i.e. linear in the *sub*sequence length.

In contrast with an equivalent feedback shift register, the DAWG of a given sequence is the most efficient means for generating the sequence if one has observed a sufficient number of characters. The equivalent FSR always requires $c(\underline{s})$ characters to generate the sequence, where $c(\underline{s})$ denotes the maximum order complexity of the sequence $\underline{s}$. By traversing the DAWG, however, at most $c(\underline{s})$ characters are required to generate the sequence, but much less may suffice. As can be seen from Figure 1, the sequence (110100) is determined by two bits in 3 out of 4 cases, yielding an average of $2\frac{1}{4}$ bits versus 3 bits for the equivalent FSR.

The described situation appears to be typical for random sequences. It can easily be seen that for DeBruijn sequences [Debr 46], however, there is no advantage in using the DAWG as $c(\underline{s})$ characters are necessary and sufficient to generate that particular sequence.

References

[Blum 83]   A. Blumer, J. Blumer, A. Ehrenfeucht, D. Haussler and R. McConnell. "Linear Size Finite Automata for the Set of all Subwords of a Word: An Outline of Results", *Bul. Eur. Assoc. Theor. Comp. Sci.*, no. 21, pp. 12–20, 1983.

[Blum 85]   A. Blumer. "A Generalization of Run-Length Coding", *IEEE Intern. Symp. on Info. Theory 1985*, Brighton, England, June 24–28, 1985.

[Debr 46]   N. G. de Bruijn. "A Combinatorial Problem", *Nederl. Akad. Wetensch. Proc.*, vol. 49, pp. 758–754, 1946.

[Jans 89]   C. J. A. Jansen. *Investigations On Nonlinear Streamcipher Systems: Construction and Evaluation Methods*, PhD. Thesis, Technical University of Delft, Delft, april 1989.

[Ruep 84]   R. A. Rueppel. *New Approaches to Stream Ciphers*, PhD. Thesis, Swiss Federal Institute of Technology, Zurich, 1984.

Solving Equations in Sequences

Kencheng Zeng *
Center for Advanced Computer Studies
University of Southwestern Louisiana
P.O. Box 4330, Lafayette, LA 70504

Minqiang Huang
DCS Center, Graduate School of Academia Sinica
P.O. Box 3908, Beijing, China

I. Introduction

Boolean combinations of LFSR sequences have long been studied in cryptology ([1], [2]). In analyzing the cryptologic characteristics of such pseudo-random sequences, one would often come across the task of solving an equation in linear sequences like

$$AX = B \tag{1}$$

where $A = (a_t)_{t \geqslant 0}$ and $B = (b_t)_{t \geqslant 0}$ are known sequences, while $X = (x_t)_{t \geqslant 0}$ is unknown in the sense that, with the initial values $x_0, x_1, \cdots, x_{n-1}$ remaining to be determined, it subjects to the known linear recurring relation

$$x_{t+n} = \sum_{i=0}^{n-1} c_i x_{t+i}, \quad t \geqslant 0.$$

The main purpose of this note is to discuss the uniqueness of the solution, and to analyze the computational complexity, as well as the solution distance, i.e. the length of the B-segment needed for solving Equation (1), under the assumption that $A \neq 0$ and $AX = B$ is solvable.

Let $g(\lambda) = \lambda^n + \sum_{i=0}^{n-1} c_i \lambda^i$, and for $t \geqslant 0$,

---

* On leave from the Graduate School of Academia Sinica, Beijing, People's Republic of China.

$$\lambda^t \equiv \sum_{i=0}^{n-1} d_{t,i} \lambda^i \quad mod\ g(x). \tag{2}$$

It can be easily derived ([3], [4]) that

$$x_t = \sum_{i=0}^{n-1} d_{t,i} x_i.$$

Let $p = [per(A), per(g(\lambda))]$, $I_A = \{0 \leqslant i < p \mid a_i = 1\}$. We have the following equivalent system of linear equations for (1):

$$\sum_{i=0}^{n-1} d_{t,i} x_i = b_t, \quad t \in I_A. \tag{3}$$

It is a matter of common sense in linear algebra that the number of solutions to (3), if any, is 2^{n-r}, where

$$r = rank\,(d_{t,i})_{p \times n}.$$

II. Some Sufficient Conditions for Uniqueness

Obviously, Equation (1), or (3), is uniquely solvable if and only if $rank\,(d_{t,i}) = n$ or $AX \neq 0$ for all $g(\lambda)$-generated $X \neq 0$. The simplest case is

Proposition 1. If A is a primitive sequence of degree m, and $g(\lambda)$ is primitive of degree n, then Equation (1) has just one solution.

Proof. We need only show that $AX \neq 0$ for all non-trivial X generated by $g(\lambda)$. This is true by [5]. It can also be easily shown as follows. Let $S = 2^m - 1$, $T = 2^n - 1$. In a period of ST consecutive bits, A contributes $T(S-1)/2$ zeros and B contributes $S(T-1)/2$ zeros, hence AX has at most

$$\frac{T(S-1)}{2} + \frac{S(T-1)}{2} = ST - \frac{(S+T)}{2} < ST$$

zeros, which implies $AX \neq 0$. $\square$

Proposition 2. If $(per(A), per(g(\lambda))) = 1$, then $rank\,(d_{t,i}) = n$.

Proof. Write $l = per(g(\lambda))$ and $A_i^{(l)} = (a_{i+lt})_{t \geqslant 0}$. First, we show that $A_i^{(l)} \neq 0$. Notice that $(A_i^{(l)})_j^{(k)} = A_{i+lj}^{(lk)}$. By taking

$$k = l^{-1} \mod per(A)$$

and

$$j = -ik \mod per(A) ,$$

we have $(A_i^{(l)})_j^{(k)} = A$, and therefore $A_i^{(l)} \neq 0$.

Now, let $\alpha = \lambda \mod g(\lambda) \in F_2[\lambda]/(g(\lambda))$,

$$G = \{\alpha^i \mid 0 \leqslant i < p = [per(A), per(g(\lambda)] , a_i = 1\} .$$

For any $i \geqslant 0$, since $A_i^{(l)} \neq 0$, there must be a t such that $a_{i+tl} = 1$, which shows $\alpha^i = \alpha^{i+tl} \in G$ and hence $G = \{a, \alpha, ..., \alpha^{l-1}\}$.

Finally, the proposition follows from the fact that

$$rank(d_{t,i}) = rank_{F_2}G = n .\qquad \square$$

Corollary. Let $f(\lambda)$ be the minimal polynomial of A, $GF(2^{k_1})$ and $GF(2^{k_2})$ be the splitting fields of $f(\lambda)$ and $g(\lambda)$ respectively. If $(k_1, k_2) = 1$, then $rank(d_{t,i}) = n$ $\qquad \square$

Particularly useful is the following result, in the case of solving equations of the form $XY = A$.

Proposition 3. Suppose X is primitive, $((per(A), per(g(\lambda))) = 1$. Let α be as above,

$$G_0 = \{\alpha^i \mid a_i = 1, x_i = 0\} ,$$

$$G_1 = \{\alpha^i \mid a_i = 1, x_i = 1\} ,$$

then

$$rank_{F_2} G_1 = n , \qquad rank_{F_2} G_0 = n - 1 .$$

Proof. By assumption that X is primitive, there exists an index i_0 such that

$$x_{i_0} = x_{i_0+1} = \cdots = x_{i_0+n-1} = 1 .$$

For each $i = i_0, i_0+1, i_0+n-1$, as shown in Proposition 2, we can choose a t with $a_{i+tl} = 1$. This implies

$$\alpha^{i_0}, \alpha^{i_0+1}, ..., \alpha^{i_0+n-1} \in G_1$$

and hence $rank_{F_2} G_1 = n$.

Similarly, the second equality holds. □

III. Required Amount of Linear Equations and Computations

The coefficient matrix $D = (d_{t,i})_{p \times n}$ of (3) has $p = [per(A), per(g(\lambda))]$ rows. In general, p is comparable to 2^{m+n}, where m is the linear complexity of A. Apparently, we need to find a much better upper bound for the number of required rows. Let

$$G_A = \{(d_{t,0}, d_{t,1}, ..., d_{t,n-1}) \mid a_t = 1\} ,$$

$$G_A(N) = \{(d_{t,0}, d_{t,1}, ..., d_{t,n-1}) \mid 0 \leqslant t < N, a_t = 1\} .$$

If there is an $N = poly(m, n)$ such that

$$rank \ G_A(N) = rank \ G_A ,$$

then the time for solving Equation (1) will be $T = O(N^2 n) = poly(m, n)$.

Proposition 4. Let $f(\lambda)$ be the minimal polynomial of A, $f * g$ be the polynomial having as roots all the distinct matching products of the roots of $f(\lambda)$ vs the roots of $g(\lambda)$ (see [5]). Then for $N = deg(f * g)$,

$$rank \ G_A(N) = rank \ G_A ,$$

and clearly,

$$N \leqslant deg \ f(x) \cdot deg \ g(\lambda) = mn .$$

Proof. Write

$$d_t = (d_{t,0}, d_{t,1}, ..., d_{t,n-1}) \in V_n(F_2) ,$$

$$g(\lambda) = \lambda^n + c_{n-1}\lambda^{n-1} + \cdots + c_1\lambda + c_0 ,$$

and

$$M = \begin{pmatrix} 1 & & & \\ & \cdot & & \\ & & \cdot & \\ & & & 1 \\ c_0 & c_1 & \cdot & c_{n-1} \end{pmatrix}.$$

We have $d_{t+1} = d_t M$.

Let L be the left shift operator of sequences, $d = (d_t)_{t \geqslant 0}$ be the sequence of vectors. It follows from Caylay-Hamilton's Theorem that

$$g(L) \cdot d = (d_t g(M))_{t \geqslant 0} = 0 .$$

For the vector sequence $Ad = (a_t d_t)_{t \geqslant 0}$, according to [5],

$$f * g(L) \cdot Ad = 0 ,$$

which implies that all the vectors $a_t d_t$ are linear combinations of $a_0 d_0$, $a_1 d_1$, $\cdots$, $a_{N-1} d_{N-1}$. Therefore,

$$rank \ \{d_t \mid a_t = 1\} = \{a_t d_t \mid t \geqslant 0\} = \{a_t d_t \mid 0 \leqslant t < N\} . \qquad \square$$

This result indicates that the amount of computations for solving $AX = B$ is polynomial in m and n. In fact,

$$Time \ (Solving \ AX = B) = O(N^2 n) = O(m^2 n^3) .$$

A slight modification of the above procedure yields

Proposition 5. Let

$$G_{A,X} = \{d_t \mid a_t = x_t = 1\} ,$$

$$G_A(N) = \{d_{t_0} \mid 0 \leqslant t < N, a_t = x_t = 1\} .$$

If $N = deg \ (f * g * g)$, then

$$rank \ G_{A,X} = rank \ G_{A,X}(N)$$

and

$$Time \ (Solving \ AX = B, \ A \ and \ X \ unknown) = O(N^2 n) = O(m^2 n^5) .$$

IV. Conclusions

The equation in linear sequences can be transformed into a system of simultaneous linear equations over F_2, and solved in a time polynomial in the sizes of the sequences. The main idea and results can be generalized to the situations with two or more unknown sequences, such as $AX + BY = C$ or $XY + AX + BY = C$.

References

[1] Beker, H.J. and Piper, F., *Cipher Systems: The Protection of Communications* Northwood Books, 1982.

[2] Rueppel, R.A., *Analysis and Design of Stream Ciphers*, Springer-Verlag, 1986.

[3] Wan, Z.X., *Algebra and Coding*, (in Chinese), Science Press, Beijing, 1976.

[4] Lidl, R. and Niederreiter, H., *Finite Field*; Encyclopedia of Mathematics and Its Applications, Addison-Wesley Publishing Company, 1983.

[5] Zieler, N. and Mills, W.H., "Products of Linear Recurring Sequences," *Journal of Algebra*, 27(1973).

SECTION 9

APPLICATIONS

The utilization of such a general network topology raises new and increasingly complex security issues which did not exist in earlier scenarios. In this paper we will briefly examine the implications and possible consequences of these security risks and propose a solution being developed within Philips as an Advanced Development Project on Security.[1] As this solution is based on the utilization of Smart Card technology as a fundamental security tool, we will first briefly describe the attributes of the Smart Card necessary for implementation in such a scenario, next describe the design and implementation of a Security Server integrating the Smart Card specifically oriented to banking applications, indicate its utilisation and limitations in typical banking scenarios and finally suggest potential improvements.

2. THE SECURITY RISKS IN BANKING NETWORKS

2.1 Banking Security Risks of Yesterday

In the original batch oriented model, tight security control was possible, as a typical banking system was concentrated in a single computer center ; data could be input on separated data entry systems under specific procedures and programs such as double entry and comparison. With the evolution towards network based systems, access control packages on the mainframe together with identity and password control in dedicated transaction handling applications became necessary. However the batch processing and dedicated nature of applications still made security a manageable process.

In the EFT environment of the early eighties, the transmission of financial transactions over data communication lines increased the risks of fraud and necessitated the introduction of cryptography to guarantee the integrity and confidentiality of data.

2.2 Security Risks Today and Tomorrow will be higher

The use of LAN and WAN in today's and tomorrow's banking infrastructure will require more security measures, as these types of networks are vulnerable and can be tapped both actively and passively so that the risk of hacking becomes realistic.

The number of IWS is increasing rapidly in typical bank organizations and as most will use standard operating systems such as UNIX, OS/2, MS/DOS, knowledge about these operating systems and potential for manipulation will become widespread. Because IWS are distributed, they are difficult to physically control from some central point, so that the risks of software corruption and in particular virus infection and propagation increase enormously. Because popular operating systems used with many IWS, such as MS/DOS, have minimal security, and thus access control and authorization are not available as standard features, passwords and user ID's are easy to steal, particularly when sent in the clear on a LAN. Additionally local file management systems can be accessed via standard PC tools, whilst monitoring of the IWS memory is also possible if somewhat more difficult.

The number of hardware and software components of a large banking system today is enormous, and monitoring of security is a continuous and resource consuming task with a large overhead. Centralized control of security in a decentralized environment is not easy, the difficulty is compounded when applications such as Telebanking and Electronic Data Interchange (EDI) are added. The hardware and software environments are being constantly updated, which leads to serious problems regarding the portability of security across changing environments. Controlling releases of a large number of software packages, from a security viewpoint, poses obvious problems of control procedures and logistics, but is a necessity in a banking environment. The volume of transactions, particularly real-time transactions are ever increasing, which implies that security failures need to be detected instantaneously if damage is to be limited. Consequently detection techniques must be automated with only minimal human intervention, that is, these automated techniques need to be highly reliable.

3. GLOBAL SECURITY SCENARIO IN BANKING NETWORKS

3.1 Generalized Network
Figure 1 illustrates a generalized banking network scenario in which IWS and Servers are connected on LAN, which in turn are interconnected via bridges to other LAN or via data communication controllers to mainframes. In each LAN, one or more Servers are configured specifically as Security Servers.

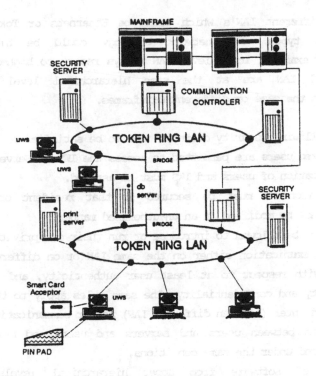

Fig. 1

Such a scenario fits neatly into an architecture where resource sharing is a basic principle. No assuptions can be made about the security of the network itself, which must be presumed to be vulnerable to attack, or about the security of the objects connected on the network, or of the operating systems and software. It must be assumed that all types of interconnections are possible.

<u>On the same LAN</u>

. Between Servers
. Between Security Servers
. Between Servers and Security Servers
. Between IWS
. Between IWS and Security Servers
. Between Servers and Mainframes
. Between IWS and Mainframes

The first five connection types must also be possible between objects connected on different LAN's which could be Ethernets or Token Rings or other Networks types. The network topology could be hierarchically structured, for example, lower level LAN always report to higher level LAN, or "flat", all LAN are at the same hierarchical level and report independently to the same or different mainframes.

Globally the following security criteria need to be satisfied :
- Only authorized users are permitted to access the IWS or Servers.
- The authentication of users and IWS must be assured.
- The IWS environment must be secured so that resident or downloaded software cannot be modified in an unauthorized manner.
- Users must not be allowed to introduce or run their own private programs.
- Inter user communication either on the same LAN or on different LAN must be secured with respect to at least user authenticity, and if necessary data integrity and confidentiality. The same rules apply to the intra (on same LAN) and inter (between different LAN) Server communications.
- Communications between users and servers and users and mainframe must also be secured under the same conditions.
- Downloading of software from upper hierarchical levels to lower hierarchical levels must be assured for integrity, typically from mainframes to Servers of from Servers to IWS.

It is debatable that all of these criteria have the same imperatives in a single network or that they can be fulfilled at acceptable cost.

4. AN OVERVIEW OF THE SMART CARD

Smart Card architecture and specifications and standardisation requirements such as physical characteristics, contacts and exchange protocols have been described elsewhere [2, 3, 4, 5]. In this section only those attributes of the Smart Card relevant to the present discussion are described.

4.1   Smart Card Security
Figure 2 is a block diagram of general Smart Card architecture showing how the different components are interconnected.

Smart Card security is based on physical protection of the chip, logical

Intergrated Chip Layout

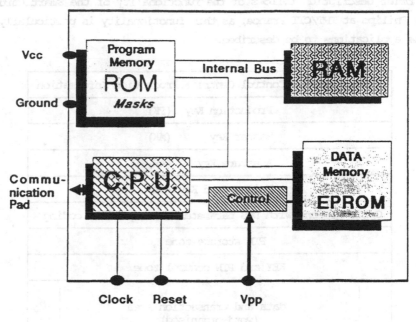

Vcc

Ground

Program Memory
ROM
Masks

Internal Bus

RAM

Commu-
nication
Pad

C.P.U.

Control

DATA
Memory
EPROM

Clock   Reset   Vpp

Figure 2

protection of access to its contents, and by control procedures during each phase of the Smart Card life cycle. Physical protection of the Smart Card chip has been developed by the chip manufacturers to provide physical security of the chip against the physical attacks likely to be perpetrated in the environment in which Smart Cards will be used. These include scanning of the card contents by electron microscopes, electromagnetic and chemical attacks. The data cells and all control and data paths are buried beneath layers of oxide. Fuse bits will blow and block further use of the card in attack scenarios.

The fundamental differences between a Smart Card and a conventional microprocessor chip are that the Smart Card chip is physically and logically protected, the Smart Card chip implements different technologies on the same chip and specialised masking allows the same Smart Card component to execute different function sets. From a security viewpoint, the Smart Card can be regarded as a Tamper Resistant Module (TRM) or a low-cost limited performance Crypto Module (CM).

4.2 Philips Smart Card based on the Data Encryption Standard (DES)

A brief description follows of the functionality of the Smart Card developed by Philips at TRT/CTI France, as this functionality is particularly suited to the applications to be described.

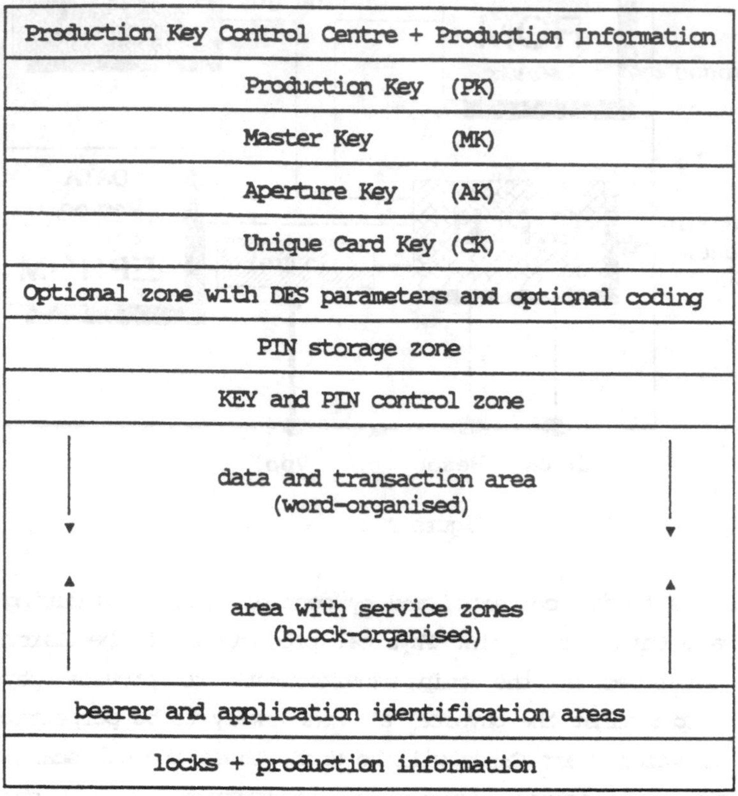

Production Key Control Centre + Production Information
Production Key (PK)
Master Key (MK)
Aperture Key (AK)
Unique Card Key (CK)
Optional zone with DES parameters and optional coding
PIN storage zone
KEY and PIN control zone
data and transaction area (word-organised)
area with service zones (block-organised)
bearer and application identification areas
locks + production information

Figure 3 : Smart Card Memory Layout

Figure 3 illustrates the Smart Card EPROM memory layout. Normally ROM is used for programs, RAM for temporary working space and EPROM or E^{2}PROM for permanent data storage. Apart from containing the microprogram for controlling the Smart Card and I/O to the outside world, the ROM storage also contains the Data Encryption Standard (DES) algorithm. Since its publication in 1974, the DES algorithm has been the preferred candidate for a cryptosystem because of its high security in withstanding cryptanalytic attacks and because it is a de facto standard. The Philips Smart Card can execute the standard DES funtions, such as encryption / decryption in

Electronic Code Book (ECB) / Cipher Block Chaining (CBC) / Cipher Feed Back (CFB) modes, generate Message Authentication Codes (MACs) and generate DES keys. Encryption performance is equivalent to a few hundred bits per second, and will be superior in new versions.

Data memory is organised in 32-bit words, each word being controlled by 4 system bits which define the type of word (secret or not, valid or not) and the type of access (read or write). Data storage and data protection can be implemented on a word (transaction area) or block (service zone) basis. Multiple service zones can be implemented with the possibility of executing cryptographic operations within them, each controlled by a different DES key.

Data and service zones can be efficiently located by a search instruction. Multiple PIN modifications by the card holder are possible under appropriate control. Besides containing an instruction set typical of many microprocessors, the Smart Card also incorporates functions of specific interest in security applications. These include :

. The possibility of storing several key types, such as master keys (for card control), service keys to control service zones, certification keys to certify the contents of a card word, and temporary encryption/decryption keys. All these keys can be invoked to execute cryptographic operations under the appropriate conditions.

. A write protect function which enables, for example, a key to be remotely loaded into a specific location in the card memory. The transfer of the key from the remote server to the card can of course occur under encryption of another key, the correctness of the key loading operation being authenticated to the remote server. This is an important capability in security applications.

. Generation of pseudo-random numbers by invoking the DES algorithm.

. Programmability of security protocols (e.g. a specific authentication protocol) in the data memory of the Smart Card, within the limits of the available memory, during the prepersonalisation phases. Such an approach results in very high security, because the authentication protocol is now essentially an interactive dialogue between an external entity (e.g. a server) and the Smart Card TRM.

5. DESIGN AND IMPLEMENTATION OF A SECURITY SERVER

5.1 Server Environment

The Security Server (SS) typically comprises a host PC with associated software and a hardware based Security Processor (SP) including a resident real-time operating system and associated management and security firmware which occupies one or more slots in the PC Cabinet. The SP is connected to the PC bus via a suitable interface such as a full duplex 10 MBPS Ethernet data link. At least a Smart Card Acceptor (SCA) can be connected (optionally an operator console) to the SP via RS 232-C interfaces. The SS itself normally includes a hard disk, and a second SCA and operator console.

5.2 Server Network Environment

For simplicity, we focus on a LAN environment interconnecting PC based IWS and Servers, in which the common operating system is OS/2, that is, the SS can be accessed on an Ethernet or Token Ring LAN. In this case the host PC acts as a Server on the LAN with the SP as a slave resource. The communication interface from IWS or other servers accessing the SS will use the standard OS/2 LAN MANAGER mechanisms (pipes, queues, etc), see figure 4.
The SS includes the basic software and application interface layers to provide a convenient programming level for applications. All IWS and servers on the LAN include a SCA, that is, have at least secure key storage and minimum cryptographic capability.

5.3 Security Server (SS) Components

The major components of the SS include :
. SP Hardware.
. SP Firmware/Software (OS, drivers, security and management).
. Software in the host PC incorporating the SS Services and Management.
. Software in the IWS necessary to access the SS and its services.
 A brief functional description and interaction of these major components follows.

5.4 Security Processor (SP) Hardware

The SP Hardware (shown in the block diagram of figure 5) is based on classical microcontroller technology and includes :

Security Server Network Environment

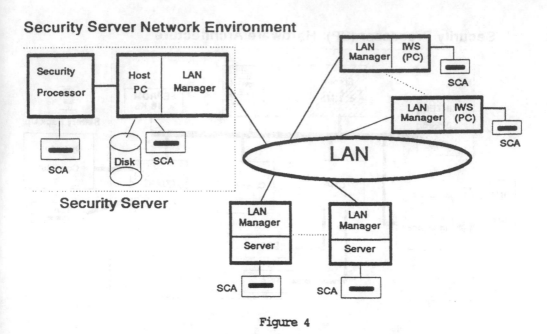

Figure 4

- A microprocessor (typically INTEL 80286/386SX) and its support circuitry.
- RAM/EPROM/EEPROM memory appropriately shared over the addressable memory space.
- Rechargeable battery backup for a specified part of CMOS memory.
- DES and RSA Hardware accessed via independent DMA channel device controllers for high performance.
- Miscellaneous components such as a Real Time Clock, etc.
- A high performance local communications controller using the Ethernet protocol links the SP to the host PC, whilst two asynchronous full duplex V24 interfaces permit the connection of a SCA and operator console locally.

5.5 Security Processor (SP) Firmware/Software

The basic SP Firmware/Software architecture is shown in figure 6 and includes the VRTX/86 real time operating system, a Starter/ Loader module and drivers for the SP hardware. Only these basic modules are permanently resident, other applications such as different cryptosystems, security protocols, tamper detection software, can be downloaded (under precise security conditions) as defined extensions. The SP operates is a multiuser, multitasking environment.

Security Processor (SP) Hardware Architecture

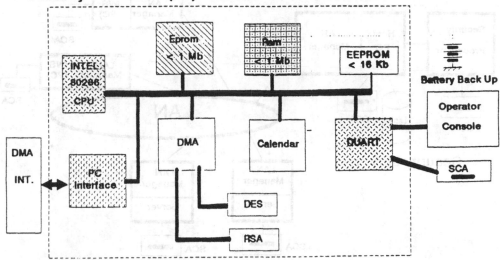

Figure 5

Security Processor (SP) Software Architecture

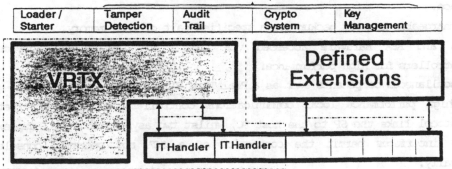

* VRTX Real Time Multitasking O.S.
* Diagnostic Tests on Hardware
* Host (PC) Data Link Driver & Task Loader
* Tamper Detection Software
* Audit Trail Management
* Drivers for Security Processor Hardware
  - DES chip, RSA chip, Serial Lines, RTC, Etc.

Figure 6

5.6 Host PC Software

The host PC Software comprises the following modules :

- The interface with the SP consisting mainly of the Ethernet driver responsible for managing data communications with the SP.
- The SS Management Service comprises the set of functions to handle user requests for the services of the SS, such as User Log-on, Authentication, User Management, Encipherment, etc, and directs these requests to the appropriate component, SP for example.
- The SS Service Dispatcher dispatches user requests to the appropriate service libraries and manages user connections (via pipes) and disconnections.
- The SS Access Management module handles SS start up and initialisation of the SS environment and user connections.

5.7 IWS Software

The IWS Software provides modules for :

- Managing the communication interface to the SS via the standard OS/2 named pipe structure.
- Managing the interface between the IWS libraries and the SS, the IWS libraries containing the set of functions required for accessing the SS Services such as user Log-On, Authentication, etc, as well as local Services. These library functions are accessed via MACROS from the application software. Other servers wishing to communicate with the SS need at least a subset of the IWS software (the SS interface, and relevant service libraries).

5.8 Functional Flow

As shown in the flow diagram of figure 7, when the user logs-on locally to an IWS, he is presented with a User Menu/Dialogue screen, which enables him to select an application which in turn invokes the necessary macros to the IWS libraries thus enabling access to the SS services.

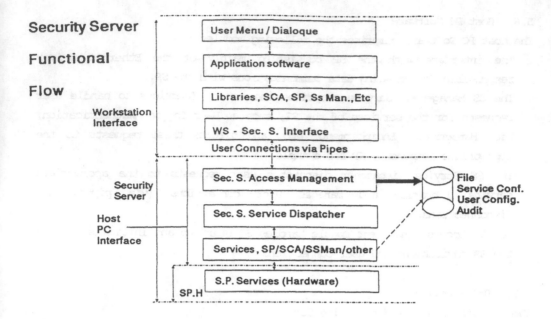

Figure 7

6. BASIC SECURITY ARCHITECTURE OF SECURITY SERVER

The security architecture of the Security Server is multilevel :
. Securing the integrity (hardware/firmware/software) of the S.P.
. Securing the access of users to the S.P. resources.
. Security implemented in the host PC acts as a primary filter.
. Security implemented in the IWS software and hardware (Smart Card).
. Secure cryptography and protocols.
. OS/2 and procedural security.
The first two aspects, integrity of and access to SP resources are
fundamental design parameters, and are described in more detail.

6.1 S.P. Integrity
Because the SP is a key component it is designed to take into account a
number of attack scenarios. Although the SP is not physically tamper
resistant , it can accomodate tamper resistance/detection if required, that
is, if physical tamper detection is implemented, the SP is able to take

appropriate security measures in such an attack scenario, erasure of sensitive memory areas under hardware control, for instance. It is assumed that the SS incorporating the SP is situated in a secure environment (secure against physical manipulation).

Security and diagnostic tests exercise the hardware (DES/RSA) and also critical software to ensure correct functioning. These tests can be run at startup and/or periodically dependant on SP load.

The security of the SP must be controlled during its entire Life Cycle if its integrity is to remain intact. This is done by creating special root Smart Cards for secure loading, modification and control during each SP life cycle phase using authentication protocols.

Thus during the Manufacturing Phase the Manufacturer Smart Card permits setting of the internal SP environment so that the manufacturer is the only authority permitted to load or update basic software and new applications.

During the Personalisation Phase the client root card is created (with the aid of the Manufacturer Card) permitting initialisation by the client of the SP set of root keys needed for eventually generating user keys (Master, Authentication, Service, Encipherment, etc).

During the Set up Phase, the client can create User Groups, initialise the internal (SP) structure of these groups (root keys, group types, protocols per group, etc) and create/delete user cards. In addition to normal users such as operators, customers, etc, the client can create/delete one or more Super User (Security Administrator) cards.

For flexible use it is desirable to be able to create additional User Groups and user cards during the Active Life Phase of the SP. This can be done by the Super User who is responsible for user management (he can create/delete/ update users at lower hierarchical levels), and SP resource management (start up/shut down/configuration/audit). Typically user groups could be classified operators, normal operators/customers and guests, those without Smart Cards for example. Such a group classification system permits definition of an access rights hierarchy for SP access and applications between users.

This separation of responsibilities during the different SP Life Cycle Phases preserves integrity. For example, loading of new applications is only possible if both the Manufacturer and Super User cards are simultaneously present. This insures that only authentic software can be loaded into the SP (implying an agreement between the Manufacture/Client regarding the development and testing of software) and avoids the problems associated by bugged software. Super Users can only be created by clients and other users by Super Users or Clients.

6.2   User Access Control

User access to the SP resources is controlled via a Validity List, which can be stored externally (on hard disc) for a large number of users, an internal structure in the SP defining users rights and internal software/application control.

The user validity list contains one record per valid user and typically has the following structure :

USER ID	GP NO	USER ID	GP NO	VALID DATE	CARD ID	OTHER INF	CURRENT RANDOM

Field 1 (clear)   Field 2 (enciphered)

Such a structure prevents replay attacks, contains redundancy for coherence and provides the right balance between security and performance. The current random is a MAC of all the current user records within a specific group. This prevents a potentially fraudulent user (for example, a user who falsely claims to have lost his Smart Card from re-inserting his old record in the validity list. The only secure method to ensure against such an eventuality is to generate a digital signature or MAC over the whole validity file which must be verified at each access (large overhead) or alternately, keeping a black list, the protection and management of which can be equally onerous. The present design involves only re-computing the MAC (the current random) over a single user group, which can be done periodically rather than for every user modification. This decreases access time, because only the specific user record must be deciphered and the current random verified, whilst maintaining security, as a stale record cannot have the correct current random. Additional information can include for example the RSA public and secret keys of the user. These public keys "notarised" by the SS could serve as the basis of a public directory for RSA key management.

The SP internal structure controls the root keys used for generating user keys. Key management is based on diversification, that is, a user key is generated by executing a DES encipherment operation on the user ID and some other appropriate information (such as the card n°), the appropriate root key serving as the encipherment key. At the Smart Card level the result of this operating is securely stored and used for communicating with the SP, that is, a key pair is created, permanently stored in the Smart Card and dynamically generated in the SP. There are as many root keys as key types (typically Authentication, Encipherment, Certification and so forth), these root keys being different for each user group. Thus the SP internal structure stores in its memory a table for each user group typically defining all root keys, the current random, validity list enciphering key, type of protocol for authenticating users associated with the specific group vis à vis the SS.

Software control must permit identification of and access to the software modules (services and applications) in the SP under the appropriate security conditions. An external file links the externally known module name to its internal ID. This file contains one record per module defining at least the Module ID and Name, User Group Access Rights, and is implemented to allow the Super User easy access to the user rights profile.

The SP stores in its internal memory a structure defining the conditions under which users can access software resources. A Module Table typically defines the Module ID, Module Type (system task/user task/procedure), Module Signature, User Group Access Rights, Link to Command Table. The Module Type could be down loadable (user tasks) or not (basic system tasks), each down loadable module has a DES MAC (signature) associated with it, this MAC being created and stored in the SP under Manufacturer control. When the module is down loaded, a MAC is computed and verified against the stored value, if no match occurs the down loaded software is considered to have undergone unauthorised modification and will be rejected. The User Group Access rights specifies whether the user can Load/Execute/Delete a module. The link to the command table points to the first module command, this table defining the command ID, User Group Access Rights (if the user can execute the command)

and Link to following command. Thus there is multilevel security (module and command access levels).

7. SECURITY SERVER CONFIGURATION SCENARIOS

We have described the basic architecture and the trusted environment of the Security Server in the preceeding sections, which provides a flexible platform for configuring the security server in differing scenarios. In addition to this basic platform, the DES and RSA cryptosystems together with the Smart Card provide the necessary tools to implement authentication, key management protocols and other security services compatible with specific security environments with differing levels of security. We outline below some of these scenarios oriented to Banking Networks.

7.1   Scenario 1 : Authentication Server in a LAN

In this scenario the network configuration of figure 4 holds. The primary function is to authenticate users (via their Smart Cards) and to authenticate their access rights to the objects connected on the LAN [6]. At a global level this implies authenticating the first five connection types defined in section 3.1, at a finer level of granularity access control to specific applications and services must be secured. In the description to follow DES is used as the cryptosystem.

In a typical application, the user logs-on to an IWS using the normal OS/2 log-on procedure. On successful termination he is presented with a Menu / Dialogue screen prompting him to insert his Smart Card. On successful termination of a local PIN verification procedure, the user is prompted to select an application. Dependent on the application, the Authentication Server might need to be accessed for specific services via appropriate Macros invoked by the application.

Assume that user A via application B wishes to execute resource C on server D, and that access to resource D is protected. Application B will first open a session with the Authentication Server (AS) to obtain the necessary access control rights to access resource C in Server D. The ID of the user A (ID_A) is first transmitted to the AS which verifies A's validity via the Validity List, next regenerates A's Authentication Key AK_A, then selects the authentication protocol to be used for A. Suppose this is a "strong authentication" protocol, in this case the AS generates a 64 bit random R, enciphers it under key AK_A and transmits it to the smart card. The smart card in turn deciphers R, replaces the least significant 32 bits of R by an internally generated random r', re-enciphers the concatenated R' = rr' under AK_A and retransmits it to the AS. The AS deciphers R' to obtain rr', verifies that r = the 32 most significant bits of R and returns R' to the smart card in clear. The smart card in turn verifies that the 32 LSB of R' matches the r' sent, in which case the mutual authentication between the AS and user A's smart card has been successfully completed, else the session is aborted.

In this case "strong authentication" implies mutual authentication, in which masquerading or replays are not possible. Notice that the protocol is executed between the AS and the Smart Card, both considered secure environments and that any modification by the IWS would be detected and cause the procedure to abort. Of course, one way authentication (Smart Card to AS) can be used if a lower level of security is acceptable.

Once the authentication phase is completed, user A's access rights to resource C in Server D are read by the AS in the Resource Rights file. Two alternatives are possible. In the first alternative the AS generates a common Session Key SK_{AD} between A and D, enciphers SK_{AD} under its shared Communication Key CK_D with server D (generated and stored in server D's (Smart Card at personalisation) and transmits it to server D. Simultaneously the AS enciphers A's access rights to C under server D's communication key CK_D, adds the Session Key SK_{AD}, enciphers the whole under the common Encipherment Key EK_A shared with A's smart card and transmits the lot to A's Smart Card. This encipherment could have the following format, where E stands for encipher :

$$EEK_A[SK_{AD},ECK_D\{ID_A,USER\ RIGHTS\ IN\ C\}]-(1)$$

A's Smart Card deciphers (1) to obtain SK_{AD}, re-enciphers the rest under SK_{AD}, and transmits the result to D.

$$ESK_{AD}[ECK_D,\{ID_A,USERS\ RIGHTS\ IN\ C\}]-(2)$$

Server D via its Smart Card deciphers (2), first with SK_{AD} then with CK_D. It is able to verify that the communication is indeed from A and is not a replay (current session key SK_{AD} and A's explicit ID_A) and can thus grant A's access rights to resource C.

If user A frequently uses resource C then addition of a validity period for the users rights in C and a Time Stamp by the AS converts (1) into a "token" or "credential" which can be used to securely access resource C in server D during a defined time period, this being the second alternative [7]. This cryptographically sealed "credential" is created the first time user A requests access to resource C and can be stored locally on secondary memory at IWS level, and can be valid for a defined time period, for example, a session, a day, a week or more. If the Smart Card has sufficient memory, several such credentials for frequently used services at different destinations can be stored in it, this having the advantage of creating "portable credentials", that is, the user is not restricted to a specific IWS. If the user smart card has EEPROM memory, the profile of "portable credentials" can be modified under appropriate security to suit a users' current needs, making the mechanism very flexible.

Such a solution provides an elegant security mechanism to optimise performance in a LAN both at the AS and IWS levels, without loss of security. Because it is desirable to control network access rights from a trusted central point such as an AS, for security and administrative reasons (flexibility to modify users rights), the validity period for credentials should be adjusted accordingly, for example, one day is a good compromise between flexibility and performance.

Apart from user authentication and access control to resources in the LAN, the AS has the inherent capability to provide additional security services at LAN level, such as Integrity and Confidentiality of sensitive long term data, particularly if users do not themselves have sufficient cryptographic capability at IWS level. The AS's ability to generate users' Encipherment and Authentication Keys, its organisation for validating users and the high performance of its DES hardware (several M bits per sec.) leads naturally to the provision of Confidentiality and Integrity services for large volume data. Users can access such services in basically the same manner as already described, except that these services are located in the AS rather than in other servers.

If users do not possess Smart Cards, an authentication protocol with the AS can use, either a user password read from a magnetic card (typical in todays banking applications), encipherment of this password occuring either in the magnetic card reader or in the IWS, or the password is manually entered by the user at the IWS keyboard and transmitted to the AS either enciphered at IWS level or in the clear. The clear or enciphered password is used to identify the user and permit access to the security services. Clearly security will be lower (strong authentication is not possible) and in the implementation proposed here, users' rights will be limited.

At AS level the RSA cryptosystem is also implemented and RSA based Authentication, Key Management and Integrity (based on RSA Digital Signatures) is an alternative or complement to the DES based approach, particularly as provision has been made for the storage of user RSA keys (secret and public) at AS level. Thus the notion of a "notarised" (by the AS) LAN level RSA Public Key Directory and an Integrity Service for users provided by the RSA hardware at AS level is implicit. It is pointed out that Smart Cards do not as yet implement the RSA (this is expected to change in the future), consequently whilst users' secret RSA keys can be stored in Smart Cards, cryptographic operations will have to be performed in the IWS with the RSA cryptosystem implemented in software, leading to reduced security. It is difficult to satisfy the criteria for secure downloading of software to IWS level or to guarantee the integrity of the IWS software environment, these security considerations being outside the scope of the present scenario. Secure downloading of IWS software implies sufficient cryptographic capability and a secure execution environment to generate and verify MAC's or digital signatures associated with secure downloading. A similar environment is required for maintaining the integrity of the IWS software environment.

Authentication of an IWS can be assured by implementing a Smart Card chip as part of the IWS hardware itself and using an authentication protocol similar to that described for a user Smart Card vis à vis the AS.

7.2   Scenario 2 : Inter LAN Gateway

In this scenario the Security Server has about the same functionality within the LAN itself as described for scenario 1, in addition it secures communications to other LAN, that is, it is integrated into and secures the gateway. This is most simply achieved if each pair of gateways in the global network share at least one common communication key in a DES environment. Thus inter LAN communications can be enciphered for confidentiality or authenticated for integrity as need be. Clearly higher performance for the cryptosystem is needed than in scenario 1. A problem arises when a user on LAN A(Domain A) needs to access another user or resource on LAN B(Domain B).

An obvious solution would be to transfer the requesting users' rights in Domain A to Domain B unmodified. It is unlikely that user profiles in Domain B would be defined exactly as in Domain A, thus the mapping of a users' rights from one Domain to another is not straight forward. In a homogeneous network controlled by a single organisation this might not be an unsolvable problem, however in a multiuser hetrogeneous network it is very unlikely that similar definitions of user rights or even the same notions of security exist.

7.3   Scenario 3 : Network Level Key Management Facility

In this scenario the basic Security Server is configured mainly as a Key Management Facility (KMF) for the global network rather than at a LAN level. Depending on the number of potential users in the global network and the expected load, one or more SP's can be configured in a single KMF with load sharing capabilities. The KMF's major task is the secure generation, distribution and updating of network wide keys at all levels (Master/Key Enciphering/Data Enciphering/User Keys), and the maintenance, updating and distribution of black lists. Sublists can also be distributed to LAN level Security Servers.

8. FUTURE EVOLUTION OF THE SECURITY SERVER CONCEPT

Implementation of the Security Server concept as described in this paper is nearing completion. The next step will be to test the concept, particularly performance, in pilot applicationis which include a realistic mix of security services.

In order to be as independant as possible of the underlying network the Security Server is basically implemented at the application layer, layer 7 in the ISO, OSI model. It is the intention to follow emerging ISO standardisation on security as closely as possible. For example if a Digital Signature Standard emerges in the future (today it is in the DP stage) it will be implemented.

Future areas for evolution and improvement could possibly include :
. Porting to other OS environments such as UNIX.
. Interfacing to other hardware environments such as VME/MCA.
. Adding of tamper resistance to the SP.

RSA as a Benchmark for Multiprocessor Machines

Rodney H. Cooper[1]   **Wayne Patterson**[2]

An Extended Abstract

The use of pipelined and parallel machines is becoming a more commonplace aspect of computing environments. As a consequence, pipelined and parallel machines will undoubtedly soon be generally available for computer security computations.

There is very little experience, however, with the implementation of standard cryptologic algorithms on multiprocessor machines.

In addition, the existence of parallelism should cause a reconsideration of algorithms used to carry out the standard algorithms in the field.

The authors have undertaken a project to develop benchmarks for RSA encryption on various pipelined and parallel machines. The initial machines selected for the analysis are the IBM 3090 Vector Computer, the Inmos T800 Transputer Development System, and the Thinking Machines CM-2 Connection Machine.

IBM 3090

The IBM 3090 with vector module attachments achieves supercomputer performance by adding vector processors to a basic mainframe design. The 3090 is availablem with 1, 2, 4, or 6 processors. Cycle time is 18.5 nanoseconds. Each CPU has 64 KB of cache storage. Code is written in IBM VS FORTRAN Version 2.

Inmos T800 Transputer Development System

A transputer is a microcomputer with its own local memory and with links for connecting one transputer to another transputer. [INMOS 1988]

The transputer architecture defines a family of programmable VLSI components. A typical member of the transputer product family is a single chip containing processor, memory, and communication links which provide point-to-point connections between transputers.

The specific transputer system (T800) for our experiment is a 40-processor, 10-board system, capable of being reconfigured in several processor topologies.

---

[1] University of New Brunswick, Fredericton, NB, Canada
[2] University of New Orleans, New Orleans, LA, USA

- Derivation of a low cost SP with or without tamper resistance for integration at IWS level. Such an evolution would greatly enhance security and add functionality (secure software downloading and software integrity control at IWS level).
- Improving performance of the Security Server by implementing higher performance hardware (RISC architecture).
- Implementing other public key cryptosystems, e.g. GF (2^n), or Key Management protocols (e.g. Diffre/Hellman).
- Design and implementation of a Smart Card able to incorporate RSA and or Zero Knowledge.

9. CONCLUSIONS

If large distributed networks will exist in the future, as market trends predict, and if their security becomes an issue, then the types of solutions proposed in this paper will be of undoubted interest.

10. BIBLIOGRAPHY

[1]   "System Requirements and Design Specifications of the SAPPHIRE Security Server", Philips TDS, Advanced Development Project on Security, 1989.
[2]   ISO/IS 7816-1 : "Identification cards, Integrated circuit(s) with contacts" (Part 1 : Physical characteristics).
[3]   ISO/IS 7816-2 : "Identification cards, Integrated circuit(s) with contacts" (Part 2 : Dimensions and location of the contacts).
[4]   ISO/IEC/DIS 7816-3 : "Identification cards, Integrated circuit(s) with contacts" (Part 3 : Electronic signals and protocols).
[5]   R. Ferreira : "On the utilisation of Smart Card Technology in High Security applications", IFIP conference SECURICOM, Monte Carlo, 1986.
[6]   R.M. Needham, M. Schroeder : "Using Encryption for Authentication in Large Networks of Computers", Comm. ACM, Vol. 21, N° 12, December 1978, pp 993-999.
[7]   R. Ferreira, J.J. Quisquater : "Towards more practical Key Management Solutions in Computer Networks using the Smart Card", Proceedings of Smart Card 2000, Vienna, October 1987.

Thinking Machines CM-2 Connection Machine

A Thinking Machines, Incorporated CM-2 Connection Machine [TMI 1988] is a computing system with facilities for SIMD parallel programs. The full CM-2 parallel processing unit contains:

> 65,536 data processors
> an interprocessor communications network
> four sequencers
> several I/O controllers and/or framebuffers.

Each processor has 64K bits of bit-addressable local memory and an ALU that can operate on variable-length operands. Each processor can access its memory at 5 Mbits per second.

When each of the processors is performing a 32-bit integer addition, the unit operates at about 2500 MIPS; with an additional floating point accelerator, the performance can increase to 3500 MFLOPS (single precision) or 2500 MFLOPS (double precision).

A most important component of the CM-2 is the *router*, which allows messages to be passed from any processor to any other. A more finely-grained communications link is the NEWS grid, which supports programmable grids with arbitrarily many dimensions. The NEWS grid allows processors to pass data in a rectangular pattern. The advantage of this over the router is that the overhead of explicitly specifying destination addresses is eliminated.

The parallel processing unit operates under the control of a front-end computer, either a Symbolics 3600 Lisp machine, or a DEC VAX 8000-series. All CM-2 programs execute on the front end; during the course of execution, instructions are sent to the parallel processing unit.

Each processor has:

> an ALU
> 64 K bits of memory
> four 1-bit flags
> optional floating point accelerator
> router interface
> NEWS grid interface
> I/O interface.

The ALU is a 3-input, 2-output logic element, reading two data bits from memory and one from a flag; then computing two output bits, one to memory, the other to a flag. The logic element can compute any 3-to-2 boolean function.

On each chip are 16 processors. One router node is also on each chip. The router nodes of all processors are connected together to form the router set. The topology of the set is a boolean n-cube. For a fully configured CM-2, it is a 12-cube with 4096 chips. Thus each router node is directly connected to 12 other router nodes.

Multiplication

At the heart of RSA encryption, and indeed, of many cryptologic computations, is the requirement to perform repeated multiplications of large integers. In addition to a comparison of machines, this project also involves the comparison of the implementation several approaches to multiplication, including divide-and-conquer, fast Fourier transform [BRASSARD 1988], and multiple-radix representation on these parallel machines.

Multiple-Radix Representation

We present here an algorithm [PATTERSON 88] that is applicable in some cases, and is easily shown to be $\mathcal{O}(n)$, where n is the length of each factor.

The key to this method is a different representation of the integers – a representation known as multiple-radix, or mixed-radix. Let $\mathbb{Z}_n$ represent the integers modulo n, with its normal ring structure using addition and multiplication. For distinct prime numbers $p_1, p_2, ..., p_m$ let $\mathfrak{R} = \mathbb{Z}_{p1} \oplus \mathbb{Z}_{p2} \oplus ... \oplus \mathbb{Z}_{pm}$ represent the ring product. There is a mapping $\emptyset$ from the interval:

$$[\, 0, \, (\prod_{i=1}^{m} p_i) - 1\,] \quad \longrightarrow \quad \mathfrak{R}$$

which is a set isomorphism. Furthermore, it is both an additive and a multiplicative homomorphism wherever the mapping on the sum or product is defined. In other words, if a, b, a +b, and a·b are in the interval described above, then $\emptyset (a + b) = \emptyset (a) + \emptyset (b)$, and $\emptyset (a \cdot b) = \emptyset (a) \cdot \emptyset (b)$. The mapping is:

$$\emptyset (a) \quad = \quad (a \bmod p_1, a \bmod p_2, ..., a \bmod p_m).$$

It is clear that multiplication of two numbers in $\mathfrak{R}$, the mixed-radix representation, is $\mathcal{O}(n)$. Suppose that numbers a and b have, say, n digits. Find a series of primes for which modular multiplication is essentially a machine level operation (in other words, choose primes $< \sqrt{\text{word length}}$ of the machine. Since there are approximately 10,000 primes $< \sqrt{2^{31}}$, this is not difficult. Then, perform the approximately $\frac{n}{4}$ machine-level multiplications of the components of $\mathfrak{R}$, giving the result of the multiplication.

Addition is implemented similarly.

A benefit to the use of the multiple-radix representation occurs when the multiplication algorithms under discussion are to be implemented on a parallel machine. The multiple-radix representation has the advantage that there are fewer "cross-products" or communications between processes.

Division

Another critical step in the RSA involves repeated reduction modulo n, for some large n. This integer division is also computationally costly. A variation on the standard algorithm, more adaptable to parallel environments, will also be compared. This method of division, called "division by multiplication" is described below.

A major step in the implementation of large-scale integer computation systems is the development of an efficient method of performing modular multiplication, which itself is a necessary step in the carrying out of integer division.

In this paper we will implement a method of carrying modular multiplication with large integers that does not involve division, at least not at any level beyond the division by individual small primes. [COOPER 1989b].

The Pascal Knapsack PKC

As a final comparison, public key encryption using the recently developed Pascal knapsack PKC [COOPER 1989a] is being subjected to the same tests.

Computations are being performed at the New Orleans Advanced Computation Laboratory of the University of New Orleans, at the University of New Brunswick's IBM 3090 Vector Computer, and at Thinking Machines, Incorporated of Cambridge, Massachusetts.

The final paper will incorporate results of all computational experiments as well as details on the programming in the more advanced environments such as the Transputer Development System (programmed in Occam) and the Connection Machine (programmed in *-LISP).

References

[BRASSARD 1988]   Gilles Brassard and Paul Bratley, *Algorithmics: Theory and Practice,* Prentice-Hall, 1988.

[COOPER 1989a]   R. H. Cooper, Ron Hunter-Duvar, and Wayne Patterson, *Pascal Knapsack Public Key Cryptosystems*, to appear in the Proceedings of ICC '89.

[COOPER 1989b]   R. H. Cooper and Wayne Patterson, *Division by Multiplication,* to appear.

[INMOS 1988]   Inmos Ltd., *Transputer Development System,* Prentice-Hall, 1988.

[PATTERSON 1988]   Wayne Patterson and R. H. Cooper, *Multiple-Radix Arithmetic,* to appear.

[TMI 1988]   Thinking Machines, Incorporated, *CM-2 Technical Summary*.

Range Equations and Range Matrices: a Study in Statistical Database Security

V.S. Alagar
Department of Computer Science
Concordia University
1455 de Maisonneuve Blvd. West
Montreal, Quebec H3G 1M8, Canada

Abstract

This paper studies range equations and matrices with range entries. Range equations and their solutions are generalizations of compositions for consecutive sets of integers. Consistency and reducibility of range equations are due to data dependant semantics and the restrictions set by the query response strategy in statistical databases. Having determined the probability that a random solution to a given range equation is reducible, this paper relates it to the extent of security affordable in statistical databases and further generalizes it to matrices with range entries.

1 Introduction

A finite set of consecutive integers taken from N, the set of all non-negative integers, will be called a range. In this paper we are interested in ranges, range equations, matrices of range equations and their applications to secure statistical databases under count queries.

In this paper we restrict ourselves to the collection $C = \{[(x-1)s, xs-1] \mid x \in P\}$, of ranges where P is the set of all positive integers and $s \in P$ is fixed. Thus C is a partition of N and every range $X \in C, X = [(x-1)s, xs-1]$, is uniquely determined by x, which we call the range indicator of X. The sum (or union) of a finite number of $d(\geq 2)$ ranges $X_1, ..., X_d$ from C is a range; but this range does not belong to X. More precisely, if x_i is the range indicator of $X_i \in C$, then $X_1 + X_2 + \cdots + X_d$ is the range $Y = [s(y-d), sy-d], y = x_1 + x_2 + \cdots + x_d$. The range $Z \in C$ with range indicator $z = y-d+1$ has a non-empty intersection with Y. Conversely, given $X \in C$ and $s, d \geq 2$, one of our interests is in determining the number of solutions to

$$Y = X_1 + X_2 + \cdots + X_d, X_i \in \mathcal{C}, i = 1, \ldots, d \qquad (1)$$

such that $X \cap Y \neq 0$. In particular, we are interested in situations where $X \subseteq Y$, as well as those when $X \nsubseteq Y, X \cap Y \neq 0$. In the former case, every integer $n \in X$ can be written, in at least one way, as $n = r_1 + r_2 + \cdots + r_d, r_i \in X_i$. Since, for $i \neq j$, the values $r_i \in X_i, r_j \in X_j$ may be equal, every integer $n \in X$ admits a composition into d parts $r_1, \ldots, r_d$ with $r_i \in X_i$. When $X \nsubseteq Y, X \cap Y \neq 0$, the ranges $X, X_1, \ldots, X_d$ can be <u>reduced</u> to $Z \subseteq X, Z_i \subseteq X_i, i = 1, \ldots, d$ such that $Z \subseteq Z_1 + \cdots + Z_d$; that is, only a subrange of integers of X admit compositions. The ratio R_x / T_x, where R_x is the number of compositions of all reducible solutions and T_x is the total number of compositions from all solutions, gives the probability that a random solution of the range equation (1) is reducible. This probability of reductions, as we shall see in Section 3, determines the extent of security afforded in a statistical database. Section 2 briefly reviews range equations and cites the results from [2] and [3] on the probability of range reduction. Section 3 is a brief introduction to matrices of range equations and the security of statistical databases. Section 4 discusses several types of reducible range equations and shows how these results can be combined to compute more accurately the probability of range reduction in the context of statistical database security.

2 Preliminary Results and a Brief Review on Range Reduction

This section provides a brief summary of the results on the first type of range reduction, called type T_1. See Alagar [2], [3] for detailed examples and proofs.

Our notation from now on is to write

$$X = X_1 + X_2 + \cdots + X_d \qquad (2)$$

for the general situation (1) and refer to (2) as a range equation. This notation is inaccurate; however this notation is used to show the generalization of the concept of compositions of a positive integer n into d parts

$$n = r_1 + r_2 + \cdots + r_d, r_i \geq 0, i = 1, \ldots, d \qquad (3)$$

Given n, d, it is known from elementary assignment problem (of balls to boxes) that the number of solutions to (3), also called the number of compositions of n into d parts is

$$J(n, d) = \binom{n+d-1}{n} \qquad (4)$$

The binomial identity

$$\sum_{t=0}^{s}\binom{r+t}{t} = \binom{r+s+1}{s} \tag{5}$$

is useful in later analysis. Example 1 illustrates reducible and irreducible range equations.

Example 1 Let $s = 5$, $d = 5$, $X = [45,49]$, $X_1=[0,4]$, $X_2=[10,14]$, $X_3=[5,9]$, $X_4=[15,19]$, $X_5=[10,14]$. It is easy to see that, for every $n \in X$, there exist (several) $r_i \in X_i, i = 1,...,5$ such that

$$n = r_1 + r_2 + r_3 + r_4 + r_5.$$

Thus

$$X = X_1 + X_1 + X_3 + X_4 + X_5$$

cannot be reduced. However, if $X = [40,44]$ and $X_1 = [0,4], X_2 = X_3 = X_4 = X_5 = [0,9]$ are given as a solution to the range equation $X = X_1 + X_2 + X_3 + X_4 + X_5$, then it follows that 40=4+9+9+9+9 is the only valid solution; that is, the equation is reducible.

Restricting to the simple case $d = 2$ we state Theorem 1 on the probability that a range equation

$$X = X_1 + X_2 \tag{6}$$

is T_1 reducible when $X, X_1, X_2 \in C, s \geq 2$. Denoting the range indicators of the ranges in (6) by x, x_1 and x_2 respectively, we seek solutions X such that

$$[(x - 1)s, xs - 1] \cap [(x_1 + x_2 - 2)s, (x_1 + x_2)s - 2] \neq 0. \tag{7}$$

Theorem 1 *The range equation (6) is T_1 reducible if $x = x_1 + x_2$. In the reducible case, the amount of reduction in each range is 1 and the reduced range equation is irreducible.*

Proof See Alagar [2].

Example 2 The equation $[25,29] = [10,14] + [15,19]$ is not reducible and $x = x_1 + x_2 - 1$ holds here. However, for the equation

$$[25, 29] = [5,9] + [15, 19], \tag{8}$$

$x = 6, x_1 = 2, x_2 = 4$ imply that $x = x_1 + x_2$ holds. Hence equation (8) is reduced to $[25,28] = [6,9] + [16,19]$.

The next theorem gives the probability of type T_1 range reduction.

Theorem 2 *Let $X = [(x-1)s, xs-1]$. The probability p_x that X is reduced is $(s-1)(x-1)/(2xs - s + 1), x \geq 1$.*

Proof The number of solutions to the equation $X = X_1 + X_2$ is the sum of the two quantities: (i) The number of reducible cases; that is, the number of compositions of x into two parts (x_1, x_2) with $x_1, x_2 \geq 1$. There are $(x - 1)$ compositions. (ii) The number of irreducible cases; that is, the number of compositions of $x + 1$ into two parts (x_1, x_2) with $x_1, x_2 \geq 1$; there are x of these. The probability p_x depends on the exact counts of writing $n = r_1 + r_2, n \in X, r_1 \in X_1$ and $r_2 \in X_2$. If $0 \leq r_1 \leq (x-1)s$, then for each value of r_1 there are s possible values for r_2 with $n = r_1 + r_2$. Hence there are $[(x-1)s+1]s$ possible (r_1, r_2) in this case. If $(x-1)s+1 \leq r_1 \leq (xs-1)$, there are $s(s-1)/2$ possible pairs (r_1, r_2) with $r_1 + r_2 = n$. Hence totally there are $s(2xs - s + 1)/2$ possible pairs (r_1, r_2) with $r_1 + r_2 = n$. There are $(x-1)s(s-1)/2$ pairs (r_1, r_2) with $r_1 + r_2 = n$ corresponding to reducible ranges. Hence the probability of reduction is p_x, as stated in the theorem.

Notice that p_x increases as x increases (for a fixed s) and as s increases (for a fixed x), $p_1 = 0$ and $p_x < \frac{1}{2}, x \geq 1, s \geq 2$.

3 Probability of T_1 Type Range Reduction - General Case

For the rest of this section we let $X = [(x-1)s, xs-1], X_i = [(x_i - 1)s, x_i s - 1], y = x_1 + \cdots + x_d, s \geq 2$, and $d = sq + r, 0 \leq r < s$.

Theorem 3 *The range equation*

$$X = X_1 + X_2 + \cdots + X_d$$

is reducible and each range can be reduced by the amount $(r - 1)$ if

$$d/s - 1 \leq y - x < (d-1)/s.$$

The equation is not reducible if

$$(d-1)/s \leq y - x \leq d - 1.$$

Proof See [2].

Corollary 1 (i) *If $d=s+1$, then for all values $y, x + 1 \leq y \leq x + d - 1$, the range equation is irreducible; otherwise $y=x+q$ is the only value of y corresponding to the reducible range equation. For $y, x + q + 1 \leq y \leq x + d - 1$, the range equation is not reducible.*

(ii) *If $y=x+q$ and $s \neq d$, each range in the range equation is reduced by the same amount $r-1$ and hence each remaining range contains only $s-r+1$ integer values. The range X is reduced to $[(x-1)s, xs-r]$ and each X_i is reduced to $[(x_i-1)s+r-1, x_i s-1]$.*

(iii) *If $y = x$ and $s = d$, each range in the range equation is reduced to a single integer.*

Example 3 Let $s=3$, $d=5$, $x=4$. We have $q=1$, $r=2$.

(i) Let $2/3 \leq y - x < 4/3$. Hence $y=x+1=5$ with $y = x_1 + x_2 + x_3 + x_4 + x_5$, $x_i \geq 1$. This forces $x_i = 1, i = 1, ..., 5$. The range equation

$$[9, 11] = [0, 2] + [0, 2] + [0, 2] + [0, 2] + [0, 2]$$

is reduced to

$$[9, 10] = [1, 2] + [1, 2] + [1, 2] + [1, 2] + [1, 2].$$

(ii) Let $4/3 \leq y - x \leq 4$. Then $y - x \in \{2, 3, 4\}$ or $y \in \{6, 7, 8\}$. Each value of y corresponds to an irreducible range equation. For example, the choice $y=7$ gives

$$y = x_1 + x_2 + x_3 + x_4 + x_5$$

and there are several solutions with $x_i \geq 1, i = 1, ..., 5$. For each choice, the range equation is irreducible.

(iii) Let $s=d=3, x=4$. We have $q=1$, $r=0$ and $0 \leq y - x < 2/3$ gives $y=x=4$. The equation $y = 4 = x_1 + x_2 + x_3$ has three sets of solutions $x_1 = 1, (1, 2), x_2 = 1, (2, 1), x_3 = 2, (1, 1)$. It is easy to verify that any one choice gives a total reduction.

The next Theorem gives the number of range equations for a given range X.

Theorem 4 *The number of solutions to a range equation $X = X_1+X_2+\cdots+X_d$ is $N(x, d, s) = \binom{x+d-1}{d} - \binom{x+q-1}{d}$.*

Proof From Corollary 1 we know $y - x \in \{q, q+1, ..., d-1\}$, where $y = x_1 + x_2 + \cdots + x_d, x_i \geq 1, i = 1, 2, .., d$. Alternately, $y - d = \sum_{i=1}^{d}(x_i - 1), (x_i - 1) \geq 0$, where $y - d \in \{x + q - d, ..., x - 1\}$. Thus the number of solutions is the sum $\sum_{j=1}^{d-q} J(x - j, d)$ where $J(x - j, d)$ is the number of compositions of $(x - j)$ into d parts. Using (4) we get

$$N(x,d,s) \;=\; \sum_{j=1}^{d-q} J(x-j,d) = \sum_{j=1}^{d-q} \binom{x-j+d-1}{d-1}$$

$$= \sum_{j=x-d+q}^{x-1} \binom{j+d-1}{j}$$

$$= \sum_{j=0}^{x-1} \binom{j+d-1}{j} - \sum_{j=0}^{x-d+q-1} \binom{j+d-1}{j} \qquad (9)$$

An application of result (5) to each sum on the right side gives the result stated in the theorem.

Theorem 5 *For a given x, the number of reducible range equations is* $R(x,d,s) = \binom{x-1+q}{d-1}$.

Proof From Theorem 3, the only solution for a reducible equation is $y = x+q, y = x_1 + \cdots + x_d$. Thus the number of solutions is the same as the number of compositions of $(x+q-d)$ into d parts.

The next theorem generalizes Theorem 2 and gives the probability of a T_1 type range reduction.

Theorem 6 *Let*

$$s_1 = \begin{cases} s-r, & s < d \\ s-d, & s \geq d \end{cases}$$

The probability $p(x,d,s)$ that the range $X = [(x-1)s, xs-1], x \geq 1$ is reducible is given by

$$p(x,d,s) = \frac{R(x,d,s)\dbinom{s_1+d}{d}}{D(x,d,s)}, \qquad (10)$$

where $D(x,d,s) = \binom{d+xs-1}{d} - \binom{d+(x-1)s-1}{d}$.

Proof For each integer $t \in [(x-1)s, xs-1]$, there are $\binom{t+d-1}{d-1}$ compositions of t into d parts of the form $t_1 + t_2 + \cdots + t_d$, where $t_i \in X_i$. Hence there are

$$D(x,d,s) = \sum_{t=(x-1)s}^{xs-1} \binom{t+d-1}{d-1}$$

compositions in all ranges. Rewriting the above expression in the form

$$D(x,d,s) = \sum_{t=0}^{xs-1} \binom{t+d-1}{d-1} - \sum_{t=0}^{(x-1)s-1} \binom{t+d-1}{d-1}$$

and using result (5) we get

$$D(x,d,s) = \binom{d+xs-1}{d} - \binom{d+(x-1)s-1}{d}.$$

From Theorem 5 we know that there are R(x,d,s) range solutions, each of which is reducible. Since the size of reduction is r-1, the number of compositions corresponding to any one reducible range is the sum

$$\sum_{t=0}^{s_1} \binom{t+d-1}{t} = \binom{s_1+d}{d}.$$

Hence the total number of compositions in all reducible range solutions is

$$R(x,d,s)\binom{s_1+d}{d}.$$

The proof follows under the assumption of uniform probability distribution for compositions. For d=2, p(x,d,s) reduces the result of Theorem 2. Notice also that $p(x,d,s) = 0, x < d \leq s$ or $s < d, x < d - q$. Table 1 shows the computed probabilities for a selected set of values.

s/d	3	5	7
	.0091	.0000	.0000
	.0189	.0041	.0004
3	.0275	.0116	.0040
	.0321	.0173	.0086
	.0350	.0215	.0128
	.0360	.0230	.0145
	.0217	.00000	.0000
	.0433	.00001	.0000
5	.0611	.00010	.0002
	.0704	.00018	.0006
	.0761	.00026	.0011
	.0780	.00029	.0013
	.0289	.00000	0
	.0566	.00005	0
7	.0789	.00041	.00000007
	.0904	.00075	.00000037
	.0973	.00103	.00000078
	.0997	.00113	.00000098

Table 1: Probability of reduction. Rows in a box are the probabilities corresponding to $x = 3, 5, 10, 20, 50, 100$.

4   Security of Statistical Databases and Matrix of Ranges

This section relates the analysis done so far to security of statistical databases. Furthermore we generalize range equations to matrices, two dimensional tables of equations that naturally correspond to contingency tables.

An important function of a statistical database is to provide statistical summaries on the stored data. Abstractly, an individual unit in the database is a record with k attributes $A_1, A_2, ..., A_k$ for some positive integer $k \geq 1$. To simplify our discussion and be consistent with our earlier analyses, we assume that each A_i can take d different values. Without loss of generality denote these values by $\{0, 1, ..., d-1\}$. A query C, in the database is a logical formula over the attributes; in particular C specifies the attributes and the type of statistics to be computed on the attribute domains. We restrict to conjunctive queries requiring a count; that is, AND is the only logical operator allowed in C and the statistic required is the number of records in the database satisfying C. If S_C denotes the set of records in the database satisfying the formula X, then the exact statistic is $|S_C|$. The formula $C = (A_1 = j)$, for a fixed $j \in [0, d-1]$, is logically equivalent to $C_1 \vee C_2 \vee \cdots \vee C_d$, where $C_i = (A_1 = j) \wedge (A_2 = i - 1), i = 1, ..., d$. Since the sets S_{C_i} are mutually disjoint,

$$|S_C| = |S_{C_1}| + \cdots + |S_{c_d}| \qquad (11)$$

It is known that a statistical database under count queries can be compromised in the sense that sensitive information is revealed when exact statistics (of aggregates of the population in the database) is revealed. Hoffman and Miller [13], Denning ([6], [7], [8], [9], [10]), Schlörer ([15], [16], [17], [18], [19], [20], [21], [22]), Achubgue and Chin [1], Fellegi ([11], [12]) and Nargundkar [14] have proposed and analysed several forms of output controls and output perturbations to prevent the deduction of any sensitive information from the output statistic.

These analyses show that it is fairly easy to devise compromising strategies for such methods. In [2] and [3] Alagar proposed "range responses" for "count queries" and showed the strength of the method through an extensive probabilistic analysis. The strategy is to output the range $X = [(x-1)s, xs] \in C$, if $|S_C| \in X$. We comment that range output is different from systematic rounding (with respect to a fixed base) which was studied by Fellegi [11] and Fellegi and Phillips [12] in the early '70s. It is probable that Statistics Canada decided to implement random rounding instead of systematic rounding, because there was some suspicion that systematic rounding is insecure. The extensive analysis in

Alagar [3] reveals that range response, a form of symmetric systematic rounding is a sound, if not absolutely secure, strategy.

From (11) and the output ranges $X, X_1, ..., X_d$, it follows that $X \cap (X_1 + \cdots + X_d) \neq 0$ is a necessary integrity constraint. Although a user may not know the distribution of values in the domains of the attributes $A_1, ..., A_k$, the user can pose $(d+1)$ AND queries, get $(d+1)$ ranges and then set up a range equation. Once the equation becomes reducible, the user is faced with less uncertainty to resolve. That is, reducible ranges reveal more information and might become a security risk in the case of total reduction. Hence, the probability of reduction R_x/T_x, defined in the introduction, is the probability that a true count may be deduced from the statistic output by the response strategy.

The analysis provided in [3] is, to our knowledge, the first mathematical analysis provided for output perturbation schemes of statistical databases. In our opinion, this analysis is still incomplete. A more complete analysis involves considerably higher dimensional tables of ranges for reducibility. Below we give a brief summary of results from Alagar [3].

We start with a simple example to illustrate reduction in a two dimensional table of ranges.

Example 4 Let SEX, MAR. STATUS and EMP. STATUS be three of the attributes in database records with their values shown below:

> SEX: Male, Female;
> MAR. STATUS: Single, married, divorced, other;
> EMP. STATUS: Professor, Manager, Programmer, Analyst.

Consider the characteristic formulas:

$$
\begin{aligned}
C &= (\text{EMP. STATUS} = \text{Professor}), \\
C_1 &= (C \ \underline{\text{AND}} \ \text{SEX} = \text{Male}), \\
C_2 &= (C \ \underline{\text{AND}} \ \text{SEX} = \text{Female}), \\
C_1' &= (C \ \underline{\text{AND}} \ \text{MAR. STATUS} = \text{Single}), \\
C_2' &= (C \ \underline{\text{AND}} \ \text{MAR. STATUS} = \text{Married}), \\
C_3' &= (C \ \underline{\text{AND}} \ \text{MAR. STATUS} = \text{Divorced}), \\
C_4' &= (C \ \underline{\text{AND}} \ \text{MAR. STATUS} = \text{Other}), \\
C_{11}' &= (C_1' \ \underline{\text{AND}} \ \text{SEX} = \text{Male}), \\
C_{12}' &= (C_1' \ \underline{\text{AND}} \ \text{SEX} = \text{Female}), \\
C_{21}' &= (C_2' \ \underline{\text{AND}} \ \text{SEX} = \text{Male}), \\
C_{22}' &= (C_2' \ \underline{\text{AND}} \ \text{SEX} = \text{Female}), \\
C_{31}' &= (C_3' \ \underline{\text{AND}} \ \text{SEX} = \text{Male}), \\
C_{32}' &= (C_3' \ \underline{\text{AND}} \ \text{SEX} = \text{Female}), \\
C_{41}' &= (C_4' \ \underline{\text{AND}} \ \text{SEX} = \text{Male}), \\
C_{42}' &= (C_4' \ \underline{\text{AND}} \ \text{SEX} = \text{Female}).
\end{aligned}
$$

These formulas are related:

$$C = (C_1 \text{ OR } C_2),$$
$$C = (C_1' \text{ OR } C_2' \text{ OR } C_3' \text{ OR } C_4'),$$
$$C_1' = (C_{11}' \text{ OR } C_{12}'),$$
$$C_2' = (C_{21}' \text{ OR } C_{22}'),$$
$$C_3' = (C_{31}' \text{ OR } C_{32}'),$$
$$C_4' = (C_{41}' \text{ OR } C_{42}'),$$
$$C_1 = (C_{11}' \text{ OR } C_{21}' \text{ OR } C_{31}' \text{ OR } C_{41}'),$$
$$C_2 = (C_{12}' \text{ OR } C_{22}' \text{ OR } C_{32}' \text{ OR } C_{42}').$$

Moreover, let us assume the following range responses:

$$\text{COUNT } (C) \in [30,32], \qquad \text{COUNT } (C_1) \in [18,20],$$
$$\text{COUNT } (C_2) \in [12,14], \qquad \text{COUNT } (C_1') \in [6,8],$$
$$\text{COUNT } (C_2') \in [6,8], \qquad \text{COUNT } (C_3') \in [9,11],$$
$$\text{COUNT } (C_4') \in [9,11], \qquad \text{COUNT } (C_{11}') \in [3,5],$$
$$\text{COUNT } (C_{12}') \in [3,5], \qquad \text{COUNT } (C_{21}') \in [0,2],$$
$$\text{COUNT } (C_{22}') \in [6,8], \qquad \text{COUNT } (C_{31}') \in [6,8],$$
$$\text{COUNT } (C_{32}') \in [0,2], \qquad \text{COUNT } (C_{41}') \in [3,5],$$
$$\text{COUNT } (C_{42}') \in [3,5].$$

These ranges are arranged in the two dimensional table Table 2 with four rows (corresponding to MAR. STATUS) and two columns (corresponding to SEX).

[3,5]	[3,5]	[9,11]
[0,2]	[6,8]	[9,11]
[6,8]	[0,2]	[6,8]
[3,5]	[3,5]	[6,8]
[12,14]	[18,20]	[30,32]

Table 2

The interpretation of the entries in Table 2 is the following: the outer row is the range equation

$$[12,14] + [18,20] = [30,32]$$

obtained from $C = (C_1 \text{ } \underline{OR} \text{ } C_2)$; the outer column is the range equation

$$[6,8] + [6,8] + [9,11] + [9,11] = [30,32]$$

obtained from $C = (C_1' \text{ } \underline{OR} \text{ } C_2' \text{ } \underline{OR} \text{ } C_3' \text{ } \underline{OR} \text{ } C_4')$; each row is a range equation corresponding to an expansion of C_i' over the attribute values of SEX; each column is a range equation corresponding to an expansion of C_i over the values of attribute MAR. STATUS. Notice that the top two rows are reducible. After reduction we get the table shown in Table 3.

[4,5]	[4,5]	[9,10]
[1,2]	[7,8]	[9,10]
[6,8]	[0,2]	[6,8]
[3,5]	[3,5]	[6,8]
[12,14]	[18,20]	[30,32]

Table 3

Now, we find the left-most column is completely reducible. After this reduction, the top two rows become reducible and as a consequence the outer row is reducible. The result shown in Table 4 cannot be reduced any further.

4	5	9
1	8	9
6	[0,2]	[6,8]
3	[3,5]	[6,8]
14	18	32

Table 4

However, the reduced table provides a lot more information and insight into the structure of data: there are 14 female professors and among them 3 are single, 6 are married, 1 is divorced and 4 are "other"; there are 4 female and 5 male divorcees; there are 8 males and 1 female whose marital status is "other". There is still some uncertainty; for, three tables (with exact counts) arise from Table 4.

If in Example 4, we change only the range entries within the table, the extent of reduction and the number of tables (with exact counts) will be different. For example, Table 5 can be reduced to Table 6 and no further. From Table 6 we get 69 tables of integer entries (possible exact counts); 2 tables corresponding to the true count 30, 16 tables corresponding to the true count 31 and 51 tables corresponding to the true count 32. Hence, a slight reduction in range size has not resulted in any information gain.

[0,2]	[6,8]	[9,11]
[0,2]	[6,8]	[9,11]
[3,5]	[3,5]	[6,8]
[6,8]	[0,2]	[6,8]
[12,14]	[18,20]	[30,32]

Table 5

[1,2]	[7,8]	[9,10]
[1,2]	[7,8]	[9,10]
[3,5]	[3,5]	[6,8]
[6,8]	[0,2]	[6,8]
[12,14]	[18,20]	[30,32]

Table 6

Next, we consider a general situation and obtain an expression for the number of consistent two dimensional tables of ranges. Our notation is the following:

$X = [(x-1)s, xs-1], x \geq 1$. A table has m rows and n columns; if

$$\begin{aligned}
X &= X_1 + \cdots + X_m, X_i = [(x_i-1)s, x_is-1] & (12)\\
X &= Z_1 + \cdots + Z_n, Z_i = [(z_i-1)s, z_is-1] & (13)
\end{aligned}$$

are the irreducible range equations in the outer column and the outer row, $y = x_1 + \cdots + x_m, (x_i \geq n), w = z_1 + \cdots + z_n, z_i \geq m, m = sq_1 + r_1, n = sq_2 + r_2$,

$$\beta_1 = \begin{cases} q_1, & m = s+1 \\ q_1 + 1, & m \neq s+1 \end{cases}$$

$$\beta_2 = \begin{cases} q_2, & n = s+1 \\ q_2 + 1, & n \neq s+1 \end{cases}$$

then from Section 2, we get

$$\begin{aligned}
\beta_1 \leq y - x \leq m - 1 & (14)\\
\beta_2 \leq w - x \leq n - 1 & (15)
\end{aligned}$$

Hence for a given x there are $J(y - mn, m)\ J(w - mn, n)$ pairs of range indicator vectors $(x_1, ..., x_m; z_1, ..., z_n)$ for the outer column and outer row.

Next, we determine the number of matrices with range entries for one pair $(x_1, ..., x_m; z_1, ..., z_n)$.

Let $(x_{ij}), i = 1, ..., m$ and $j = 1, ..., n$ denote the range indicators of range entries in a table. Denote the row sums of the range indicators by $y_i, i = 1, ..., m$ and the column sums of the range indicators by $w_j, j = 1, ..., n$. Then from Section 2 we get

$$\begin{aligned}
\beta_2 \leq y_i - x_i \leq n - 1, i = 1, ..., m & (16)\\
\beta_1 \leq w_j - z_j \leq m - 1, j = 1, ..., n & (17)
\end{aligned}$$

Fixing the range indicators of the outer column and outer row at $(x_1, ..., x_m; z_1, ..., z_n)$ we get

$$y_i \in [x_i + \beta_2, x_i + n - 1], i = 1, ..., m$$

and

$$w_j \in [z_j + \beta_1, z_j + m - 1], j = 1, ..., n.$$

Since $(y_1 + \cdots + y_m) \in [m\beta_2 + y, m(n-1) + y], (w_1 + \cdots + w_n) \in [n\beta_1 + w, n(m-1) + w]$, and $y_1 + \cdots + y_m = w_1 + \cdots + w_n (= t$, say), we must have $t = [L, U]$, where

$$\begin{aligned}
L &= \max\{m\beta_2 + y, n\beta_1 + w\}\\
U &= \min\{m(n-1) + y, n(m-1) + w\}.
\end{aligned}$$

The number of pairs $(y_1, ..., y_m; w_1, ..., w_n)$ satisfying these constraints is

$$W(L, U) = \sum_{t=L}^{U} Q_1(t, m)\, Q_2(t, n),$$

where

$$Q_1(t, m) = \text{coefficient } of\ h^{t-m\beta_2-y} \text{ in } (1 + h + \cdots + h^e),\ e = m - \beta_2 - 1$$

and

$$Q_2(t, n) = \text{coefficient } of\ h^{t-n\beta_1-w}\ \in \text{ in } (1 + h + \cdots + h^f),\ f = m - \beta_1 - 1;$$

see Alagar [3], for a proof of these results.

Having determined $W(L, U)$, the number of pairs $(y_1, ..., y_m; w_1, ..., w_n)$ of outer column and outer row range indicators, next we determine $T(y_1, ..., y_m; w_1, ..., w$ the number of matrices for the pairs $(y_1, ..., y_m; w_1, ..., w_n)$ of row sums; this is given in the next theorem.

Theorem 7 *Let $(r_1, ..., r_m)$ be the row sums and $(c_1, ..., c_n)$ be the column sums with $r_i \geq n, i = 1, ..., m$ and $c_j \geq m, j = 1, ...,$. Assume $r_1 \geq r_2 \cdots \geq r_m$ and $c_1 \geq c_2 \cdots \geq c_n$. The number of $m \times n$ matrices with positive integral entries is*

$$T(r_1, \cdots, r_m; c_1, \cdots, c_n) = \text{ the coefficient of } t_{t_1}^{c_1} t_{t_2}^{c_2} \cdots t_{t_{n-1}}^{c_{n-1}} \text{ in } F(t_1, \cdots, t_{n-1}),$$

where $F(t_1, \cdots, t_{n-1}) = \displaystyle\prod_{i=1}^{m} F_i,\ F_i = \displaystyle\sum_{j_1=1}^{e_1(r_i)} t_1^{j_1} \sum_{j_2=1}^{e_2(r_i)} t_2^{j_2} \cdots \sum_{j_{n-2}=1}^{e_{n-2}(r_i)} t_{n-2}^{j_{n-2}} G_i(t_{n-1}),$

$$G_i(t_{n-1}) = t_{n-1} + t_{n-1}^2 + \cdots + t_{n-1}^{e_{n-1}(r_i)}$$

and $e_b(r_i) = r_i - (j_1 + j_2 + \cdots + j_{b-1}) - n + b,\ b = 1, 2, \cdots, n - 1.$

Proof Let $j_1, 1 \leq j_1 \leq r_i - n + 1$ be an entry in the ith row of column 2. Then the minimum and maximum possible values of an entry in the ith row of column 2 are 1 and $r_i - j_i - n + 2$. That is, an entry j_2 in the ith row of column 1 must be in the range $1 \leq j_2 \leq r_i - j_1 - n + 2$.

Hence,

$$
\begin{aligned}
e_1(r_i) &= r_i - n + 1, \\
e_2(r_i) &= r_i - j_1 - n + 2, \\
e_3(r_i) &= r_i - j_1 - j_2 - n + 3,
\end{aligned}
$$

etc. The generating function F_i for the ith row is

$$F_i = \sum_1^{e_1(r_i)} t_1^{j_1} * \qquad \text{(a function for the rest of the columns in that row}$$
$$= \sum_{j_1=1}^{e_1(r_i)} t_1^{j_1} \sum_{j_2=1}^{e_2(r_i)} t_2^{j_2} \quad \text{(a function for the rest of the columns in that row}$$

Proceeding by one column at a time for i^{th} row, we get F_i. Clearly then, the product of F_i's is the generating function for all the rows and columns, i.e., for all tables. In this product it is easy to see that the terms $t_{t_1}^{c_1} t_{t_2}^{c_2} \cdots t_{t_{n-1}}^{c_{n-1}}$ are those with given column sums $(c_1, c_2, \cdots, c_n)$. Hence the coefficient $t_{t_1}^{c_1} t_{t_2}^{c_2} \cdots t_{t_{n-1}}^{c_{n-1}}$ must give the number of $m \times n$ tables with row sums $(r_1, \cdots, r_m)$ and column sums $c_1, \cdots, c_n$). Hence the theorem is proved.

The next example illustrates the result of Theorem 7.

Example 5 Let $m = n = 4$, $r_1 = r_2 = 4$, $r_3 = r_4 = 6$. We have $e_1(r_1) = e_1(r_2) = 1$, $e_1(r_3) = e_1(r_4) = 3$, $e_2(r_1) = e_2(r_2) = 2 - j_1$, $e_2(r_3) = e_2(r_4) = 4 - j_1$, $e_3(r_1) = e_3(r_2) = 3 - j_1 - j_2$, $e_3(r_3) = e_3(r_4) = 5 - j_1 - j_2$.

$$F_1 = F_2 = t_1 t_2 t_3.$$
$$F_3 = F_4 = \sum_{j_1=1}^{3} t_2^{j_2}(t_3 + \cdots + t_3^{5-j_1-j_2}).$$

So, the generating function is

$$F(t_1, t_2, t_3) = F_1 F_2 F_3 F_4 = (t_1 t_2 t_3)^4 [t_1^2 + t_1 g + t_2 g + h]^2$$

where

$$g(t_2, t_3) = 1 + t_2 + t_3,$$
$$h(t_2, t_3) = 1 + t_3 + t_3^2.$$

If $c_1 = c_2 = c_3 = c_4 = 5$, the number of tables is the coefficient of $t_1^5 t_2^5 t_3^5$ in $F(t_1, t_2, t_3)$. It is straightforward to find this to be 6. The six tables are shown in Table 7:

1	1	2	2	6		2	2	1	1	6		1	2	2	1	6
2	2	1	1	6		1	1	2	2	6		2	1	1	2	6
1	1	1	1	4		1	1	1	1	4		1	1	1	1	4
1	1	1	1	4		1	1	1	1	4		1	1	1	1	4
5	5	5	5			5	5	5	5			5	5	5	5	

2	1	2	1	6		1	2	1	2	6		2	1	1	2	6
1	2	1	2	6		2	1	2	1	6		1	2	2	1	1
1	1	1	1	4		1	1	1	1	4		1	1	1	1	4
1	1	1	1	4		1	1	1	1	4		1	1	1	1	4
5	5	5	5			5	5	5	5			5	5	5	5	

Table 7

For $c_1 = c_2 = 6$, $c_3 = c_4 = 4$, the coefficient of $t_1^6 t_2^6 t_3^6$ is 3 and the three tables are shown in Table 8.

1	1	1	3	6
1	1	3	1	6
1	1	1	1	4
1	1	1	1	4
4	4	6	6	

1	1	3	1	6
1	1	3	2	6
1	1	1	1	4
1	1	1	1	4
4	4	6	6	

1	1	2	2	6
1	1	2	2	6
1	1	1	1	4
1	1	1	1	4
4	4	6	6	

Table 8

Since a range indicator uniquely determines its range, Theorem 7 enables us to compute all the tables with range entries. The total number $T(x, m, n, s)$ of $m \times n$ tables for the given range $X \in C$ is

$$T(x, m, n, s) = \sum \sum T(y_1, \cdots, y_m; w_1, \cdots, w_n),$$

where the inner sum is taken over $W(L, U)$ pairs of $(y_1, \cdots, y_n; w_1, \cdots, w_n)$ and the outer sum is taken over $J(y - mn, n)$ $J(w - mn, n)$ pairs of outer column, outer row range indicator vectors $(x_1, \cdots, x_n; z_1, \cdots, z_n)$ determined by x, m, n, s. In a similar fashion we can determine the number of matrices in which one or more rows (or/and columns) are reducible.

It is the purpose of the next section to study indirect reductions, as illustrated by Example 4, in a general setting and indicate how the probability of a reducible matrix can be computed.

5 Indirect Reductions

Example 4 is a simple instance of indirect reduction. In general, indirect reductions occur due to data dependency and integrity constraints in a database. However, our discussions here are held at the level of equations and not at the level of data that gave rise to the equations. The general situation is the following: A range equation is not T_1 reducible (i.e., not directly reducible). However, one or more of the ranges in this equation are reduced due to their participation in other range equations. This in turn may cause a reduction in the original equation. It is the purpose of this section to investigate two types of such indirect reductions.

5.1 Indirect Reduction of Type T_2

This is a general situation of Example 4 where one or more rows of a matrix are reducible causing reductions in one or more ranges of the outer column range equation. That is, the equation

$$X = X_1 + X_2 + \cdots + X_d \qquad (18)$$

is not T_1 reducible and t, $1 \le t \le d$ ranges X_i are T_1 reduced. We say X is T_2 reducible when these t reduced ranges force a reduction in equation (18).

Without loss of generality assume that $X_1, X_2, \cdots X_t$ are T_1 reducible. Let

$$X_i = X_{i1} + X_{i2} + \cdots + X_{id}, \; i = 1, 2, \cdots, d,$$

be the range equations that caused reductions in X_i. If x_{ij} is the range indicator of X_{ij} and w_i is their sum, then from Theorem 3 we have

$$q + 1 \le y - x \le d - 1$$

$$d/s - 1 \le w_i - x_i < (d-1)/s, \; i = 1, \cdots, t$$

$$\text{and} \quad (d-1)/s \le w_1 - x_1 \le d - 1, \; i = t + 1, \cdots, d.$$

Moreover, the amount of reduction in each X_i, $i = 1, \cdots, t$ is

$$s_1 = \begin{cases} r, & s < d \\ d, & s \ge d \end{cases} \qquad (19)$$

and the reduced ranges are

$$X_i = [(x_i - 1)\, s, \; x_i\, s - s_1], \; i = 1, \cdots, t.$$

The right side of (18), after substitution of these reduced ranges and simplification, becomes $[s(y-d), sy - d']$, where $d' = ts_1 + d - t$. The integrity constraints in a database and the query-answering strategy demand the consistency of equation (18); that is,

$$X \cap [s(y-d), \; sy - d'] \ne \phi.$$

This implies

$$d'/s - 1 \le y - x \le d - 1/s.$$

Similar to our discussions on Theorem 3, it can be shown that X is reduced if

$$d'/s - 1 \le y - x \le (d' - 1)/s$$

and X is not reduced if

$$(d' - 1)/s \le y - x \le d - 1.$$

A complete analysis of T_2 reducibility is done below: by examining four situations arising out of the intersection of the range $[q + 1, \; d - 1]$ with the ranges $[d'/s - 1, \; (d' - 1)/s]$ and $[(d' - 1)/s, \; d - 1]$.

Case 1

$$\text{Let } q+1 < \frac{d'}{s} - 1 \le y - x \le \frac{d'-1}{s} < d-1. \tag{20}$$

This arises when t satisfies the inequality

$$\frac{(q+2)s - d}{s_1 - 1} < t < \frac{(s-1)(d-1)}{s_1 - 1}.$$

Let $d' = s\alpha + \beta$, $0 \le \beta \le s - 1$. Substitution in (20) gives

$$\alpha - 1 + \frac{p}{s} \le y - x \le \alpha + \frac{\beta - 1}{s}. \tag{21}$$

If $\beta = 0$, then $y - x = \alpha - 1$ is the only solution and in this case X is reduced to the single point $(x-1)s$. If $\beta > 0$, $y - x = \alpha$ is the only solution given by (21); in this case X is reduced by the amount $\beta - 1$ and the reduced range is $[(x-1)s, \, xs - \beta]$.

Case 2

$$\text{Let } q+1 < \frac{d'}{s} - 1 \le y - x \le d-1 < \frac{d'-1}{s}. \tag{22}$$

It is easy to verify that no reduction in X is possible for $d = 2$. So, it is sufficient to consider $d \ge 3$. From (22) it follows that

$$d' \ge \max\{A, B\}, \text{where } 2A = (q+2)s, \ B = s(d-1) + 1.$$

2.1 Let $s \le d$. This implies that $q = 0$ and $B > A$. Consequently, $t \ge s - 1$. Since $0 \le t \le d$, $s \ge d$, $t \ge s - 1$ there are three possibilities:
$s = d$, $t = d - 1$. Now $d' = s^2 - s + 1$ and hence (22) becomes

$$s - 2 + \frac{1}{s} \le y - x \le s - 1.$$

So, $y - x = s - 1$ is the only solution and X remains irreducible.
$s = d$, $t = d$. Now $d' = s^2$, $y - x = s - 1$ and once again $y - x = s - 1$ is the only solution; but X reduces to the single point $(x-1)s$.
$s > d$, $t = d$. It is easy to verify there is no reduction in X.

2.2 Let $s < d$. Since $d' = t(r-1) + d$, $0 \le r \le s - 1$, $0 \le t \le d$, it follows that $d' < dr$ and $\frac{d'-1}{s} < d - 1$ which contradicts the earlier inequality (22). That is, this situation cannot arise.

Case 3

$$\text{Let } \frac{d'}{s} - 1 < q + 1 \le y - x \le \frac{d'-1}{s} < d - 1. \tag{23}$$

This arises when $t < \min\{A, B\}$, where

$$A = \frac{(q+2)s - d}{s_1 - 1}, \quad B = \frac{(s-1)(d-1)}{s_1 - 1}.$$

3.1 Let $s \geq d$. Now, $q = 0$, $s_1 = d$. It is easy to prove that $A < B$ and hence $t \leq A$, which in turn gives

$$1 \leq t \leq \frac{2s - d}{d - 1}. \tag{24}$$

Moreover, $1 \leq y - x \leq \frac{d'-1}{s}$ implies that $y - x$ lies in an interval of length

$$\frac{d' - 1}{s} - 1 - \frac{t(d-1) + d - 1 - 1}{s}.$$

Using (24) we can show that this length is < 1 and this forces $y - x = 1$. Hence X is reduced to

$$\begin{aligned}
[(x-1)s, \; sy - d'] &= [(x-1)s, \; sx + s - d'] \\
&= [(x-1)s, \; xs - \beta],
\end{aligned}$$

where $\beta = d' - s - 1$.

Note that even when $s = d$, (24) gives $t = 1$ and $\beta = d' - s = 2d - 1 - s = s - 1$. Hence the reduced range is $[(x-1)s, \; xs - s + 1] = [(x-1)s, \; (x-1)s + 1]$ and reduction to a single point cannot occur.

3.2 Let $s < d$. Now we have $s_1 = r$ and

$$A = \frac{2s - r}{r - 1}, \quad B = \frac{(s-1)(d-1)}{r - 1}.$$

For $d \geq 3$ we can show $A \leq B$; consequently, $t \leq A$ and $1 \leq t \leq \frac{2s-r}{r-1}$. Now we have

$$\begin{aligned}
\frac{d'-1}{s} &= \frac{t(r-1) + d - 1}{s} \\
&< \frac{s(q+2) - 1}{s} \\
&= q + 2 - \frac{1}{s}.
\end{aligned}$$

Since $q + 1 \leq y - x \leq \frac{d'-1}{s} < (q+2) - \frac{1}{s}$ we conclude that $y - x = q + 1$ is the only integer solution. Moreover, we must have

$$d' = s(q+1) + \beta, \; 0 \leq \beta \leq s - 1.$$

Therefore

$$sy - d' = s(x + q + 1) - d' = sx - \beta$$

Hence X is reduced to $[(x-1)s, xs - \beta]$, and $\beta - 1$ is the amount of reduction in X.

Case 4

$$\text{Let } d'/s < q+1 \le y-x \le d-1 < (d'-1)/s \tag{25}$$

Since the length of the interval $[d'/s-1,(d'-1)/s]$ is less than 1, $q+1 = d-1 = y-x$ must hold. If $s \ge d$, then $q = 0$ and this forces $d = 2$, and $y-x = 1$.

4.1 $s = 2$, $d = 2$. From (25) we get

$$(s-1) \le t \le 2(s-1) \tag{26}$$

That is, $1 \le t \le 2$.

For $t = 1$, X is reduced to $[2(x-1), 2x-1]$ and for $t = 2$, X is reduced to the single point $2(x-1)$.

4.2 $s = 3$, $d = 2$. It follows from (26) that $2 \le t \le 6$ and hence $t = 2$ is the only possibility. Since $d' = 4$, there is total reduction in X. Notice $s > 3$ does not arise.

This concludes the analysis for T_2 type reduction.

Example 6 Let $d = 3$, $s = 5$, $x = 7$, $y = 8$. The equation

$$[30,34] = [10,14] + [10,14] + [5,9]$$

is not T_1 reducible. Let the ranges $[10,14]$, $[10,14]$ be both T_1 reducible.

Since $s > d$, we have

$$q = 0, \ d' = t(d-1) + d = 7$$

Hence

$$d'-1 = 2/5, \ q+1 = 1, \ (d'-1)/s = 6/5 \text{ and } d-1 = 2.$$

It is straightforward to verify that all conditions of Case 3 are satisfied. The amount of reduction in $[30,34]$ is $d'-s-1 = 1$. This is indeed the amount of reduction achieved in $[30,34]$ when $[10,14]$ reduces to $[10,12]$.

Example 7 Let $d = 3$, $s = 6$, $x = 3$. For this value of x, consider the three range equations that are not T_1 reducible:

$$[12,17] = [0,5] + [0,5] + [6,11] \tag{27}$$

$$[12,17] = [0,5] + [0,5] + [12,17] \tag{28}$$

$$[12,17] = [0,5] + [6,11] + [6,11] \tag{29}$$

We shall now discuss the possible reductions in $[12,17]$ due to one or more T_1 reducible ranges that occur in the right hand side of the above equations.

According to our analysis, no reduction in $[12, 17]$ can occur when only one of the ranges on the right side of the above equations is T_1 reducible. Suppose two of the ranges are T_1 reducible. Now $d' = 7$.

For all three equations

$$1 \le t \le (2s - d)/(d - 1), \quad s \ge d$$

holds. For equation (27) $y - x = 1$ and hence the amount of reduction is $d' - s - 1 = 0$. For the other two equations $y - x = 2$ and hence there is no reduction. That is, no equation is reducible for $t = 2$. This agrees with Case 3 above.

Next consider $t = 3$. Now $d' = 9$. Once again, Case 3 applies. Since $y - x = 2$ for equations (28) and (29), they are not reducible. Equation (27) is reducible and the amount of reduction in each range is $d' - s - 1 = 2$. For, due to T_1 reductions of all the ranges on the right hand side of (27) we must have $[6, 11] \rightarrow [6, 9]$, $[0, 5] \rightarrow [0, 3]$, $[0, 5] \rightarrow [0, 3]$. So (27) becomes

$$[12, 17] = [0, 3] + [0, 3] + [6, 9].$$

Since the maximum sum of true counts from the right hand side is only 15, the range $[12, 17]$ is reduced to $[12, 15]$. Hence only one range equation which was not T_1 reducible has become reducible and the amount of reduction is 2.

Example 8 Let $s = 5$, $d = 8$ and $x = 6$. For any value $y - x \in \{2, 3, 4, 5, 6, 7\}$, a range equation $X = X_1 + \cdots + X_8$ with $y = \sum_{i=1}^{8} x_i$ is not T_1 reducible. For our purposes here we take $y = 9$ and consider the range equation

$$[25, 29] = 7 \star [0, 4] + [5, 9], \tag{30}$$

where $7 \star [0, 4]$ means the sum of $[0, 4]$ taken seven times.

If $1 \le t \le 4$, there is no reduction in $[25, 29]$. Let $t = 5$. Now $d' = 18q + 1 = 2$, $d'/s - 1 = 13/5$, $y - x = 3$, $(d' - 1)/s = 17/5$ and hence Case 1 applies. Since

$$(2s - r)/(r - 1) = 7/2$$

and

$$(s - 1)(d - 1)/(r - 1) = 28/3$$

all the conditions of Case 1 are satisfied. The amount of achievable reduction in the range $[25, 29]$ is $d' \bmod s - 1 = 2$.

Next, it is appropriate to ask for the probability that range X which is not T_1 reducible is T_2 reducible. Although an elegant closed form expression for this probability cannot be obtained, we shall show how such a probability can be computed.

We split our discussions into two parts. First we compute the probability that a range X is T_2 reducible, given there are t $(1 \le t \le d)$ ranges X_i in that equation which are T_1 reducible. Next we shall calculate the probability that given an X which is not T_1 reducible there are t $(1 \le t \le d)$ X_i's that are T_1 reducible. Clearly, the product of these two probabilities is what we want.

Let $P_t[X \mid X_1, \cdots, X_t]$ denote the probability that X is T_2 reducible when $X_1, \cdots, X_t$ are T_1 reducible. The following theorem is similar to the one in Section 2 which computed the probability of a T_1 reduction.

Theorem 8 *Let s_1, q and d' be as defined in this section and $d' = s\alpha + \beta$, $0 \le \beta \le s - 1$. Define*

$$
q_t = \begin{cases} q + 1, \ 1 \le t \le (2s - s_1)/(s_1 - 1), \\ (\alpha s - 1) \\ (2s - s_1)/(s_1 - 1) < t < \min\{d, (s-1)(d-1)/s_1 - 1)\}, \end{cases}
$$
$$
s_t = s - \beta
$$

Then the probability that a range $X = [(x-1)s, xs - 1]$ may be T_2 reducible given t of the X_i's in its range equation are T_1 reducible is

$$
P_t[X \mid X_1, \cdots, X_t] = \frac{\binom{s_t + d}{d}\binom{x + q_t - 1}{d - 1}}{D(x, d, s)}
$$

where

$$
D(x, d, s) = \binom{d + xs - 1}{d} - \binom{d + (x-1)s - 1}{d}.
$$

Proof See Alagar [3].

Next we develop an expression for the probability $P(X_1, \cdots, X_t)$ that a range $X = [(x-1)s, xs - 1]$ which is not T_1 reducible has a range equation

$$
X = X_1, \cdots, X_d
$$

and there are t, $1 \le t \le d$ ranges X_i that are T_1 reducible. Recall that $y = x + j$, $q + 1 \le j \le d - 1$, holds for a range X that is not T reducible. Let w_j denote the number of compositions of $x + j$ into d parts in which each summand is at least 1. This number w_j is the number of range solutions to $x + j$ and each summand in the decomposition is the range indicator of a range. Consider one composition of $(x + j)$ and call its summands $a_1, a_2, \cdots, a_d$. Let Q_t denote all t element subsets of $a_1, a_2, \cdots, a_d$. Now the sum

$$
\sigma_j = \left[\sum_{Q_t} \prod_{a_i \in Q_t} p(a_i, d, s) \prod_{a_i \notin Q_t} (1 - p(a_i, d, s)) \right]
$$

taken over all t element subsets Q_t is the probability that there are exactly t a_i's in the composition of $(x+j)$ such that the ranges with these a_i's as indicators are T_1 reducible. But then there are w_j solutions for $(x+j)$. Hence the probability that t ranges are T_1 reducible for a given $x+j$ is

$$r_j = (\Sigma \sigma_j)/w_j$$

where the sum is over all w_j solutions. When x is given, and $x+j$ occurs with probability $1/(d-q-1)$. Hence the probability that t ranges are T_1 reducible is

$$P(X = X_1, \cdots, X_t) = \left(\sum_{q+1}^{d-1} r_j \right) /(d-q-1). \tag{31}$$

Finally, we derive the probability that a range X is T_2 reducible.

Theorem 9 *The probability $p_2(x,d,s)$ that a range $X = [(x-1)s, xs-1]$ is T_2 reducible is*

$$p_2(x,d,s) = \sum_{t=1}^{d} P(X \mid X_1, \cdots, X_t) \cdot P(X_1, \cdots, X_t).$$

Proof The proof follows from the concept of conditional probability, Theorem 8 and the discussions following that theorem.

We have not computed $p_2(x,d,s)$ and leave the details of its computation as well as its comparison with $p(x,d,s)$ to the reader.

5.2 Indirect Reduction - Type T_3

Assume that the range equation

$$X = X_1 + X_2 + \cdots + X_d$$

is not T_1 reducible. There are d other range equations and the X_is occur in different equations. Assume that c out of these d equations are T_1 reducible. These reduced equations are then substituted and if this causes a reduction in X we say X is T_3 reducible. This generalizes the column reduction discussed in Example 4. The analysis can be carried out similar to the previous section. We omit the details and summarize the results in the next theorem.

Theorem 10 *Let the range equation $X = X_1 + \cdots + X_d$ be not T_1 reducible. There are other range equations so that X_i occurs in the ith equation and c, $1 \le c \le d$, of these range equations are T_1 reducible. The range X is T_3 reducible iff one of the following conditions is true:*

1) $s = d = 2$ and $y = x + 1$. The range X is reduced to a single point for $c = 1$.

2) $s = 3$, $d = 2$ and $y = x + 1$. The range X is reduced to $[3x - 2, 3x - 1]$ for $c = 1$ and is reduced to a single point for $c = 2$.

3) $s \geq 4$, $d = 2$ and $y = x + 1$. The range X is reduced to $[(x-1)s+c, sx - 1]$ and c is the amount of reduction in X.

4) $d \geq 3$, $1 \leq c \leq (s-1)(s_1 - 1)$ and $y = x + d - 1$. The range is reduced to $[(x-1)s + s_c, xs - 1]$ where the amount of reduction is $s_c = c(s_1 - 1)$.

5) $d \geq 3$, $(s-1)/(s_1 - 1) < c < s(d - q - 2)/(s_1 - 1)$ and $y = x + d - \alpha - 1$ where $\alpha = c(s_1 - 1)\underline{divs}$. The range X is reduced to $[(x-1)s + s_c, xs - 1]$ where the amount of reduction is $s_c = c(s_1 - 1)\underline{mods}$.

6) $d \geq 3$, $s(d - q - 2)/(s_1 - 1) \leq c \leq d$ and $y = x + q + 1$. The range X is reduced to $[(x-1)s + s_c, xs - 1]$ where the amount of reduction is $s_c = c(s_1 - 1)\underline{mods}$.

Example 9 Let $d = 4$, $s = 3$ and $x = 5$. Consider the range equation ($y = 8$)

$$[12, 14] = [3, 5] + [0, 2] + [6, 8] + [3, 5]. \tag{32}$$

Suppose $c = 2$ and the following equations are known.

$$\begin{aligned}
[18, 20] &= [0, 2] + [15, 17], \\
[30, 32] &= [6, 8] + [21, 23].
\end{aligned}$$

These equations are T_1 reducible and the ranges $[0, 2]$ and $[6, 8]$ are reduced to $[1, 2]$ and $[7, 8]$. Substituting these in (32) we get

$$[12, 14] = [3, 5] + [1, 2] + [7, 8] + [3, 5].$$

It follows at once that there is an exact disclosure since the sum ranges on the right hand side is $[14, 20]$ and

$$14 = 3 + 1 + 7 + 3$$

is the only solution. It is easy to verify that there is exact disclosure when any two of the ranges on the right hand side of (32) are reduced by 1. Notice that this is Case 4 of Theorem 10.

6 Concluding Remarks

A range equation is a consequence of the mutually exclusive ranges in $\mathcal{C}$, the integrity constraints of a database and the additive property of the statistic (count). Interval analysis and additive interval functions are well known methods applied to problems in numerical analysis and probability theory. The interesting and significant part of our work is to combine range arithmetic with probabilistic analysis for studying the security of statistical databases.

As remarked earlier, the analysis done in this paper is purely quantitative and probabilistic. The assumption of uniform probability distribution is both simple and natural, although it may not be accurate for real-life databases. Even with this simple assumption, the analysis is non-trivial and incomplete. An important unresolved issue is the determination of the probability distribution $P_{ij}(x, m, n, s)$, the probability that i rows and j columns, $1 \leq i \leq m, 1, \leq j \leq n$, are completely reduced in a $m \times n$ range matrix for a given x and s. Yet, the results reported in the paper should convince the reader that the range response strategy protects small counts all the time and large counts most of the time.

Finally, we emphasize that the analyses of range equations must be carried out and interpreted with the semantics of data. For, two identical ranges occurring in a range equation are the outputs for different queries. Since different queries have different query expansions, these identical ranges may behave differently when their associated queries are expanded further. Hence in indirect reductions identical ranges must be treated as distinct - in the sense that in any mathematical analysis formal mathematical substitutions must be done with caution. Due to two important reasons, our analysis does not suffer from this context sensitive semantics: i) We perform the analysis at the level of range equations and not at the level of queries that gave rise to the equations; and ii) All count queries are equiprobable, for each query all ranges are equiprobable and for a given range all the consecutive integers are equally likely to be true counts. For a given range and m, n, s, we have computed the number of $m \times n$ matrices with range entries. Based on this enumeration and the number of reducible matrices for a selected set of test data, we conclude that there are a large number of true counts that are not revealed. Hence both the formal analysis and computational experience support the view that range output for count queries is a secure strategy for statistical databases.

References

[1] Achugbue, J.O. and Chin, F.Y., "The Effectiveness of Output Modification by Rounding for Protection of Statistical Databases", *INFOR* 17(3), pp. 209-218 (1979).

[2] Alagar, V.S., "Probabilistic Analysis of Range Equations with Applications to Statistical Database Security", Submitted for publication (1988).

[3] Alagar, V.S., "Range Response: an Output Modification Technique and an Analysis of its Effectiveness for Security of Statistical Databases", Technical

Report, Department of Computer Science, Concordia University, Montreal, Canada (1983).

[4] Chin, F.Y. and Ozsoyoglu, G., "Statistical Database Design", *ACM Trans. on Database Systems* 6(1), pp. 113-139 (1981).

[5] DeMillo, R., Dobkin, D. and Lipton, R., "Combinatorial Inference", in *Foundations of Secure Computations*. Ed: Richard A. DeMillo, David P. Dobkin, Anita K. Jones and Richard J. Lipton, Academic Press (1978).

[6] Denning, D.E., Schlörer, J. and Wehrle, E., "Memoryless Inference Controls for Statistical Databases", Computer Science Department, Purdue University, West Lafayette, IN 47907, USA and Universität Ulm, *Klinische Dokumentation*, Eythstr. 2, D-7900 Ulm, West Germany (1982).

[7] Denning, D.E. and Denning, P.J., "Data Security", *Computing Surveys.* 11,3, pp. 227-249 (1979).

[8] Denning, D.E., and Denning, P.J., and Schwartz, M.D., "The Tracker: a Threat to Statistical Database Security". *ACM Trans. Database Syst.* 4(1), pp. 76-96 (1979).

[9] Denning, D.E., and Schlörer, J., "A Fast Procedure for Finding a Tracker in a Statistical Database", *ACM Trans. Database Syst.* 5(1), pp. 88-102 (1980).

[10] Denning, D.E., "Secure Statistical Databases with Random Sample Queries", *ACM Trans. Database Syst.* 5(3), pp. 291-315 (1980).

[11] Felligi, I.P., "On the Question of Statistical Confidentiality", *J. Amer. Stat. Assoc.* 67(337), pp. 7-18 (1972).

[12] Feligi, I.P. and Philips, J.L., "Statistical Confidentiality: Some Theory and Applications to Data Dissemination", *Annals Econ. Soc'l. Measurment* 3(2), pp. 399-409 (1974).

[13] Hoffman, L.J. and Miller, W.F., "Getting a Personal Dossier from a Statistical Data Bank", *Datamation* 16(5), pp. 74-75 (1970).

[14] Nargundkar, M.S. and Saveland, W., "Random Rounding to Prevent Statistical Disclousure", *Proc. Amer. Stat. Assoc., Soc. Stat. Sec.*, pp. 382-385 (1972).

[15] Schlörer, J., "Identification and Retrieval of Personal Records from a Statistical Data Bank", *Methods Inform. Med.* 14(1),pp. 7-13 (1975).

[16] Schlörer, J., "Confidentiality of Statistical Records: a Threat Monitoring Scheme for On Line Dialogue", *Methods Inf. Med.* 15(1), pp. 36-42 (1976).

[17] Schlörer, J., "Confidentiality of Security in Statistical Data Banks", pp. 101-123, in *Data Documentation: Some Principles and Applications in Science and Industry*; Proc. Workshop on Documentation, Ed: W. Guas and R. Henzler, Verlag Dokumentation, Munich, Germany (1977).

[18] Schlörer, J., "Disclosure from Statistical Databases: Quantitative Aspects of Trackers", *ACM Trans. on Database Syst.* 5(4), pp. 467-492 (1980).

[19] Schlörer, J., "Security of Statistical Databases: Multi-dimensional Transformations", *ACM Trans. on Database Syst.* 6(1), pp. 95-112 (1981).

[20] Schlörer, J., "Security of Statistical Databases: Ranges and Trackers", *Klinische Dokumentation*, Universität Ulm, W. Germany (1981).

[21] Schlörer, J., "Empirical Investigations on the Identification Risk in Statistical Databases", *Klinische Dokumentation*, Universität Ulm, W. Germany (1982).

[22] Schlörer, J., "Query Based Output Pertubations to Protect Statistical Databases", *Klinische Dokumentation*, Universität Ulm,W. Germany (1982).

[23] Sicherman, G.L., De Jonge, W. and DeRiet, Riend P. Van, "Answering Queries without Revealing Secrets", *ACM Trans. Database Syst.* 8(1), pp. 41-59 (1983).

Record Encryption in Distributed Databases

Thomas Hardjono

Department of Computer Science
University College
University of New South Wales
Australian Defence Force Academy
Canberra, ACT 2600, AUSTRALIA

Abstract

This paper investigates a possible application of the Rivest-Shamir-Adleman cryptosystem for the encryption of records in a distributed database system. A simple scheme is presented in which sites of the distributed database share a common modulus, while keeping another local secret modulus.

Keywords: Data encryption, Security, Database systems, Distributed database systems.

CR Categories: E.3, H.2.0, H.3

1 Introduction

With the increasing developments in distributed database systems the security of the data stored and exchanged between sites becomes an increasingly important issue. In this paper a simple application of the Rivest-Shamir-Adleman cryptosystem (RSA) [1] is presented. Each site S_i of the distributed database system share a common modulus N_G and keeps a local secret modulus N_i. Records are encrypted using a combination of the two, allowing the records to be exchanged between sites without too much overhead in their preparation. The simple scheme can be applied as an extension of the work in [2], which will be discussed briefly in the paper.

The work here is related to several other efforts to use encryption as a method to secure records in database systems. One possible context in which this work can be placed is that of the idea of "security filters" [3,4], where all the interaction between the user and the database management system passes through a security module which filters out data of higher security level than the user's security clearance. The security filter idea is also realized in the integrity-lock approach to database security [5,6,7], whereby records are kept in plaintext form but checksums are used to detect indirect illegal tampering of the records in the database. The work in [2] presents a possible multilevel organization of keys for the encryption of records via the security filter idea, and can be viewed as being related to the work in [8] and in [3]. In this paper we extend the ideas in [2] to distributed database systems. The unit of encryption used in the following discussions is a single record, however, the ideas can be applied to smaller units such as fields within records. The use of checksums is also assumed, allowing the detection of illegal tampering. The checksums are for local use, and are not included as answers to remote queries.

In this paper we assume that at each site the commercial database system is a relational one, also commonly referred to as an "off-the-shelf" database system. Due to the complex nature and management of distributed databases we will assume a homogeneous distributed database system belonging to one given organization. Furthermore, it is assumed that some form of secure storage exists at each site to hold the cryptographic information at each site. One form of such storage is a "tamper-proof" device, such as a smartcard. With a smartcard a small amount of cryptographic information, such as keys, can be maintained securely and loaded by the trusted database administrator of each site into main memory of the local system. No copies of cryptographic information should be stored on (unsecure) secondary storage.

An important additional assumption is that user and site authentication is performed whenever a query site S_j commences interaction with a remote site S_i $(i \neq j)$. One example of a user authentication mechanism that can also be used for site authentication is that of an authenticating "certificate", given in [9] and [10]. Information regarding the user is placed inside the user's smartcard which is used in conjunction with an access control terminal. In [10] the access terminals are connected to a public network, hence it is not unreasonable to assume that such access terminals exists at each site in the distributed database system and that sites can perform key exchanges, such as that in [11] and [12].

2 Encrypted Records in Distributed Databases

An important theorem regarding the Euler Totient function which can be of use in conjuction with the RSA cryptosystem and the Diffie-Hellman [11] key exchange system was given by Carmichael in [13], and is presented in the following.

Theorem 1 (Carmichael) *Except when n and $n/2$ are the only numbers whose (Euler) totient is the same as that of n, the congruence $x^{\phi(n)} \equiv 1$ holds for a modulus which is some multiple of n.*

In practical terms this means that if the equation $\phi(n) = a$ has more than two solutions n and $n/2$, then

$$M^{\phi(n)} \equiv 1 \bmod (c.n) \tag{1}$$

holds for any constant c $(c > 0)$, where $(M, (c.n)) = 1$.

Using this theorem let $N_G = P_G.Q_G$ and let $\alpha = lcm(P_G - 1, Q_G - 1)$ as in the RSA cryptosystem. Here P_G and Q_G are suitable primes. Both N_G and α are secrets shared among all the sites, however primes P_G and Q_G are not known except to a trusted centre which has the responsibility of issuing sites with their cryptographic information. We assume that the trusted site behaves similar to those described in identity-based cryptosystems, an example of which is presented in [12]. Here we also assume that the sites of the distributed database system which share the common information can be trusted to some extent, and one site should not be treated as an "opponent" with respect to the other sites. Note, however, that site authentication is still required to prevent masquerading.

Knowing the shared values N_G and α each site S_i then chooses a sufficiently large secret integer N_i where $N_i = P_i.Q_i$. Here P_i and Q_i are large enough primes or integers containing large enough prime factors. Note that N_i should satisfy Carmichael's theorem stated previously. For ease of discussion, let each site S_i have two pairs of encryption-decryption keys (K_i^e, K_G^d) and (K_i^e, K_i^d) satisfying:

$$K_i^e . K_G^d \equiv 1 \bmod \alpha \qquad (2)$$

and

$$K_i^e . K_i^d \equiv 1 \bmod \beta \qquad (3)$$

where $\beta = \phi(N_i)$ (in this case ϕ is the Euler Totient function).

Encryption of the v-th record R_i^v at site S_i is performed using the encryption key K_i^e modulo $N = N_G.N_i$ as follows:

$$(R_i^v)^{K_i^e} \bmod N \equiv C \qquad (4)$$

Depending on the use of the encrypted records, the record cryptogram C can be converted into two smaller cryptograms. First, for local use, C can be reduced modulus N_i as:

$$C \bmod N_i \equiv C_i^v \qquad (5)$$

which is decryptable using K_i^d. For use with other sites, C can be reduced modulus N_G as:

$$C \bmod N_G \equiv C_G^v \qquad (6)$$

which is decryptable using K_G^d.

When a given record R_i^v is kept in the local database, it is stored in the form of the cryptogram $\overline{C}$ where

$$\overline{C} = C \oplus r_i^v \qquad (7)$$

Here r_i^v is a sufficiently large secret random number kept by site S_i. This random number should be unique for each record R_i^v in the database at site S_i, and should be easily generated using some unique property of the record, such as the unique record identifier or record search key (search field). Equation 7 is done so that if an opponent compromises N_G at site S_j, and decides to attack site S_i ($i \neq j$), he or she cannot simply access the encrypted records C_G^v at site S_i by reducing C modulo the compromised N_G.

When site S_i wants to answer queries from a remote query site S_j ($i \neq j$) it must agree upon a working key with the query site. Site S_i can recover C from $\overline{C}$ since it knows r_i^v and further reduce it modulus N_G to get the cryptogram C_G^v. Using some form of key exchange system (such as that in [11] and [12]), site S_i and the query site S_j determine a common working encryption and decryption key (W_{ij}^e, W_{ij}^d) where

$$W_{ij}^e . W_{ij}^d \equiv 1 \bmod \alpha$$

Before site S_i sends the encrypted record to the query site, it "prepares" the record by further encrypting it using key

$$W_G \equiv W_{ij}^e . K_G^d \bmod \alpha \tag{8}$$

as follows:

$$(C_G^v)^{W_G} \equiv (M^{K_i^e})^{W_{ij}^e . K_G^d} \equiv M^{W_{ij}^e} \equiv C_{ij} \bmod N_G \tag{9}$$

C_{ij} is sent to the query site who decrypts it using W_{ij}^d. Note that the records are never in plaintext form except when it used for answers to local queries via the filter. This implies that some method of (encrypted) record searching exists. In this paper we consider the issue of record searching in encrypted databases as a separate problem, and it will only be briefly discussed in section 6.

3 Multilevel Encryption

In [2] a multilevel encryption scheme was proposed which was applied to centralized database systems, via the security filter. The notation in this section is chosen for convenience and should not be confused with those in other sections. In brief, the scheme uses a number of pairs of encryption and decryption keys, derived from:

$$D(N_i) = (k.\phi(N_i)) + 1 \tag{10}$$

where k is an integer greater than zero. Any two factors of $D(N_i)$ whose product equals $D(N_i)$ can be used as the encryption and decryption key. Here N_i should be composed of two or more large secret primes and $\phi(N_i)$ in this case is the Euler Totient function. In the original scheme all the keys and modulus N_i were kept secret, hence the effort of solving the discrete logarithm and factorization problems are harder to start with.

The multilevel property stems from the fact that given a pair of divisors of a given $D(N_i)$ (for a suitable N_i and k) whose product equals $D(N_i)$, these keys can encrypt and decrypt data which have already been encrypted using some of the other keys generated from the same $D(N_i)$. Say $D(N_i)$ has n suitable divisors. Let the pairs of these suitable divisors of $D(N_i)$ be

$$(d_1, d_n), (d_2, d_{n-1}), \ldots, (d_k, d_{n-k+1})$$

where the product of each of the pairs equal $D(N_i)$.

To each of the security levels we assign a pair of these divisors. If we encrypt data using d_{n-j} and we find that $d_{j+1} \mid d_k$, then we can decrypt it using d_k because $d_k = \delta.d_{j+1}$ for some integer $\delta > 0$ and positive $j < n$. Thus we have:

$$d_{n-j}.d_k \equiv d_{n-j}.(\delta.d_{j+1}) \equiv \delta \bmod \phi(N_i) \tag{11}$$

and we proceed to calculate

$$d_k/\delta = d_{j+1} = d_{decrypt} \tag{12}$$

where the encrypted data is finally decrypted using $d_{decrypt}$.

Assuming we have a three level security classification, labelled A-level, B-level and C-level (in decreasing security classification) we select from the divisors of $D(N_i)$ three pairs (d_1, d_n), (d_2, d_{n-1}) and (d_3, d_{n-2}) where $d_1 \mid d_3$, $d_1 \mid d_2$ and $d_2 \mid d_3$. We assign (d_3, d_{n-2}) to the A-level and use d_{n-2} to encrypt all A-level data (or records). Similarly we assign (d_2, d_{n-1}) to the B-level data and use d_{n-1} for encryption, and (d_1, d_n) to the C-level and use d_n for encryption.

Using equations 11 and 12 a user's clearance can be determined, and access to a record of a given security level is permitted by the filter only when $d_{decrypt}$ is greater than zero and matches one of d_1, d_2 or d_3. Note that if $d_{decrypt}$ does not match any of the three keys, decryption will not result in the correct plaintext message. Further information on the multilevel idea can be found in [2].

4 Application to Distributed Databases

In this section we will apply the idea in [2] to distributed databases. Given that the distributed database system is homogenous, controlled under one organization it is reasonable to assume that a trusted centre exists, and that some common agreement and the sharing of secret cryptographic information occurs between the sites. Furthermore, it is also reasonable to assume that different sites may have a different arrangement of security classifications. Hence, a user may have a different (lower or higher) security clearance at a remote site to that of his or her home site. For illustration purposes we will assume that there are five security levels available to each site, ranging from the A-level to E-level in decreasing security classification. The notation in this section will follow that in section 2 with minor modifications to denote the security clasification.

Previously the shared secret between the different sites were the modulus N_G and α. Now each site will also share a set of keys following the description in section 3. First, the trusted centre issues the group modulus N_G, α and five group secret decryption keys, namely K_{GA}^d, K_{GB}^d, K_{GC}^d, K_{GD}^d and K_{GE}^d corresponding to the five security levels. Note that the five keys follows the multilevel idea in in section 3. Each site then chooses its own secret modulus N_i. Using the five shared secret keys, each site S_i calculates two sets of its own secret keys, each containing five keys corresponding to the five security levels. The first set of keys are of the form $K_{i\chi}^e$ and are used for local queries, while the second set of keys are of the form $K_{i\chi}^d$ and are used to answer remote queries ($\chi = A, \ldots, E$). For a given security level, the corresponding keys from the two sets must satisfy equation 2 and 3. Thus, for the key $K_{G\chi}^d$ the key $K_{i\chi}^e$ calculated by site S_i must satisfy:

$$K_{i\chi}^e \cdot K_{G\chi}^d \equiv 1 \bmod \beta$$

as in equation 3 and the key $K_{i\chi}^d$ must satisfy:

$$K_{i\chi}^e \cdot K_{i\chi}^d \equiv 1 \bmod \alpha$$

as in equation 2 ($\chi = A, \ldots, E$). Record encryption uses key $K^e_{i\chi}$ modulo N as in equation 4.

The trusted centre issues to each user information necessary to authenticate the user, which is stored in a "tamper-proof" device such as a smart-card. In addition, the decryption key $K^d_{G\chi}$ is also placed inside the smart-card, for a user with security clearance χ ($\chi = A, \ldots, E$) at his or her home site. Following the description of an authenticating certificate in [9] and [10], this key is included in the certificate that is used to authenticate a user at a local site. Furthermore, the certificate is sent to the remote site by an access control terminal so that the remote database system can determine the user's security clearance with respect to the security clearance configuration at that remote site. A particular user may have clearance for B-level data at a site, yet have a lower security clearance at another site. This idea of differing security clearance configurations between sites in a distributed database system is not unreasonable, and may be tailored by the needs of the organization.

When a certificate arrives at a remote site S_i, and is decrypted and authenticated, the user's security level is checked by way of solving equations 11 and 12 using key $K^d_{G\chi}$ in the certificate and key $K^e_{i\chi}$ of level χ ($\chi = A, \ldots, E$) [2]. Once the security level has been determined, records in the form of $\overline{C}$ that have security level equal or lower than that of the user's clearance can be retrieved, converted to C and then reduced modulo N_G to become C^v_G. As explained in section 2, after the two sites agree on a new working key modulo the shared N_G, the (encrypted) records of security level χ are "prepared" by site S_i by way of encrypting them using

$$W_G \equiv W^e_{ij} . K^d_{i\chi} \bmod \alpha$$

as in equation 8, which is decryptable using W^d_{ij}. An example using small numbers is given in the following section.

5   Example

In this section we will consider a small example, assuming that five security levels exists. Each site has three security levels which are chosen from the five security levels.

First, let the trusted centre choose the following five group shared keys $K^d_{GA} = 3125$, $K^d_{GB} = 625$, $K^d_{GC} = 125$, $K^d_{GD} = 25$ and $K^d_{GE} = 5$. Let $P_G = 149$ and $Q_G = 167$, $N_G = 24883$ and $\alpha = lcm(149 - 1, 167 - 1) = 12284$.

Site S_i then chooses a secret modulus $N_i = 15317 = 17^2.53$, where $\beta = \phi(N_i) = 14144$. (Note that here $\phi(N_i)$ satisfies Carmichael's theorem, with $\phi(15317) = \phi(21783) = \phi(22593) = \phi(28352) = \phi(29044) = \phi(30124) = \phi(30634) = \phi(35440) = \phi(42528) = \phi(43566) = \phi(45186) = \phi(53160) = \phi(N_i) = 14144$). The site then proceeds to calculate $K^e_{i\chi}$ ($\chi = A, \ldots, E$) using β. This results in the five keys $K^e_{iA} = 3037$, $K^e_{iB} = 1041$, $K^e_{iC} = 5205$, $K^e_{iD} = 11881$ and $K^e_{iE} = 2829$. Then for each these five key and α the next set of keys are calculated: $K^d_{iA} = 12013$, $K^d_{iB} = 12225$, $K^d_{iC} = 2445$, $K^d_{iD} = 11461$ and $K^d_{iE} = 5241$.

Consider the record or data $M = 31$ of A-level security classification stored at site S_i. The record M is encrypted using $K_{iA}^e = 3037$ modulo

$$N = N_G.N_i = 24883.15317 = 381132911$$

as follows:

$$M^{K_{iA}^e} \bmod N \equiv C$$

$$(31)^{3037} \bmod 381132911 \equiv 103919594$$

This record is stored as $\overline{C}$ as in equation 7 using a random r_i^v. When the record is to be used for local purposes it is reduced modulus N_i as:

$$C \bmod N_i \equiv C_i^v$$

$$103919594 \bmod 15317 \equiv 9066$$

which is decryptable using $K_{GA}^d = 3125$ as follows:

$$(C_i^v)^{K_{GA}^d} \bmod N_i \equiv M$$

$$(9066)^{3125} \bmod 15317 \equiv 31 \equiv M$$

For use with remote query sites, the record is reduced modulus N_G as:

$$C \bmod N_G \equiv C_G^v$$

$$103919594 \bmod 24883 \equiv 8186$$

Let the working keys between the site S_i and the query site S_j $(i \neq j)$ be the pair $(W_{ij}^e, W_{ij}^d) = (1249, 5665)$ where $1249.5665 \equiv 1 \bmod \alpha$. Then the key to convert C_G^v to C_{ij} is calculated as follows:

$$W_G \equiv W_{ij}^e.K_{iA}^d \bmod \alpha$$

$$W_G \equiv 1249.12013 \equiv 5473 \bmod \alpha$$

and W_G is used to encrypt C_G^v as:

$$(C_G^v)^{W_G} \equiv (8186)^{5473} \equiv 15375 \equiv C_{ij} \bmod N_G$$

which is sent as the query answer by site S_i to site S_j.

The query site S_j can decrypt C_{ij} using $W_{ij}^d = 5665$ as:

$$(C_{ij})^{W_{ij}^d} \equiv (15375)^{5665} \equiv 31 \equiv M \bmod N_G$$

6 Comments and Remarks

From the previous sections it is clear that the security of the shared N_G, α and the keys $K_{G\chi}^d$ ($\chi = A, \ldots, E$) among all the sites is imperative. These cryptographic information should always be kept in secure storage, or in a smartcard carried by the trusted database administrator of each site. The only guard against an opponent that knows N_G and α is the fact that the records are encrypted modulus N and then further modified via the random r_i^v. Note that the keys $K_{G\chi}^d$ ($\chi = A, \ldots, E$) can be discarded by each site after the local keys are calculated. The encryption of the records modulus N sets them ready to be accessed locally or sent as answers to querying sites. This eliminates the need to decrypt the records into plaintext modulo N_i and encrypting it again using a common working key modulo N_G.

The use of a local secret modulus N_i reflects the need for each site to be autonomous to some extent, although it must be recognized that in a distributed database system a trade-off may have to be found between security and autonomy. A site can "reconfigure" the security of its database system by choosing a new and suitable modulus N_i and the corresponding new encryption keys.

The immediate disadvantage of using N_G and N_i together is the large size of the resulting cryptogram records. Given that N_i is secret, it may be of smaller size compared to N_G. However, a trade-off must be found between speed, security and storage space, and with the increasingly cheaper cost of space this problem may be tolerated. It is also envisaged that for an implementation of this scheme a hardware encryption module is used instead of software. With the recent advances in the area of hardware development for cryptographic purposes, this is not an unreasonable requirement.

One of the main difficulties with the encryption of records is that record searching becomes severely limited. Work in solving this problem can be found in [14], in which a low level approach has been taken in encrypting records before they are written to the disk storage. This solution, however, requires specialized database architectures, in which access to low level routines of the database systems is possible. Such low level access is rarely possible in commercial database systems.

Given that a security filter is used on a commmercial database system, one possible way around the problem is to use an external hashing mechanism within the filter that maintains an indicator of the records stored in the database. Assuming that each record has a unique plaintext identifier, a hashing mechanism can be use which will hash records to an appropriate *Record Identifier Set* rather than to buckets. One suitable hashing scheme is *multiattribute hashing* [15]. In multiattribute hashing each field type of the records in the database is assigned a hash function. The output of the hash function depends on the division of the range of the possible values for the field. To find out in which bucket (or in this case, which record identifier set) a record should be placed, the output of the hash functions are concatenated forming the "address" of the bucket.

By using the record identifier sets, a query can be satisfied by calculating its query signature which is hashed to the most appropriate identifier set. This set should contain the unique record identifiers of those records in the database that have the highest probability of satisfying the query. Searches into the database is based on the identifiers in the record identifier sets. The use of such a hashing mechanism may provide a temporary solution to the problem of record searching in encrypted databases. It also points to the fact that in the long term specialized database architectures are needed to secure records and to perform searching of encrypted records.

7 Conclusion

In this paper a simple scheme based on the RSA cryptosystem for the encryption of records in distributed databases has been presented. Each site keeps a local secret modulus, while sharing another modulus with the other sites of the distributed database system. This shared modulus is kept secret among the sites, and the combination of the two values allow encrypted records to be converted easily for use between sites. The security of the scheme depends on the ability of each site in keeping secret the shared cryptographic information. The main disadvantage of the scheme is the large record cryptograms stored at each site. However, with the increasingly cheaper cost of storage this may not present a problem. The scheme can be used with the "security filter" idea and a possible solution to the problem of record searching in encrypted database systems is suggested.

Acknowledgements

The author would like to thank Dr. Joseph Pieprzyk, Prof. Jennifer Seberry, Assoc. Prof. Svein Knapskog and Dr. Chris Lokan for the many important comments and suggestions.

References

[1] R. L. Rivest, A. Shamir, and L. Adleman, "A method for obtaining digital signatures and public-key cryptosystems," *CACM*, vol. 21, pp. 120–128, Feburary 1978.

[2] T. Hardjono and J. Seberry, "A multilevel encryption scheme for database security," in *Proceedings of the 12th Australian Computer Science Conference*, pp. 209–218, Feburary 1989.

[3] D. E. Denning, "Cryptographic checksums for multilevel database security," in *Proceedings of the 1984 IEEE Symposium on Security and Privacy*, pp. 52–61, IEEE Computer Society, 1984.

[4] D. E. Denning, "Commutative filters for reducing inference threats in multilevel database systems," in *Proceedings of the 1985 Symposium on Security and Privacy*, pp. 134–146, IEEE Computer Society Press, Apr 1985.

[5] R. D. Graubart, "The integrity-lock approach to secure database management," in *Proceedings of the 1984 IEEE Symposium on Security and Privacy*, pp. 62–74, IEEE Computer Society Press, Apr 1984.

[6] R. D. Graubart and S. Kramer, "The integrity lock support environment," in *Proceedings of the 2nd IFIP International Conference on Computer Security, IFIP Sec/'84* (J. H. Finch and E. G. Dougall, eds.), pp. 249–268, North-Holland, 1984.

[7] R. D. Graubart and K. J. Duffy, "Design overview for retrofitting integrity-lock architecture onto a commercial dbms," in *Proceedings of the 1985 IEEE Symposium on Security and Privacy*, pp. 147–159, IEEE Computer Society, Apr 1985.

[8] D. E. Denning, "Field encryption and authentication," in *Advances in Cryptology: Proceedings of CRYPTO 83* (D. Chaum, ed.), pp. 231–247, Plenum Press, 1983.

[9] J. K. Omura, "A computer dial access system based on public-key techniques," *IEEE Communications Magazine*, vol. 25, pp. 73–79, Jul 1987.

[10] P. J. Lee, "Secure user access control for public networks," in *Proceedings of AUSCRYPT '90*, Jan 1989.

[11] W. Diffie and M. Hellman, "New directions in cryptography," *IEEE Transactions on Information Theory*, vol. IT-22, pp. 397–427, June 1976.

[12] K. Koyama and K. Ohta, "Security of improved identity-based conference key distribution systems," in *Proceedings of EUROCRYPT '88 (Lecture Notes in Computer Science No. 330)* (C. G. Gunther, ed.), pp. 11–19, Springer-Verlag, 1988.

[13] R. D. Carmichael, "Notes on the simplex theory of numbers," *Bulletin of the American Mathematical Society*, vol. 15, pp. 217–223, 1909.

[14] R. Bayer and J. K. Metzger, "On the encipherment of search trees and random access files," *ACM Transactions on Database Systems*, vol. 1, pp. 37–52, Mar 1976.

[15] J. B. Rothnie and T. Lozano, "Attribute based file organization in a paged memory environment," *Communications of the ACM*, vol. 17, pp. 63–69, Feb 1974.

SECTION 10

IMPLEMENTATIONS

VLSI Design for Exponentiation in $GF(2^n)$

W. Geiselmann
D. Gollmann

Universität Karlsruhe

1 Introduction

The security of cryptographic mechanisms like the Diffie-Hellman key exchange system, the ElGamal system, or some zero-knowledge schemes [3], [6] is based on the assumed difficulty of the discrete logarithm problem. To make these schemes efficient fast exponentiation algorithms have to be provided. Very fast algorithms exist in particular for the finite field $GF(2^n)$. It should be noted that computing discrete logarithms is easier in $GF(2^n)$ than in $GF(p)$, p prime [4]. This aspect has to be considered in any comprehensive evaluation of an implementation of one of the mechanisms listed above. However, latest developments in number theory again blur this difference. In this paper we will compare two efficient VLSI designs for exponentiation in $GF(2^n)$, efficiency being defined both in terms of area requirements and time performance.

Of course, the way elements in $GF(2^n)$ are represented bears on the implementation of arithmetic operations. We will consider only polynomial basis and normal basis architectures. Dual basis multipliers have some advantages when circuits are designed from certain standard cells (e.g. TTL), but there seems to be no reason to prefer them in full custom VLSI-design, thus they will not be considered further. Exponentiation can be decomposed into square-and-multiply. Multiplication in a polynomial basis usually is performed by feedback shift registers. Squaring, although a linear operation, can in general not be implemented much more efficiently than multiplication. In a normal basis squaring becomes a cyclic shift. This advantage has to be paid for by larger and often quite cumbersome layouts for multiplication. Hence, in both cases we have to look for special classes of bases that allow efficient exponentiation and are of sufficient generality to be of practical interest.

2   Polynomial Basis

The standard way of representing elements in $GF(2^n)$ makes use of the polynomial basis $(1, \alpha, \alpha^2, \ldots, \alpha^{n-1})$ where α is a root of an irreducible polynomial $p(x) = x^n - \sum_{i=0}^{n-1} p_i x^i$. $u \in GF(2^n)$ can be written as $u = \sum_{i=0}^{n-1} u_i \alpha^i$. The product of two elements u and v is the product of the two corresponding polynomials reduced modulo $\alpha^n = \sum_{i=0}^{n-1} p_i \alpha^i$. The Horner scheme for multiplication

$$u \cdot v = u \cdot \sum_{i=0}^{n-1} v_i \alpha^i = (\ldots ((v_{n-1} u \alpha + v_{n-2} u) \alpha + \ldots + v_1 u) \alpha + v_0 u$$

gives rise to the serial input / parallel output architecture of figure 1 where the intermediate result is multiplied by α at each step.

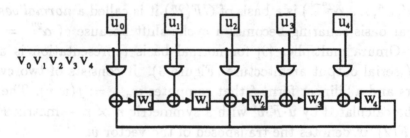

Figure 1: A serial input / parallel output polynomial basis multiplier for $GF(2^5)$ with $p(x) = x^5 + x^2 + 1$.

There seems to be no general way to obtain circuits for squaring that are distinctly better than multipliers. However, for special feedback polynomials we may do better. Consider the trinomials $p(x) = x^n + x^k + 1$, with n odd, $k < \frac{n}{2}$ even. Figure 2 shows a parallel / input serial output squarer.

This squarer produces two output bits at a time. If we adapt the above multiplier so that it accepts two input bits simultaneously we arrive at a design for "serial" square-and-multiply, i.e. square-and-multiply starting at the most significant bit of the exponent, that requires only three shift registers and performs an exponentiation in $(n+1)^2/2$ steps. Irreducible trinomials over $GF(2)$ are quite abundant, a list up to $n = 1000$ can be found in [11]. It should be noted that there exist no irreducible trinomials over $GF(2)$ if $n \equiv 0 \mod 8$.

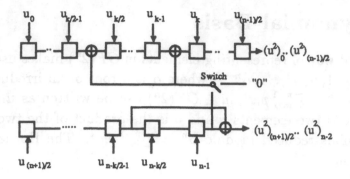

Figure 2: A parallel input / serial output squarer.

3   Normal Basis

If $(\alpha, \alpha^2, \alpha^4, \ldots, \alpha^{2^{n-1}})$ is a basis of $GF(2^n)$ it is called a *normal basis* [8]. In a normal basis squaring becomes a cyclic shift because of $\alpha^{2^n} = \alpha$. The Massey-Omura multiplier [9] for normal basis representation is a parallel input / serial output architecture (Figure 3). It consists of two cyclic shift registers and a bilinear form f that computes $w_{n-1} =: f(u, v)$. The function f can be calculated by $u \cdot F \cdot v^t$ with a symmetric $n \times n$ − matrix $F = (\varphi_{i,j})$ over $GF(2)$, v^t denotes the transposed of the vector v.

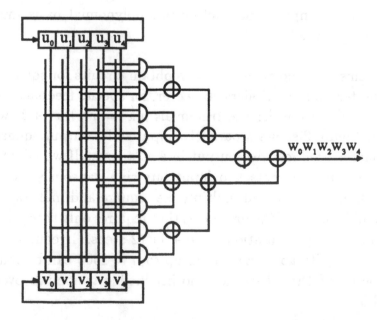

Figure 3: A parallel input / serial output normal basis multiplier for $GF(2^5)$ with $p(x) = x^5 + x^4 + x^3 + x^2 + 1$.

There is an "$AND(u_i, v_j)$" in the layout iff $\varphi_{i,j} = 1$. So $\#\{(i,j)|\varphi_{i,j} = 1\}$ is a convenient measure for the complexity of a normal basis multiplier. The matrix F and with it the complexity of the multiplier depend on the choice of the normal basis. A "random" normal basis of $GF(2^n)$ has complexity $\approx \frac{1}{2}n^2$, hence the area of an VLSI-implementation of such a multiplier will become very large. Thus it is essential to find low complexity normal bases. There are two important ways known to optimize normal basis multiplication:

- Choose a "good" normal basis. This problem was solved in [2], [10] for most finite fields of characteristic 2 by presenting an algorithm for finding good normal bases for $GF(2^n)$ for $n \not\equiv 0 \mod 8$.

- Reduce the layout by taking advantage of some symmetry in the normal basis multiplication.

We obtain a serial input / parallel output normal basis multiplier by writing the product $u \cdot v$ in a way similar to Horners rule, i.e.

$$
\begin{aligned}
u \cdot v &= v_0 u\alpha + v_1 (u^{\frac{1}{2}}\alpha)^2 + v_2 (u^{\frac{1}{4}}\alpha)^4 + \ldots + v_{n-1}(u^{(\frac{1}{2})^{n-1}}\alpha)^{2^{n-1}} \\
&= (\ldots((v_{n-1}u^{(\frac{1}{2})^{n-1}}\alpha)^2 + v_{n-2}u^{(\frac{1}{2})^{n-2}}\alpha)^2 + \ldots + v_1 u^{\frac{1}{2}}\alpha)^2 + v_0 u\alpha.
\end{aligned}
$$

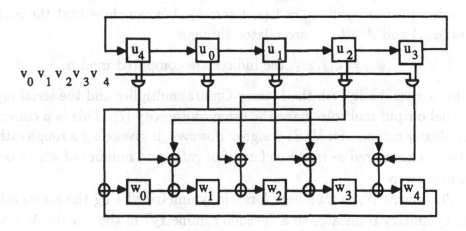

Figure 4: A serial - parallel input / parallel output normal basis multiplier for $GF(2^5)$ with $p(x) = x^5 + x^4 + x^3 + x^2 + 1$.

Multiplication is reduced to squaring, drawing square roots (which again is a cyclic shift), and multiplication by α. The multiplication $\alpha \cdot \tilde{u}$ can be represented by the multiplication of the vector $\tilde{u}$ by a $n \times n$ - matrix $A = (a_{i,j})$. Corresponding to the Massey-Omura-multiplier each "1" in the

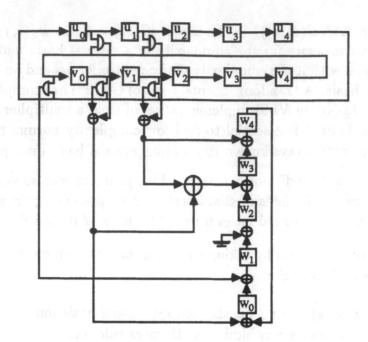

Figure 5: A parallel input / parallel output normal basis multiplier for $GF(2^5)$ with $p(x) = x^5 + x^4 + x^3 + x^2 + 1$.

matrix A corresponds to one gate in the layout of this normal basis multiplier. $u_i \cdot v$ is added to w_j iff $a_{i,j} = 1$ (s. figure 4). We can show that the matrices $F = (\varphi_{i,j})$ and $A = (a_{i,j})$ are related through

$$\varphi_{i,j} = a_{j,i+j-1}, \text{ the indices are computed mod } n.$$

So for a given basis both the Massey-Omura multiplier and the serial input / parallel output multipler have the same complexity [7]. This is a convenient complexity measure for VLSI designs. However, it gives only a rough estimate of the area required as issues as fanout of gates and number of wire crossings are not covered.

The matrix F, as a bilinear form, is symmetric. Using the above relation this symmetry translates to a "pseudo symmetry" in the matrix A: Column i is column $n - i$ shifted i–times ($0 \le i < n$). This fact can be used to reduce the layout of the serial input / parallel output normal basis multiplier (see figure 5).

The number of steps for exponentiation using a normal basis implementation will depend on the weight of the exponent. When we assume that all exponents are equally probable we require on average $n/2$ multiplications and n shifts, i.e. $n \cdot \frac{n}{2} + n$ steps.

4 CMOS-Layouts

In a hierarchical VLSI-CAD system one of the most important design issues during layout is the choice of an architecture that allows to place basic cells (e.g. NAND, XOR, shift register cells, ...) on the chip in such a way that the space for routing between cells becomes as small as possible.

The basic cell of our polynomial basis design is sketched in figure 6. v is the serial input, during exponentiation the output of the squarer will play this rôle. A particular feature of this design is the communication structure. Cells are abutted and while squaring cells communicate with their next neighbours only, multiplier cells skip one stage in an array to get their data. In this layout, as well as in the layout for the normal basis multiplier presented below, only dynamic shift registers were employed. The area of a current 2μ – CMOS layout for $n = 333$ is less than $5mm \times 5mm$ where only $3mm \times 3mm$ are occupied by the exponentiator proper.

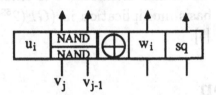

(Size of the cell: $65\mu \times 375\mu$)

Figure 6: The floorplan for the basic cell of a polynomial basis exponentiator.

For any multiplier for large n it is impossible to perform all inputs and outputs in parallel on the chip. Therefore extra shift registers u_in, v_in, w_out, for input and output have been added to the design of the normal basis multiplier from figure 5. To minimize the space for routing between basic cells the shift registers u_i, v_i, 2 *NANDs* and 1 *XOR* should be close together (for calculating $u_i \cdot v_{n-1} \oplus u_{n-1} \cdot v_i$).

The most important application for normal basis architectures is *exponentiation*. So it should be possible to transfer the result of the multiplication (w) in one time cycle into one of the multiplication registers (for example v). Therefor w_i with one *XOR* should be close to v_i. This leads to the normal basis cell shown in figure 7.

The inputs and outputs on the right hand side of all the cells have to be connected corresponding to the matrix A defined above. This routing and placement of the n "normal basis cells" can be performed by automatic

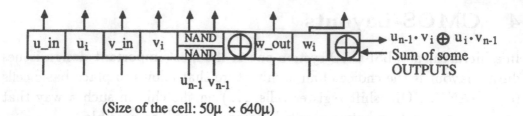

(Size of the cell: $50\mu \times 640\mu$)

Figure 7: The floorplan for the basic cell of the Normal Basis Multiplier shown in figure 5.

placement and routing tools on the VLSI graphic system. Using this architecture, a 2μ – CMOS process, and a normal basis of the field $GF(2^{332})$ with complexity 1321 ($< 4 \cdot 332$) a multiplier for $GF(2^{332})$ was realized on a chip of about $8.5mm \times 2.8mm \approx 24mm^2$. A normal basis of the Mersenne prime field $GF(2^{607})$ with complexity 3621 ($< 6 \cdot 607$) leads to a normal basis multiplier of about $1cm^2$. A similar architecture is employed in an encryption chip based on normal basis multiplication in ($GF(2^{593})$) that is available as a commercial product [5].

5 Conclusion

The fact that in a normal basis representation squaring is a cyclic shift is by itself not sufficient to make normal basis exponentiators superior to polynomial basis architectures. Good normal bases are an essential prerequisite. At the same time, "good polynomial bases exist, e.g. bases over trinomials. The basic cell of the resulting polynomial basis exponentiator is slightly smaller than the normal basis basic cell, if we omit the extra cells added to the normal basis design to enhance communication. Additional routing area, however, can almost be neglected in the polynomial basis design, so we suggest that the use of normal basis designs is indicated in particular in situations where there are substantially less multiplications than squarings. This can be achieved either by limiting the weight of the exponent or by employing special exponentiation algorithms (see e.g. [1]). Otherwise we expect that the simple and very regular structure of polynomial basis devices will give smaller and faster VLSI designs.

References

[1] G.B.Agnew, R.C.Mullin, S.A.Vanstone; *Fast Exponentiation in $GF(2^n)$*; in Proc. Eurocrypt 88, ed. C.G.Günther, Springer LNCS 330, pp.251-255, 1988.

[2] D.W.Ash, I.F.Blake, S.A.Vanstone; *Low Complexity Normal Bases*; Discrete applied Math. 25, 1989, pp.191-210.

[3] T.Beth; *Efficient Zero-Knowledge Identification Scheme for Smart Cards*; in Proc. Eurocrypt 88, ed. C.G.Günther, Springer LNCS 330, pp.77-84, 1988.

[4] I.F.Blake, P.C. van Oorschot, S.A.Vanstone; *Complexity Issues in Public Key Cryptography*; University of Waterloo, preprint.

[5] Calmos Semiconductor Inc., *CA34C168 DEP Datasheet*, Document #02-34168-001-8808, Kanata, Ontario, 1988.

[6] D.Chaum, J.-H.Evertse, J.van de Graaf, R.Peralta; *Demonstrating Possession of a Discrete Logarithm Without Revealing It*; Proc. Crypto 86, ed. A.M.Odlyzko, Springer LNCS 263, pp.200-212, 1987.

[7] W.Geiselmann, D.Gollmann; *Symmetry and Duality in Normal Basis Multiplication*; in Proceedings of AAECC-6 (Rome), ed. T.Mora, Springer LNCS 357, pp.230-238, 1989.

[8] R.Lidl, H.Niederreiter; *Finite Fields*; Cambridge University Press, 1984.

[9] J.L.Massey, J.K.Omura; *Computational method and apparatus for finite field arithmetic*; U.S.Patent application, submitted 1981.

[10] R.C.Mullin, I.M.Onyszchuk, S.A.Vanstone, R.M.Wilson; *Optimal Normal Bases in $GF(p^n)$*; Discrete applied Math. 22, 1988/89, pp.149-161.

[11] N.Zierler, J.Brillhart; *On primitive trinomials (mod 2)*; Inform.Control, vol.13, pp.541-554, 1968.

A Fast Modular-multiplication Module for Smart Cards

Hikaru Morita

NTT Communications and Information Processing Laboratories
Nippon Telegraph and Telephone Corporation
3-9-11 Midori-cho Musashino-shi Tokyo, 180 Japan

Background

Public-key cryptosystems such as the Rivest-Shamir-Adleman scheme (RSA) [1] are the most effective systems for secure communications and digital signatures. To guarantee security, the cryptosystems have to use significantly greater word-length modular exponentiation than conventional computer word-lengths. Accordingly, the cryptosystems are generally implemented with special hardware, and their cost is prohibitive for small-scale systems using smart cards or IC cards on which an LSI chip enciphers and deciphers important data. Therefore, modular multiplication, which is indispensable to modular exponentiation, must be speeded up and the amount of hardware must be reduced. The algorithm presented here fulfills both these requirements.

New Algorithm based on Higher Radix

If "n" is the bit length of the modulus N, then modular multiplication is represented as $A \times B$ *modulo* N where A, B, and N are three n-bit binary integers related by A, $B \in [0, N-1]$.

A conventional modular multiplication algorithm as shown in *Figure 1(a)* multiplies first and then divides the $2n$-bit product by the modulus N. This algorithm cannot accomplish high-speed computation and needs a massive amount of hardware because data transmission between processors and memories occurs frequently.

To simplify the operation, Baker [2] designed an algorithm which combines modular subtraction based on radix 4 and multiply-addition based on radix 2. This algorithm is shown in *Figure 1(b)*. However, until now, the problem of overflowing a finite computation range has prevented the development of an algorithm that uses a higher radix to speed-up modular multiplication.

The new algorithm [3] was designed to speed-up modular multiplication by using the same higher radix throughout for both multiply-addition and modular subtraction. In practice, the number of addition and subtraction steps can be reduced to half or less of that for other algorithms. This algorithm is shown in *Figure 1 (c)*.

The proposed algorithm is executed by using the following equation repeatedly.

$$R^{(k-1)} \leftarrow rR^{(k)} + b(k)A - c(k)N, \tag{1}$$

where "k" is the step number of the repeated processing, "r" is the radix number, $R^{(k)}$ and $R^{(k-1)}$ are partial remainders, $b(k)A$ is a partial product, and $c(k)N$ is a modular subtracter. To overcome the overflow problem, this algorithm uses two approaches:

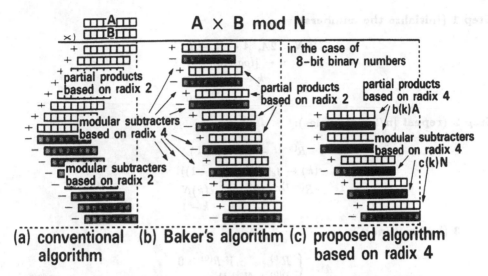

Figure 1 Comparison of modular multiplicaiton algorithms

(1) To prevent the absolute value of the next partial remainder $R^{(k-1)}$ from overflowing the upper limit N, the modular subtracter $c(k)N$ is determined using the next partial product $b(k-1)A$ in advance.

(2) To reduce the absolute range of the partial product $b(k)A$, the multiplicand A is modified from the range $[0, N-1]$ to the range $[-N/2, N/2]$.

The algorithm procedure based on radix 4 is shown in *Figure 2*.

Hardware Implementation for Smart Cards

An LSI chip on a smart card must contain the modular-multiplication module for cryptosystems as well as a microprocessor for general use. Therefore, the module cannot be allocated much hardware.

The module consists of RAM and an operation unit. The RAM stores not only variables A, B, and N, but also temporary variables such as R. The operation unit can be constructed by arranging cells as shown in *Figure 3*, into an array. However, since each cell contains about 60 gates, an operation unit having n cells may exceed the allowable number of gates. Therefore, each addition and subtraction of the modular multiplication is divided into a number of parts.

We developed a new division method which can reduce the number of memory access events. For example, if m-bit-length operation unit is used, then $\lfloor \frac{m-7}{4} \rfloor$ coefficients $c(k)$ can be calculated beforehand and the operation unit adds and subtracts $(2\lfloor \frac{m-7}{4} \rfloor n)/m$ times for $\lfloor \frac{m-7}{4} \rfloor$ addition and subtraction steps. However, the number of memory access events can be reduced to n/m. The 32-bit-length operation unit can be made with about 3 Kgates. The 512-bit modular exponentiation using a module with 3 Kbits of RAM and a 32-bit-length operation unit will be finished in about 1.7 seconds at a 4-MHz clock.

Step 1 (initialize the numbers):

$$\text{If } N < 2A, \ A \leftarrow A - N.$$
$$n \leftarrow \lfloor log_2(N) \rfloor + 1$$
$$k \leftarrow \lfloor n/2 \rfloor + 1$$
$$R^{(k)} \leftarrow 0$$

Step 2 (repeat $\lfloor n/2 \rfloor + 1$ cycles):

$$R^{(k-1)} \leftarrow 4R^{(k)} + b(k)A$$
$$c(k) \leftarrow f_c(R^{(k-1)}, b(k-1))$$
$$R^{(k-1)} \leftarrow R^{(k-1)} - c(k)N$$
$$k \leftarrow k - 1$$

Step 3 (last routine):

$$A \leftarrow \begin{cases} R^{(0)} & \text{if } R^{(0)} \geq 0 \\ R^{(0)} + N & \text{otherwise} \end{cases}$$

Notes:
• function $b(k)$ for the multiplier B: $b(k) \leftarrow -2B[2k] + B[2k-1] + B[2k-2]$ where $B[0] = 0$. Consequently, $b(k) \in \{-2, -1, 0, 1, 2\}$.

• function $f_c(R, b)$:

$$f_c \leftarrow \begin{cases} i & \text{if } L_i N_{top} < R_{top} \leq L_{i+1} N_{top} \\ 0 & \text{if } L_{-1} N_{top} \leq R_{top} \leq L_1 N_{top} \\ -i & \text{if } L_{-i-1} N_{top} \leq R_{top} < L_{-i} N_{top} \end{cases}$$

where $L_i \equiv i - 1/2 - bA_{top}/(4N_{top})$, $L_{-i} \equiv -i + 1/2 - bA_{top}/(4N_{top})$, $R_{top} \equiv \{\text{top 9 bits of } N\}$, $N_{top} \equiv \{\text{top 7 bits of } N\}$, and $A_{top} \equiv \{\text{top 7 bits of } A\}$.

The number of top bits is derived from an approximate method. L_i and L_{-i} are prepared beforehand for all b.

Figure 2 Procedure for Radix 4

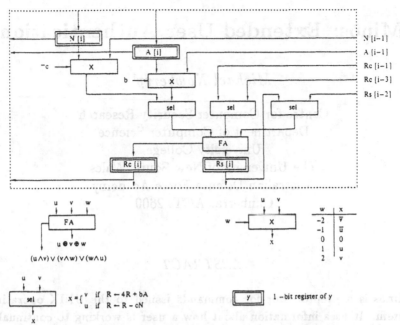

Let R represent the usual binary remainder in carry-save redundant form.
R has two components: the sum R_S and the carry R_C where $R \equiv R_S + 2R_C$.

Figure 3   Cell of "$R^{(k-1)} \leftarrow 4R^{(k)} + b(k)A - c(k)N$"

Conclusion

A new compact modular multiplication algorithm based on a higher radix can reduce the amount of processing to half or less that of conventional algorithms. When applied to smart cards, the new division method can reduce the number of memory access events. Consequently, a module containing 3 Kbits of RAM and a 3-Kgate operation unit will calculate a 512-bit modular exponentiation in under two seconds at a 4-MHz clock.

Acknowledgment

The author wishes to express his thanks to Dr Kunio Murakami who provided him with the opportunity to pursue this research. He would also like to thank Tsutomu Ishikawa and the other members of NTT LABS who discussed the research.

References

[1] Rivest, R.L., Shamir, A., and Adleman, L., "A Method for Obtaining Digital Signatures and Public-Key Cryptosystems," Comm.ACM, Vol.21(2), pp.120-126, Feb. 1978.

[2] Baker, P.W., "Fast Computation of A * B Modulo N," Electron.Lett., Vol.23, No.15, pp.794-795, July 1987.

[3] Morita, H., "A Fast Modular-multiplication Algorithm based on a Higher Radix," CRYPTO'89 Abstracts, pp.363-375, Aug. 1989.

Minòs: Extended User Authentication

Michael Newberry[1]

Centre for Computer Security Research
Department of Computer Science
University College
The University of New South Wales
Australian Defence Force Academy
Canberra. ACT. 2600

ABSTRACT

Minòs is a preprocessor for commands issued to the UNIX operating system. It uses information about how a user is working to continually authenticate them. Minòs learns with the user as they develop new habits with the computer. This paper describes in detail the plans for Minòs, and their realisation to date.

1   Introduction

> This way I went, descending from the first
>   into the second circle, that holds less space
>   but much more pain — stinging the soul to wailing.
> There stands Minòs grotesquely, and he snarls,
>   examining the the guilty at the entrance;
>   he judges and dispatches, tail in coils.
> By this I mean that when the evil soul
>   appears before him, it confesses all,
>   and he, who is the expert judge of sins,
> knows to what place in Hell the soul belongs;
>   the times he wraps his tail around himself
>   tells just how far the sinner must go down.
> The damned keep crowding up in front of him:
>   they pass along to judgement one by one;
>   they speak, they hear, and then are hurled below.   [2], p. 109

---
[1]Michael Newberry is supported in part by the Australian Telecommunications and Electronics Research Board

In Dante's Hell, Minòs was the magistrate to the damned. He examined the actions of those condemned, and accordingly assigned each a place for eternity. Unlike Dante's magistrate, computers frequently assess the character of users from their words, not from their actions. The idea of using the behaviour of a user as a guide to their identity has found little application in computer security. This paper describes the Minòs system, developed by the author since 1987. This Minòs uses information about how a user operates a computer, to identify breaches in security.

There have been three other efforts at analysing user actions in order to identify intrusions (these are summarised in [11], p. 378). Denning's *Intrusion Detection System* uses elaborate database techniques to monitor computer resource usage (see [3] for the theoretical model). To date it has used "connect time, shift of login, and location of login" ([6], p. 63) as detection metrics. Its emphasis has been on identifying an intruder from exceptional behaviour. A similar approach is reported in [4]. There, the authors looked for features that were typical of intruder sessions, in order to detect future intrusions. Their emphasis was on identifying typical "intruder behaviour". A different approach was used in [12]. Here, the author developed models of typical user behaviour, and then identified potential intrusions from their deviation from this behaviour. The emphasis was on modelling the user, and identifying intruders by their deviation from the user's normal behaviour, rather than seeking characteristic intruder behaviours. This is also the approach used by the computer system Minòs.

Minòs acts as a filter between the user and the operating system (in this case UNIX[2]). Minòs continually examines the actions of the user, looking for uncharacteristic behaviours that may indicate an intruder at work. Upon finding such behaviour, Minòs will be capable of a variety of responses, depending on the individual site's requirements (see [8], p. 171).

Minòs is capable of using many metrics to identify users. It is planned to use four in the current implementation:

- The user's choice of commands

- The user's sequencing of commands

- The user's location in the computer (That is, which directories/folders are accessed by a user)

- The way a user types (This is measured by examining the intervals between the keystrokes of the user)

---

[2]UNIX is a trademark of AT&T Bell Laboratories

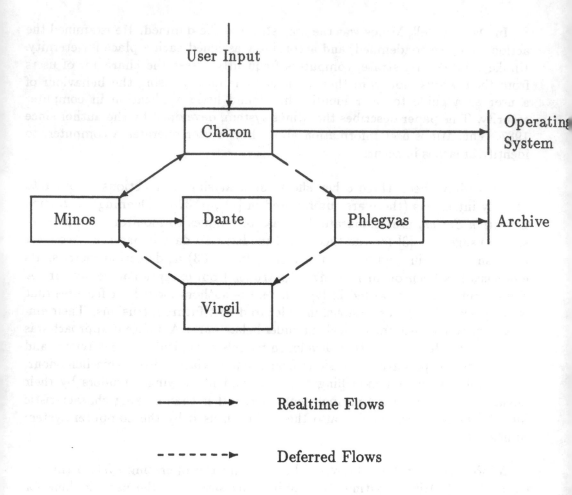

Figure 1: Process Diagram for each Metric

The first metric has already been implemented. The others are currently being implemented.

2 Minòs Overview

Figure 1 presents an overview of the principle data flows within Minòs. The Minòs system involves five processes. (All these processes are named after characters in Dante's *Inferno*.[3])

---

[3] *Dante* was the pilgrim, travelling through Hell. *Charon* was the ferryman between the world of the living and the world ruled by Minòs. *Minòs* was the magistrate of Hell. *Phlegyas*

Charon is a filter that sits between the user and the operating system. Each time the user types a command, Charon intercepts it, records it for archiving purposes, and passes it onto Minòs. Each command is also passed to the operating system for execution.

Dante is a system monitor that runs on a supervisor's console. Whenever Minòs detects any possible intruders, this information is reported to Dante, along with details about terminal location and machine.

Minòs decides the authenticity of the user. This information is then passed onto Dante. It could also be passed back to Charon, if the site management wanted Charon to take direct action (such as logging the user off, or disallowing some commands).

Virgil supplies the artificial intelligence for this system. Virgil uses the data supplied by Charon to learn the typical behaviour of the user. It combines this with information about the particular operating system to provide Minòs with the information needed to assess authenticity. For reasons of efficiency, in this implementation Virgil runs in batch mode. That is, Virgil operates periodically, updating the information provided to Minòs at regular intervals.

Phlegyas is the final component of this system. It acts as a ferry between Charon and Virgil, transferring the required data.

Minòs and Virgil are the heart of the system. It is in these processes that the key decision making and learning is performed. These two processes will thus be discussed in some detail. Charon, Phlegyas and Dante are of necessity closely linked to the operating system. In contrast, Minòs and Virgil are designed to be almost completely independent of the operating system they are running under. Most information required by these two systems about the working environment comes from a series of rules provided at system initialisation. Should Minòs and Virgil ever be transported to an operating system other than UNIX™, then only these rules, and the command parsing, need be re-written.

3 Minòs

Once fully installed, Minòs will use information from four metrics to decide the authenticity of computer users. Currently only one metric, command choice, is operational.

---

was the ferryman between Minòs's kingdom and the lower parts of Hell. *Virgil* was Dante's guide through Hell. These names were all suggested by Cathy Newberry.

3.1 Command Choice

This metric relies on the user's habits in choosing particular commands to distinguish between the authentic user and a masquerading intruder.

The choice of technique for making this decision was one of:

- neural nets;

- bayesian statistics; or

- decision trees.

After experimenting with decision trees and various statistical methods (see [8]), the author decided to use bayesian statistics. These were the most compact, and the best understood of the available options. Neural nets, although technologically appealing, were rejected as being too immature a technology. (It was judged that to attempt to use a neural net as decision maker would involve diverting too many resources away from the central thrust of this work.)

3.1.1 Bayes's Rule

Bayes's rule states:

$$P[h_i|e] = \frac{P[h_i]P[e|h_i]}{\sum_j P[h_j]P[e|h_j]}$$

where h_i are hypotheses, and e is some evidence, and $P[h|e]$ is the probability of that hypothesis being true given evidence e. For this work, the hypotheses are the authenticity of the user u_i, and the evidence is the presence of a command c_j. That is, $P[u|c]$ is the probability of a user u being authentic given a command c having been typed.

Ultimately this analysis must assign a value to $P[u|c_1 \& c_2 \& ... \& c_n]$ (the probability of user u being authentic after typing n commands). To calculate this using Bayes's rule requires the calculation of intermediate probabilities $P[c_1 \& c_2]$, $P[c_1 \& c_2 \& c_3]$, and all other combinations of the c_i's (see [9], p. 4). This is impractical for even a small number of commands, as the storage requirements would be enormous. Consequently, an assumption of conditional independence is used.

Two pieces of evidence e_i and e_2 are conditionally independent if the likelihood of e_2 occurring does not change if e_1 occurs. That is:

$$P[e_2|e_1] = P[e_2]$$

This is arguably untrue in this case — the probability of a user typing a command c_2 may well increase if command c_1 was just typed. However, because of the intractability of attempting to model these relationships, this issue of sequencing commands is treated separately, in the command sequence analyser (see section 3.2).

For a population of k known users, all of whom are assumed to be equally likely, Bayes's rule reduces to:

$$P[u_i|c] = \frac{1}{k} \frac{P[c|u_i]}{\sum_j P[c|u_j]}$$

Thus, combining this with the assumption of conditional independence, Bayes's rule gives:

$$P[u_i|c_1 \& c_2 \& ... \& c_n] = \frac{1}{k} \frac{P[c_1|u_i]P[c_2|u_i]...P[c_n|u_i]}{\sum_j P[c_1|u_j]P[c_2|u_j]...P[c_n|u_j]}$$

This expression, although involved, can be calculated. $P[c|u]$ is the probability of a user u typing a command c, which are be modelled by the relative frequency of c in the previous login sessions of u.

In this way, the command choice analysis produces a probability of the unauthenticated user being any of the known users, after n commands have been typed.

This metric has been implemented. The results are given in section 5.

3.2 Command Sequencing

This metric analyses the sequencing of commands. To analyse every possible sequence of commands would require enormous quantities of storage. To eliminate this problem, sequences are handled by assigning all commands into classes. For example, all the commands for changing directory {cd, up, down, d+, d-, pop, push, popd, pushd} belong to the class cd. The number of classes are less than twenty.

Processing here will be exactly the same as for command choice (see section 3.1). Decisions will be made using Bayesian analysis producing another conditional probability:

$$P[u|z_1\&z_2\&...\&z_n] = \frac{1}{k}\frac{P[z_1|u_i]P[z_2|u_i]...P[z_n|u_i]}{\sum_j P[z_1|u_j]P[z_2|u_j]...P[z_n|u_j]}$$

where z_i is a sequence of command classes.

The assumption of conditional independence is even more questionable here than for the command choice. For example, let $z_1 = \{cd, ls\}$, and $z_2 = \{edit, cc, edit\}$. For most programmers, the sequence $\{z_1, z_2\}$ would be common. Thus the occurrence of z_1 would increase the likelihood of z_2. However, because the storage requirements required for a full treatment would be prohibitive, the current treatment will stand.

3.3 User Location

This metric involves studying which sections of a computer a user typically accesses, and recording an exception when they go into areas they usually avoid. An excess of such behaviour could indicate an intruder.

Once again, a bayesian statistic will be used here. From data assembled by Virgil, Minòs will observe the user, continually tracking their current location, and the location of the files they are accessing.

3.4 Typing Styles

The analysis here is the most complex of any required by Minòs. Preliminary work has been performed using several metrics:

1. **Summary Methods.** Information about the intervals between successive keystrokes will be analysed using summary statistics in order to authenticate users. The summary statistics will include measures such as *skewness* and *kurtosis* (see [5], p. 154). These techniques summarise the intervals between keystrokes to produce a single number, which will be used in verifying the user's identity.

2. **Interval Methods.** Interval methods will compare keystroke intervals typed by the (unauthenticated) user, with the same intervals typed by the known user in order to arrive at a decision. The comparisons will consider the effects of context (at least one character before and after the digraph in question). The actual metric for comparisons is still under development.

The alternatives being considered include the mean/standard-deviation combination used in [8], p. 164, the use of speech techniques such as deterministic time warping (see [10]) or the use of confidence intervals (see [7], p. 312).

4 Learning

In order for any behavioural metric to be effective, it must be capable of adapting to the changing patterns of the authenticated user. That is, it must be capable of learning new patterns, and discarding old ones. Machine learning is very important for Minòs, as it is impractical for any person to deduce all the characteristics of a user's behaviour. Moreover, Minòs will be more secure if the characteristics are assembled by an "impartial" authority, rather than one user in the user population. Hence the need for Virgil, the learning intelligence behind Minòs.

The entire Minòs system must exhibit two different forms of limited intelligence, varying in their time frames. This system must be capable of recognising different user modes, and understanding which ones are meaningful to authenticate, and which ones are not. Also, Minòs must detect when a user has created a new command, at the time of creation. Both these actions must be performed in realtime. In addition it must be capable of learning long term drifts in a user's overall behaviour. The first actions need to be performed by Minòs as it authenticates the user's behaviour. The second form of learning can be relegated to batch processing, and Virgil.

4.1 Minòs

4.1.1 When to Authenticate

Because Charon operates between the user and the operating system, it can have difficulty in distinguishing keystrokes typed for execution by the operating system, and those typed for a particular command (say an editor). Because these two actions may represent two very different forms of behaviour, Minòs needs to be able to distinguish between these occasions. This is accomplished in two ways.

1. When Charon detects a change in the way the operating system is interpreting keystrokes, it stops authenticating, until the operating system

starts treating keystrokes as normal commands.[4]

2. Minòs maintains a list of commands that require further input from the user. When one of these commands is typed, Minòs stops the authentication process until it recognises a familiar operating system command.

4.1.2 New Commands

When a user creates a new command (by using a command such as a compiler) it is not unreasonable that they be allowed to use it without this reducing Minòs's estimate of their authenticity. This is achieved by maintaining within Minòs a list of those commands (*meta-commands*) that generate new commands. When a meta-command is encountered Minòs should determine the new command being created, and allow its further use without the user being penalised.

4.2 Virgil

There are several different types of mechanised learning, varying in the amount that the computer must infer (see [1], p. 8). These are listed here in order of increasing inference.

- Rote Learning

- Learning from Instruction

- Learning by Analogy

- Learning from Examples

- Unsupervised Learning

Learning from examples (archived login sessions) is most suitable for Virgil, as it is the only form of learning that allows Virgil to be selectively trained on only those login sessions found to be authentic.

For Virgil to produce a useful system, the examples (login sessions) must be chosen carefully. Virgil must only learn from sessions which belong to authentic users, otherwise it will "learn" to mis-classify intruders as authentic. Thus it is not appropriate to allow Virgil to learn from all login sessions that belong to a

---

[4]In UNIX™ this is done by watching the terminal mode. If it changes into *cbreak* or *raw* modes, or *echo* is disabled, then then Charon stops authenticating, until these are restored.

user claiming to be genuine. However, if Virgil is allowed to select the examples for learning itself, according to some predefined rules, then accurate learning will not occur. For example, if all login sessions that are authenticated by Virgil are selected for learning, then this will enforce Virgil's existing behaviour. Any new behaviour by a user would be rejected by Virgil as being atypical, not authenticated, and rejected from learning sets. In this way Virgil would only learn from examples that conformed to the rules it had already formulated. This method would hamper Virgil learning new rules as the user developed new habits.

Thus Virgil will require a teacher to select its training examples. The most likely strategy is for the security administrator to provide Virgil with examples from the audit (archived) records, if those records are proven genuine by the passage of time demonstrating an absence of any intruders. This method has clear shortcomings. An intruder penetrating a system over a long period of time would not be detected as the gradual incursions would probably not be detected by an administrator before intrusion examples are provided to Virgil. If a period of a week passes before an administrator provides Virgil with training examples, then Virgil's expectations will be a week out of date.[5] This lag may cause Virgil to incorrectly classify login sessions. Conversely, if training examples are provided to Virgil promptly (say at the end of each day), then insufficient time may have passed for security breaches to be noticed. A compromise would be for all breaches of security identified by Virgil (or any other agent) to be investigated manually. If they prove to be genuine intrusions, then the login sessions could be eliminated from Virgil's training set. Otherwise (in the case of a false rejection error) the example would be included in the training set.

Virgil provides the Bayesian incarnation of Minòs with the various probabilities required, by recording the relative frequencies of, say, the commands in the archived sessions of a particular user.

5 Results

To date Minòs has been tested over three months with two users, using the choice of command as a metric. The small user population was selected as Minòs was still under development. During these initial trials, Minòs took no action when an intruder was suspected. Instead it simply informed Dante of its suspicions.

---

[5] The issue of time lags between training and authentication is discussed further in section 5.

miken	Sept.	Oct.	Nov1.1.	Nov1.2.	Nov2.1.	Nov2.2.	Dec.	Total
Oct.	6.6%	—	2.8%	3.8%	2.0%	3.5%	3.2%	3.4%
Nov1.	1.9%	—	—	—	1.6%	1.0%	3.7%	2.2%
Nov2.	2.1%	—	—	—	—	—	3.8%	3.2%
Dec.	2.3%	—	—	—	—	—	—	2.3%

Table 1: False Rejection Errors for User miken

cathyn	Oct.	Nov1.1.	Nov1.2.	Nov2.1.	Nov2.2.	Dec.	Total
Oct.	—	8.7%	0.5%	1.3%	1.4%	1.3%	4.1%
Nov1.	—	—	—	1.2%	1.4%	0.9%	1.2%
Nov2.	—	—	—	—	—	1.3%	1.3%

Table 2: False Rejection Errors for User cathyn

Thoroughly testing Minòs proved to be more difficult than was originally anticipated. While testing for false rejections (that is instances where Minòs incorrectly labels a valid user as an intruder) was relatively simple, testing for false acceptances (occasions where an intruder is incorrectly authenticated) was more difficult, as instances of an intruder's behaviour have not been recorded.

In the discussion that follows, the error rates are the percentage of commands that were incorrectly authenticated ("command" error rates), as Minòs makes an authentication decision every time a command is typed.

5.1   False Rejections

The tables 1 and 2 give the percentage of commands that Minòs failed to authenticate for valid users. In these tests Virgil was trained on four, increasingly larger, training sets, consisting of the user's login sessions. The sessions included in the training sets are denoted by a dashes in the tables. Minòs was then tested on the sessions not included in the training sets. When Minòs was tested on its training set, their was always a 0% error.

For the purposes of these tests, November was divided into four equal parts, each approximately one week long, that were analysed separately.

cathyn	Sept.	Oct.	Nov1.1.	Nov1.2.	Nov2.1.	Nov2.2.	Dec.	Total
Oct.	41%	—	46%	48%	47%	43%	48%	46%
Nov1.	41%	—	—	—	43%	46%	48%	46%
Nov2.	41%	—	—	—	—	—	48%	46%
Dec.	41%	—	—	—	—	—	—	41%

Table 3: Percentage of Commands Rejected for User cathyn authenticating "Intruder" Login Sessions (see text for explanation).

miken	Oct.	Nov1.1.	Nov1.2.	Nov2.1.	Nov2.2.	Dec.	Total
Oct.	—	37%	54%	51%	54%	50%	46%
Nov1.	—	—	—	51%	54%	51%	52%
Nov2.	—	—	—	—	—	51%	51%

Table 4: Percentage of Commands Rejected for User miken authenticating "Intruder" Login Sessions (see text for explanation).

5.2 False Acceptances

During this trial phase several informal experiments were conducted when people attempted to use the accounts of the two registered users. In these cases, as long as the "intruder" used standard UNIX commands (such as changing directories, or listing directory contents) they were not identified by Minòs. However, as soon as they attempted to go beyond this (to say viewing a file's contents, or editing it) Minòs would record an "intruder". From these informal tests it became clear that many of the commands typed by an intruder would not actually generate a warning (as many UNIX commands are common to all users. On average 18% of commands typed in these trials were for changing directory (cd) or for listing their contents (ls)). Thus to to consider the percentage of commands issued by an intruder that were accepted by Minòs as the false acceptance error would lead to misleadingly high error rates. A more meaningful error may the number of "damaging" commands that had been misclassified. However, such analysis was too complex to be completed in time for this report.

The tables 4 and 3 lists the percentages of commands that Minòs refused to authenticate when, for various training sets, Minòs was presented with the login sessions of one user, and told that they were typed by the other user. Clearly Minòs authenticated far fewer commands here than for login sessions by the authentic user, and was thus able to distinguish between the authentic user, and the intruder.

5.3 Training Sets

From these experiments, it became clear that Minòs could still acceptably function with a delay of up to a fortnight between being trained, and carrying out the authentication. In fact, the size of the training set appeared to make little difference to the effectiveness of Minòs. Thus the learning strategies discussed in section 4 are certainly workable.

6 Conclusion

In summary, the Minòs project aims to produce an active agent, capable of a wide range of responses to intruders. It will detect security violations normally missed by existing systems because it will examine every command typed by a user. It will use a Bayesian decision maker to decide the authenticity of computer users on the basis of four metrics. The first, command choice, has already been implemented.

The Minòs project has been designed to be useful in the widest possible range of environments. In its range of metrics and responses to intruders it will be possible to provide a unique form of security when Minòs is completed in December 1990.

7 Acknowledgements

I would like to thank the many people who have contributed to the success of this project. The financial support of the Australian Telecommunications and Electronics Research Board, Telecom Australia, Prime RD&E Australia, and the Commonwealth Department of Employment, Education has been invaluable. The assistance of the other members of the Centre for Computer Security Research has also been important. Finally I would like to thank Cathy Newberry for her assistance and willingness to be part of this project.

References

[1] J. G. CARBONELL, R. S. MICHALSKI, AND T. M. MITCHELL, *An overview of machine learning*, in Machine Learning: An Artificial Intelligence Ap-

proach, R. S. Michalski, J. G. Carbonell, and T. M. Mitchell, eds., Springer–Verlag, 1983, p. 3.

[2] A. DANTE, *Inferno*, in The Divine Comedy, vol. 1, Penguin Books, 1984. Translated by Mark Musa.

[3] D. E. DENNING, *An intrusion-detection model*, in Proceedings of the 1986 IEEE Symposium on Security and Privacy, 1986.

[4] L. HALME AND J. V. HORNE, *Rep. tr-86007*, tech. rep., Sytek, Inc., 1986. Original unseen — cited in [11].

[5] J. LONG, I. NIMMO-SMITH, AND A. WHITEFIELD, *Skilled typing: A characterization based on the distribution of times between responses*, in Cognitive Aspects of Skilled Typewriting, W. E. Cooper, ed., Springer-Verlag, New York, 1983.

[6] T. F. LUNT AND R. JAGANNATHAN, *A prototype real-time intrusion detection system*, in Proceedings of the 1988 IEEE Symposium on Security and Privacy, 1988, p. 59.

[7] W. MENDENHALL, R. L. SCHEAFFER, AND D. D. WACKERLY, *Mathematical Statistics with Applications*, PWS Publishers, Boston, 1986.

[8] M. NEWBERRY AND J. SEBERRY, *User unique identification*, in Proceedings of the Twelfth Australian Computer Science Conference, P. Nickolas, ed., University of Wollongong, February 1989, Department of Computer Science, p. 163.

[9] J. O'NEILL, *Plausible reasoning*, The Australian Computer Journal, 19 (1987), p. 2.

[10] H. SAKOE AND S. CHIBA, *Dynamic programming algorithm optimization for spoken word recognition*, IEEE Transactions on Acoustics, Speech and Signal Processing, 26 (1978), p. 43.

[11] R. C. SUMMERS AND S. A. KURZBAN, *Potential applications of knowledge-based methods to computer security*, Computers and Security, 7 (1988), p. 373.

[12] W. T. TENER, *Discovery: an expert system in the commercial data security environment*, in Proc. IFIP/Sec '86, 1986. Original unseen — cited in [11].

SECTION 11

RUMP SESSION

UNIVERSAL LOGIC SEQUENCES

Ed Dawson*   Bruce Goldburg**

* School of Mathematics
Queensland University of Technology
GPO Box 2434
Brisbane Q 4001

** School of Electrical and Electronic Systems Engineering
Queensland University of Technology
GPO Box 2434
Brisbane Q 4001

ABSTRACT

A sequence defined to be a generalised multiplexed sequence using several linear feedback shift registers (LFSR) to select from the stages of another LFSR can be shown to be zero-order correlation immune. By introducing one bit of memory it is shown how one can use a generalised multiplexed sequence in the construction of a sequence which is correlation immune. This new sequence is classified as a universal logic sequence. Several properties of universal logic sequences will be developed and empirical results on other properties will be given based on the results of simulation.

1. INTRODUCTION

A stream cipher is the process of encryption where a binary sequence (z_t) is added modulo two to binary plaintext where the first term of this sequence is z_0. If (z_t) is produced by a nonlinear combination of LFSRs sequences there are several properties described in [1] that the sequence (z_t) should satisfy in order to be secure from cryptanalytic attack.

Property 1.1
The period of (z_t) should be large.

Property 1.2
The linear complexity of (z_t) should be large.

Property 1.3
The noiselike characteristics of (z_t) should be good.

Property 1.4
The order of correlation immunity of (z_t) should be large.

Generalised multiplexed sequences were introduced and several of their properties were developed in [2]. In Section 2 a definition of a generalised multiplexed sequence will be given along with its standard properties. It will be shown that generalised multiplexed sequences satisfy Properties 1.1 - 1.3. However, as will be described a generalised multiplexed sequence fails Property 1.4 in that a generalised mutliplexed sequence is zero-order correlation immune. As was shown in [3] a sequence which is zero-order correlation immune can be attacked by using ciphertext only, provided this sequence is used directly in a stream cipher to encrypt binary plaintext.

In Section 3 it is shown how one can avoid this correlation attack on a generalised mutliplexed sequence by using one bit of memory. This new sequence is defined to be a universal logic sequence. Several properties of universal logic sequences will be developed in Section 4. In Section 5 empirical results on Properties 1.2 and 1.3 for universal logic sequences will be given on the basis of simulation.

2. GENERALISED MULTIPLEXED SEQUENCES

A special case of a generalised multiplexed sequence (u_t) from [2] is defined as follows. Let $(a_{1t}), \ldots, (a_{kt})$ be binary sequences from linear feedback shift registers LFSR(1), ..., LFSR(k) whose characteristic polynomials are primitive of degree ℓ_i for $i = 1, \ldots, k$ where the ℓ_i are pairwise relatively prime integers. Let n be an integer which is pairwise relatively prime to ℓ_i for $i = 1, \ldots, k$ and which is greater than or equal to 2^k. Let A_t represent a binary k-tuple defined by $A_t = (a_{1t}, \ldots, a_{kt})$.

A mapping γ is selected which denotes a one to one mappling between binary k-tuples and the set of integers $\{0, 1, 2, \ldots, n-1\}$. Let (b_t) be a binary sequence defined by shift register LFSR(n) of length n whose characteristic polynomial is primitive. Define a sequence (u_t) by

$$u_t = b_{t + \gamma(A_t)} \tag{1}$$

which is classified as a generalised multiplexed sequence.

Three properties of the sequence (u_t) are given in [2].

Property 2.1
The linear complexity of (u_t) is $n\, (\ell_1 + 1)\, (\ell_2 + 1) \ldots (\ell_k + 1)$.

Property 2.2
The period length of (u_t) is $(2^n - 1)\, (2^{\ell_1} - 1) \ldots (2^{\ell_k} - 1)$.

Property 2.3
Let (a_{it}) and (a'_{it}) be output sequences of LFSR(i) for $i = 1, \ldots, k$. Let (b_t) and (b'_t) be output sequences of LFSR(n). For a given scrambling function γ let (u_t) and (u'_t) be defined as in (1) by $a_{1t}, \ldots, a_{kt}, b_t$ and $a'_{1t}, \ldots, a'_{kt}, b'_t$ respectively. Then the sequences (u_t) and (u'_t) are translates of one another.

Let p denote the period of (u_t) which is $(2^n - 1)\, (2^{\ell_1} - 1) \ldots (2^{\ell_k} - 1)$ as shown in Property 2.2. The number of ones in every p terms of (u_t) can be determined as shown by Property 2.4.

Property 2.4
The number of ones in a period of (u_t) is $2^{n-1}\, (2^{\ell_1} - 1) \ldots (2^{\ell_k} - 1)$.

Proof.

Let $e = (2^{\ell_1} - 1) \ldots (2^{\ell_k} - 1)$.

For $j = 0, \ldots, e - 1$ the sequences (u_{j+t}) define m-sequences of period $2^n - 1$ formed by decimating the sequence (b_t) as shown in [1] since e is relatively prime to $2^n - 1$. Each of these e m-sequences contains 2^{n-1} ones in every $2^n - 1$ terms. Hence the sequence (u_t) contains $e2^{n-1}$ ones in every p terms. ◊

In [4] it is demonstrated that a multiplexed sequence as defined in [5] is zero-order correlation immune. By applying the same technique one can show that a generalised multiplexed sequence is zero-order correlation immune as given by Property 2.5. Hence a generalised multiplexed sequence can be attacked by using the correlation attack from [3]. In this case the attack consists of correlating the ciphertext with the sequence (b_t) in order to derive the tap settings and the initial state vector for this sequence.

Property 2.5

The sequence (u_t) is zero-order correlation immune.

3. DEFINITION OF UNIVERSAL LOGIC SEQUENCES

In order to avoid the correlation attack on generalised multiplexed sequences a sequence (z_t) will be constructed by using one bit of memory. In this construction let $(a_{1t}), \ldots, (a_{kt})$ be the sequences as defined in the previous section. As shown in [1] a sequence (z_t) defined by

$$z_t = \sum_{i=1}^{k} a_{it} + s_{t-1} \tag{2}$$

has correlation immunity of order $k-1$ where the memory bit sequence (s_t) is defined in terms of $a_{1t}, \ldots, a_{kt}, s_{t-1}$ given by a Boolean function f as

$$s_t = f(a_{1t}, \ldots, a_{kt}, s_{t-1}). \tag{3}$$

There is no restriction on the choice of the Boolean function f. The initial value s_{-1} of the memory sequence (s_t) is usually selected to be zero.

The generalised multiplexed sequence from (1) will be used to define the sequence (s_t) in (2) and (3). Let (b_t) be the sequence defined in the previous section except in this case the integer n is greater than or equal to 2^{k+1}. Let $X_t = (a_{1t}, \ldots, a_{kt}, s_{t-1})$ be a binary $k+1$-tuple. Let α define a one to one mapping between binary $k+1$-tuples and $\{0, 1, \ldots, n-1\}$. Define the sequence (s_t) by

$$s_t = b_{t+\alpha(X_t)} \tag{4}$$

The sequence (z_t) in (2) formed by using (4) to define the sequence (s_t) will be called a universal logic sequence.

4. PROPERTIES OF UNIVERSAL LOGIC SEQUENCES

As described in Section 3 and stated in Property 4.1 below a universal logic sequence has a correlation immunity order of $k-1$. Hence, this sequence is secure from the ciphertext only attack from [3].

Property 4.1
A universal logic sequence as defined in (2) and (4) has a correlation immunity order of $k-1$.

Let $p = (2^n - 1)(2^{\ell_1} - 1) \ldots (2^{\ell_k} - 1)$. In Property 4.2 it will be shown that universal logic sequences as defined in (2) and (4) have a period length of p provided where necessary a few terms at the start of some sequences are deleted. In Property 4.3 it will be shown that universal logic sequences satisfy a translate property similar to Property 2.3 for generalised multiplexed sequences.

Property 4.2
The period of a universal logic sequence as defined in (2) and (4) is p provided that one deletes a few terms when necessary at the start of the sequence.

Proof
From Property 3.2 the smallest value of h such that $a_{i,\,j+h} = a_{ij}$ and $b_{j+h} = b_j$ for $i = 1, \ldots, k$ and all j is $h = p$. Hence by (2) and (4) $z_j = z_{j+p}$ if and only if $\alpha(X_j) = \alpha(X_{j+p})$ if and only if $s_{j-1} = s_{j-1+p}$.

Suppose that $s_{t-1} = s_{t-1+p}$ then $s_t = s_{t+p}$ as shown by (3) and (4) so that by mathematical induction $s_{j-1} = s_{j-1+p}$ for all $j \geq t$. This implies that $z_j = z_{j+p}$ for all $j \geq t$.

If $s_{-1} = s_{p-1}$ then from above $z_j = z_{j+p}$ for all j. As defined in Section 3 we have selected s_{-1} to be zero so that the probability that $s_{-1} = s_{p-1}$ depends on the probability that s_{p-1} is zero. This is equivalent to the probability that b_t is zero which is $\dfrac{2^{n-1} - 1}{2^n - 1}$.

This is approximately $\frac{1}{2}$ for sufficiently large n. Similarly the probability that $s_{-1} \neq s_{p-1}, s_0 \neq s_p, \ldots, s_{i-2} \neq s_{p+i-2}$ is approximately $(\frac{1}{2})^i$. Hence the probability that $s_{p+i-1} = s_{i-1}$ and $s_{p+j} \neq s_j$ for $j = -1, \ldots, i-2$ is approximately $1 - (\frac{1}{2})^i$. Clearly we can find such a value of i after a small number of steps.

From above if $s_{p+i-1} = s_{i-1}$ then $z_{p+j} = z_p$ for all $j \geq i$. Hence the period of (z_t) is p provided one deletes if necessary the first i bits of the sequence where $s_{-1} \neq s_p, s_0 \neq s_{p+1}, \ldots, s_{i-2} \neq s_{p+i-2}$ and $s_{i-1} = s_{p+i+1}$. $\lozenge$

Property 4.3
Let (a_{it}) and (a'_{it}) be sequences produced by LFSR(i) for $i = 1, \ldots, k$. Let (b_t) and (b'_t) be sequences produced by LFSR(n). Then for a given scrambling function γ the universal logic sequences (z_t) and (z'_t) defined by $(a_{1t}), \ldots, (a_{kt}), (b_t)$ and $(a'_{1t}), \ldots, (a'_{kt}), (b'_t)$ respectively are translates of one another provided a few initial bits of (z'_t) are deleted in certain cases.

Proof
Since the characteristic polynomials of LFSR(i) for $i = 1, \ldots, k$ and LFSR(n) are primitive, the sequences (a_{it}) and (a'_{it}) for $i = 1, \ldots, k$, and (b_t) and (b'_t) are translates of one another. By the Chinese Remainder Theorem there exists an integer r

where $0 \le r \le p-1$ such that
$$a'_{it} = a_{i,t+r} \text{ and } b'_t = b_{t+r}$$
for $i = 1, \ldots, k$ and all t.

Suppose that $s'_{j-1} = s_{j-1+r}$ then using similar reasoning to that in Property 4.2 $s'_{t-1} = s_{t-1+r}$ for all $t \ge j$ which implies that $z'_t = z_{t+r}$ for $t \ge j$. As shown in Property 4.2 the probability that $s'_{-1} = s_{-1+r}$ is approximately $\frac{1}{2}$. Hence the probability that $s'_{j-1} \ne s_{j-1+r}$ for $j = 0, \ldots, i-1$ and that $s'_{i-1} = s_{i-1+r}$ is approximately $1 - (\frac{1}{2})^i$. In which case from above $z'_t = z_{t+r}$ for all $t \ge i$. $\Diamond$

5. EXAMPLES OF UNIVERSAL LOGIC SEQUENCES

Several examples of universal logic sequences were studied in order to derive empirical results on the linear complexity and the number of ones in a period. The Berlekamp-Massey algorithm from [6] was used to find the linear complexity of each sequence. In each case examined the natural map was used to define the mapping α.

In the case where k is one the universal logic sequence as defined by (2) and (4) takes the form

$$z_t = a_{1t} + s_{t-1} \tag{5}$$

$$s_t = b_{t+\alpha (X_t)} \tag{6}$$

where $X_t = (a_{1t}, s_{t-1})$. All primitive polynomials of degrees three, five and seven were used to define the sequence (a_{1t}) and all primitive polynomials of degree four were used to define the sequence (b_t). Table 1 below gives lower and upper bounds on the linear complexity and the number of ones in a period from the universal logic sequences generated using (5) and (6).

ℓ_1	Period	Minimum Linear Complexity	Maximum Linear Complexity	Minimum Number of Ones	Maximum Number of Ones
3	105	101	105	48	53
5	465	458	464	228	234
7	1905	1792	1799	942	958

Table 1

In the case where k is two, one universal logic sequence was examined using shift register of lengths three, five and eight respectively for sequences (a_{1t}), (a_{2t}) and (b_t) whose characteristic polynomials were $x^3 + x + 1$, $x^5 + x^2 + 1$ and $x^8 + x^4 + x^3 + x^2 + 1$. This defined a universal logic sequence of period 55335. The linear complexity of this sequence was 55332 and the number of ones in a period was 27440.

From [1] the linear complexity profile ℓ of a pseudorandom sequence to be used in a stream cipher should follow closely the $\frac{\ell}{2}$ line. Figure 1 is the linear complexity profile of a universal logic sequence (z_t) as defined by (5) and (6) where shift registers of lengths three and four were used to define the sequences (a_{1t}) and (b_t) respectively whose characteristic polynomials were $x^3 + x + 1$ and $x^4 + x + 1$.

As shown in this figure the linear complexity profile for this sequence follows closely the $\frac{\ell}{2}$ line.

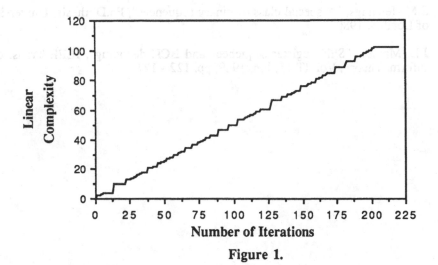

Figure 1.

6. CONCLUSION

As shown by Properties 4.1 and 4.2, and the empirical results from Section 5 on linear complexity and the number of ones in a period universal logic sequences seem to satisfy Properties 1.1 - 1.4 for a sequence to be used in a stream cipher secure from cryptanalytic attack provided sufficiently large shift registers are used. The results obtained on universal logic sequence in regards to these properties indicate that these sequences offer a similar level of security to integer addition sequences described in [1].

Furthermore the one to one function α described in Section 3 provides further security for universal logic sequences since there are

$$2^{k+1}! \binom{n}{2^{k+1}} \text{ choices for } \alpha.$$

In relation to generalised multiplexed sequences, universal logic sequences offer better security from attack in terms of the order of correlation immunity and the larger linear complexity. As shown by Property 2.1 the linear complexity of a generalised multiplexed sequence is a polynomial function of the linear complexities of the input sequences. On the other hand the linear complexity of a universal logic sequence as indicated by the results from Section 5 is an exponetial function of the linear complexities of the input sequences.

REFERENCES

1. R.A. Rueppel, "Analysis and Design of Stream Ciphers", Springer-Verlag, 1986.

2. M. Liu and Z. Wan, "Generalised Multiplexed Sequences", Advances in Cryptology, Eurocrypt 85, pp. 135 - 141.

3. T. Siegenthaler, "Decrypting a Class of Stream Ciphers Using Ciphertext Only", IEEE Trans. Comp., Vol. C-34, No. 1, 1985, pp. 81 - 85.

4. S. Mund, D. Gollmann, T. Beth, "Some Remarks on the Cross Correlation Analysis of Pseudo Random Generators", Advances in Cryptology, Eurocrypt 87, pp. 25 - 33.

5. S.M. Jennings, "A special class of binary sequences", Ph.D. thesis, University of London, 1980.

6. J.L. Massey, "Shift register sequences and BCH decoding", IEEE Trans. on Inform. Theory, Vol. IT-15, Jan. 1969, pp. 122 - 127.

The Three Faces of Information Security

John M. Carroll
Department of Computer Science
The University of Western Ontario
London, ON, CANADA
currently with:
Information Security Research Centre
Queensland University of Technology
Brisbane, QLD, AUSTRALIA

Abstract

CONFIDENTIALITY, INTEGRITY, and AVAILABILITY are the three faces of information security. These goals can all be achieved by use of cryptography. This paper will discuss:

1. Initial authentication of user and host using:

    (a) asymmetric ciphers; or
    (b) symmetric ciphers.

2. Exchange of cryptographic keys for:

    (a) privacy transformation; and
    (b) message authentication.

3. Continuous re-authentication to:

    (a) test user and host presence; and
    (b) assure channel integrity.

4. Implementation in all comunication modes:

    (a) two-party (one-to-one);
    (b) broadcast (one-to-many); and
    (c) conference (many-to-many).

The existance of umpires, network security officers or certificating authorities (CA) is not presumed, nor is it ruled out.

INTRODUCTION

There are at least a dozen attacks that can be mounted against data communications, against which authentication protocols can be effective. The following list gives brief definitions and suggests countermeasures:

1. **Masquerading.** Falsely assuming the identity of another. Countermeasure (CM) is mutual host/user authentication.

2. **Forgery.** Fraudulently altering message contents between originator and recipient. CM is message authentication/integrity codes (MAC or MIC).

3. **Active Wiretapping.** Usurping an active host, user or channel. CM is mutual host/user authentication and continuous re-authentication.

4. **Passive Wiretapping.** Obtaining information from a channel without authorization. CM is encrypted host/user authentication and end-to-end message encryption.

5. **Deletion.** Improperly removing a message from a channel. CM is encrypted serialization.

6. **Delay.** Improperly re-ordering message arrivals. CM is encrypted timestamping.

7. **Service Denial.** Intentionally making facilities unavailable. CM is continuous re-authentication.

8. **Replay.** Infiltrating previously sent messages. CM is encrypted timestamping and serialization.

9. **Source Repudiation.** Originator falsely denies sending a message. CM is digitally signed messages.

10. **Delivery Repudiation.** Recipient falsely denies receiving a message. CM is digitally signed receipts.

11. **Traffic Analysis.** Inferring message contents from plaintext headers. CM is link encryption of headers.

12. **Cryptanalysis.** Obtaining information from encrypted material without having been invested with the key. CM are:

    (a) Use strong ciphers.

    (b) Keys must never appear as plaintext.

    (c) Deny data for known-plaintext attacks.

(d) Deny data for cryptotext-only attacks.

This analysis does not address:

1. Physical intrusion.

2. Waste retrieval.

3. Covert channel interpretation.

4. Undesired signal emanation (USDE) interpretation.

5. Electronic eavesdropping.

6. Remanent magnetic flux recovery.

7. Personnel defection or agent insertion.

It is assumed that at the lowest OSI level there exist provisions for electrical handshaking, bit and burst error detection and correction, and line-quality monitoring. It is also assumed that conversions are made betweeen native language (eg. ASCII), transmission language (eg. HEX) and binary codes as required for encryption/decryption operations.

Every host or user is assumed to have at least one microprocessor with RAM/ROM and a large EPROM or equivalent for key storage. The microprocessors must be able to run pseudo random number generators, counters to maintain synchronism, binary search of EPROM, and symmetric and/or asymmetric encryption/decryption algorithms.

Every data channel or group of them is assumed to have a subchannel available for continuous re-authentication. This could be a guard fibre, time or frequency division multiplex channel, or interpacket burst arrangement.

INITIAL HOST/USER MUTUAL AUTHENTICATION

Initial host or user mutual authentication consists of the secure exchange of keys to a symmetric cipher (sometimes called IK for Interchange Keys) to be used to exchange other symmetric-cipher keys (sometimes called DEK for Data Encrypting Keys) to protect subsequent operations.

The host or user authentication can be performed recursively: the user's token (eg. Smart Card) and workstation can authenticate themselves, the workstation and host computer can authenticate themselves, and so on. Only one pair will be shown here.

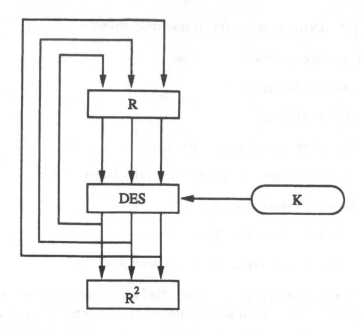

Figure 1: OBFB Encryption

The symmetric cipher is assumed to be a block product cipher, such as DES, and to be used in the output block feedback (OBFB) mode. This is a secure mode in which a random bit string is continually re-enciphered while bits selected from each encipherment are used to form a long stream-cipher key. See Figure 1. The keys exchanged by host and user are the block cipher key (K) and the initial random bit string (R).

Asymmetric Cipher Exchange

When the RSA cipher is used to exchange keys, User 1 encrypts K in User 2's public encrypting key E2 and User 2 decrypts it using its secret decrypting key D2. User 2 encrypts R in E1 and User 1 decrypts it using D1. See Figure 2. Here K is a symmetric cipher key; R is initial random fill.

The exchange proves that each user has had its public encrypting key deposited in a directory to which the other user has access, and that each has been invested with the proper algorithms to calculate secret decrypting keys.

Participants can contract with each other only in the asymmetric mode, that is by using the digital signature and receipt technique. See Figure 3.

Key exchange messages may have to be randomly padded to prevent a successful discrete logarithm attack.

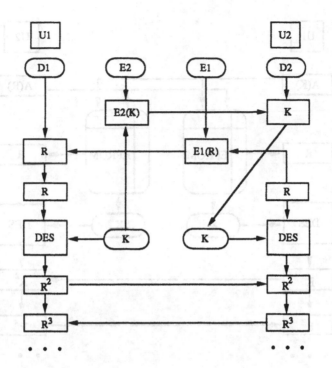

Figure 2: Asymmetric Handshake and Ping-Pong Protocol

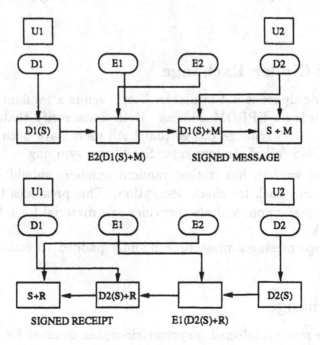

Figure 3: Digital Signatures

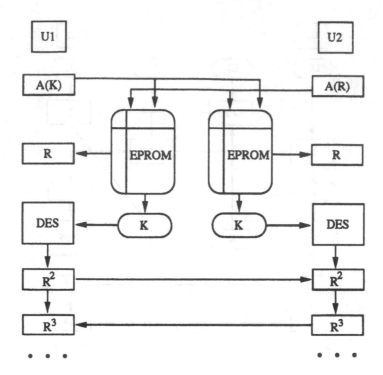

Figure 4: Symmetric Handshake and Ping-Pong Protocol

Symmetric Cipher Exchange

If an asymmetric cipher is not available, User 1 sends a random number $A(K)$ to User 2. This is an EPROM address. Both users verify that the addresses are legitimate. This exchange proves that both users have been invested with identical EPROMS full of random keys. See Figure symping

Conventional wisdom has it that random numbers should be exchanged, encrypted and sent back for check decryption. This presumes that keys have already been agreed upon, and also provides raw material for a hostile known-plaintext attack.

Key exchange messages must be randomly padded to conceal the size of EPROM.

Key Interchange

The users have now established a symmetric-cipher channel for the secure exchange of keys for:

1. Exchange of secret identifying information.

2. Continuous user-presence and channel-integrity testing.

3. Computation of message authenticating codes (MAC/MIC).

4. Message privacy encryption, if desired.

CONTINUOUS TESTING

Continuous user-presence and channel-integrity testing has been used to assure against hostile overrun of, or interference with military forward observers and treaty-compliance observers.

The problem of an active terminal or channel falling into hostile hands, known variously as active wiretapping, "spoofing", piggybacking, or between-the-lines attack, has been recognised in resource-sharing computer systems. However, conventional countermeasures such as time-out disconnection or host or user requests for re-authentication have not proved to be effective.

Continuous testing requires each user to continuously and synchronously re-encrypt the same random bit string. Each user alternately sends output over the guard channel to the other for comparison. If a comparison fails, the channel is shut down and the intial authentication must be performed again. This can be regarded as a "ping-pong" reauthentication mode. See Figure 2 and Figure 4.

To deny a large work sample to hostile cryptanalysts, the guard channel should be opened only for randomly selected encryption cycles, and only a randomly selected number of bits (¡R) should be allowed to pass during an exchange.

MESSAGE AUTHENTICATION

A message authenticating (or integrity) code (MAC/MIC) is calculated by regarding the message as a bit stream, enciphering each block in the cipher-block-chaining mode(CBC), and adding each successive block without carry(XOR). See Figure cbc. The MAC is appended to the message before encryption and the MAC keys R and K are exchanged using the IK cipher. Observe that the R used for CBC initial fill can result from any re-encryption of R.

It is feared that the 64-bit DES block size may provide insufficient protection against a "birthday" attack on message integrity — concocting a spurious message having a MAC identical to that of the target message.

For this reason and others, such as the need to increase key space and the vulnerability of 8-byte blocks to cryptanalysis, at least one 128-bit block product

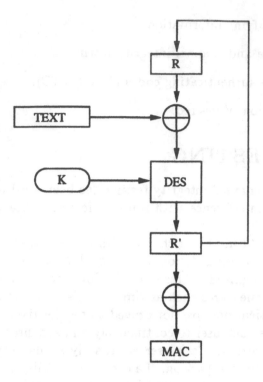

Figure 5: CBC Message Authentication

cipher is under development. Figure 6 shows how two DES chips in a triple-pass application can be used to simulate a 128-bit block cipher.

MESSAGE PRIVACY

For many users, message encryption is their least important consideration, just as it may be the most important consideration for others. In either case, the integrity and availability protection of continuous and mutual authentication is available whether messages are encrypted or not.

The encrypted MAC can be appended to a plaintext message and the MAC keys exchanged by the IK protocols.

BROADCAST MODE

The broadcast mode is implemented in a star network of N stations with the host (station #1) in the centre. The principal differences between the two-party and broadcast modes concern initialisation and presence testing. There are two sub-modes of the broadcast mode: (1) when the broadcaster (host) cares about

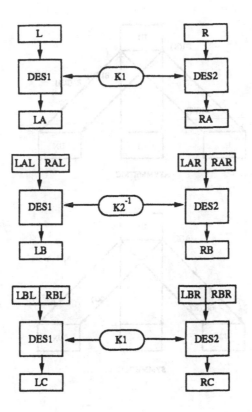

Figure 6: 128-Bit DES

user presence, and (2) when it doesn't.

Initialization

In both sub-modes 1 and 2, the host assumes that only members of the audience in good standing either know the public key E1 (asymmetric initialisation); or possess the key-storage EPROM (symmetric initialisation). In the asymmetric case, the host broadcasts the universal key K encrypted under private key D1; see Figure 7, bottom.

In sub-mode 1, each user must supply a unique R. In asymmetric initial-isation, this can be accomplished by having each user send its R component encrypted under E1. See Figure 7, top. In symmetric initialisation, this is accomplished by partitioning the key address space among users. See Figure 7, bottom.

Observe that in the asymmetrically initialised broadcast mode, users have been authenticated only weakly. The users have not been forced to use their private keys. This was done to spare the host the trouble of decrypting initiali-sation messages by using a large number of user public keys. This is a situation in which the system designer has to trade off between security and efficiency.

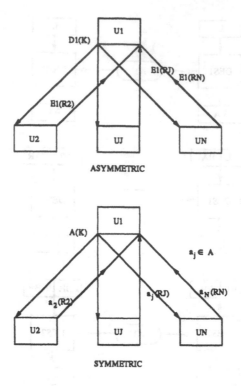

Figure 7: Broadcast Handshakes: asymmetric and symmetric

Presence-Testing

Continuous re-authentication is carried out according to a "round-robin" resource-sharing model. It is initiated by the host to avoid difficult synchronization problems. See Figure 8. Consider the i-th re-authentication of User j. The host loads $Rj\hat{\ }(i-1)$ from an (N-1)-component storage vector. That is, the (i-1)-th, or previous re-encryption of Rj, the random bit string contributed by the j-th user, under K.

The host re-encrypts $Rj\hat{\ }(i-1)$ and sends $Rj\hat{\ }i$ to User j. User j re-encrypts its copy of $Rj\hat{\ }(i-1)$ and checks the result with the copy of $Rj\hat{\ }i$ received from the host. If the two copies are identical, User j re-encrypts $Rj\hat{\ }i$ and sends $Rj\hat{\ }(i+1)$ to the host.

CONFERENCE MODE

The conference mode is implemented in a ring network. There is no designated host. The host is whoever is addressing the conference at any given time.

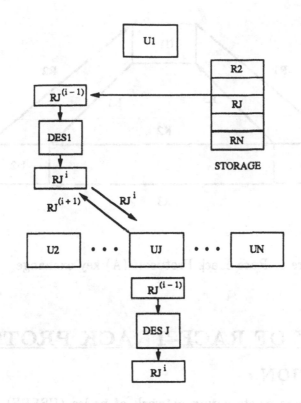

Figure 8: Round-Robin Protocol

Initialization

The conference ring is made up of N overlapping pairs of users: 1-2; 2-3; ... N-1, N. At each node, equipment is duplicated or shared such that each pair establishes a symmetric channel as in the two-party mode. The pairs, however, are interlocked to form an integral ring:

User 1 sends K1 to User 2 and User 2 sends R2 to User 1.
User 2 sends K2 to User 3 and User 3 sends R3 to User2.
...
User N sends KN to User 1 and User 1 sends R1 to UserN.

Each user is a member of two pairs. In general, User J belongs to both pair (J-1) and pair J. See Figure 9A.

Presence Testing

Continuous re-authentication can be carried out pair-wise using the Ping-Pong protocol. There is, however, a stronger test for ring integrity. We call it the "Race-Track" protocol.

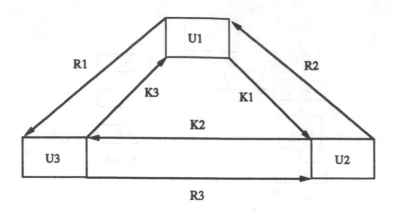

Figure 9: Race-Track Protocol: (A) key exchange

SUMMARY OF RACE-TRACK PROTOCOL

INITIALISATION

We assume that there exists a ring network of nodes (USERS) who are intermittently authenticating themselves pair-wise using the Ping-Pong protocol in which each pair continually and simultaneously re-encrypts and alternately exchanges for comparison the same pseudo-random number (TOKEN).

Figure 9A shows that the network consists of users U1, U2 and U3. At the time we invoke the Race-Track protocol to test the integrity and availability of the ring:

User U1 has just generated tokens R1' and R2';
User U2 has just generated tokens R2' and R3'; and
User U3 has just generated tokens R3' and r1'.

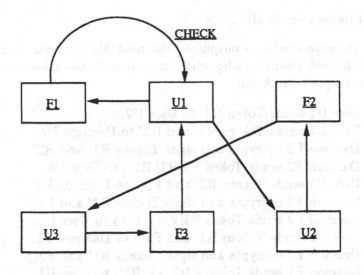

Figure 9: Race-Track Protocol: (B) Challenge

Each User is accompanied by a function-box or doppelganger (DAEMON). The purpose of the Daemon is to accept two tokens from a User, perform a secret and irreversible transformation on them, sign the result, and pass it to another User. Daemons are independent of all Users.

Execution of the Ping-Pong protocol is frozen for the duration of the Race-Track protocol, which consists of a Challenge and a Response.

CHALLENGE

User 1 is about to intervene in a network conference. User 1 challenges the integrity of the network by placing the network in CHALLENGE mode, and sending the Token of its own cardinality, Token R1', to its nearest neighbour in the clock-wise sense, User 2. See Figure 9B.

The following rules govern the Challenge mode:

1. Every User, except the Challenger, sends the monoidal token of its own cardinality, and the highest-order transformation it possesses, to the Dae-

mon of its own cardinality.

2. Every Daemon sends its output to the next highest user in cardinality, except the last Daemon, who sends its output to the Challenger to serve as an encrypted checksum.

EXAMPLE User U1 sends Token R1' to User U2.

User U2 sends Tokens R1' and R2' to Daemon F2.

Daemon F2 encrypts and signs Tokens R1' and R2'.

Daemon F2 sends Token F2(R1' R2') to User U3.

User U3 sends Tokens R3' and F2() to Daemon F3.

Daemon F3 encrypts and signs Tokens R3' and F2().

Daemon F3 sends Token F3(F2(), R3') to User U1.

User U1 sends Tokens R1' and F3() to Daemon F1.

Daemon F1 encrypts and signs Tokens R1' and F3().

Daemon F1 sends Token F1(F3(), R1') to User U1.

User U1 holds Token F1() as an encrypted checksum.

RESPONSE

The RESPONSE MODE, illustrated in Figure 9C, is invoked automatically. The following rules govern the response mode:

1. Every User sends a monoidal token not of its own cardinality, and the highest order transformation it possesses, to the Daemon of the next higher cardinality.

2. Every Daemon sends its output to the User of its own cardinality.

EXAMPLE User U1 sends Tokens R1' and R2' to Daemon F2.

Daemon F2 encrypts and signs Tokens R1' and R2'

Daemon F2 sends Token F2() to User U2.

User U2 sends Tokens R3' and F2() to Daemon F3.

Daemon F3 encrypts and signs Tokens R3' and F2().

Daemon F3 sends Token F3() to User U3.

User U3 sends Tokens R1' and F3() to Daemon F1.

Daemon F1 encrypts and signs Tokens R1' and F3().

Daemon F1 sends Token F1() to User U1.

User U1 compares this representation of Token F1() with the one it is holding.

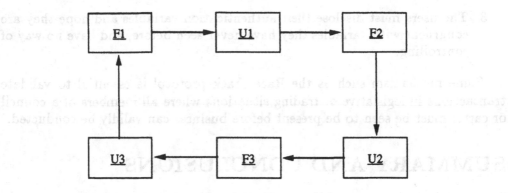

Figure 9: Race-Track Protocol: (C) response

Essentially, two independently derived versions of the checksum chase one another around the ring with the Response being out of "phase" with the Challenge by one node.

If the two representations of token F1() are found to be equal, User 1 concludes that all Users are present and authentic, and proceeds with its intervention. Otherwise, the ring must be dissolved and reconstituted with new pairings randomly selected.

EVALUATION

This protocol enables every participant in a network conference to test the integrity and availability of the network before making an intervention.

1. The protocol forces each user to disclose all the current Ping-Pong variables it possesses.

2. It computes a one-way, encrypted function of all authentication variables over two independent paths.

3. The users must disclose their authentication variables and hope they are congruent with variables they have never seen before, and have no way of controlling.

Some mechanism such as the Race-Track protocol is essential to validate transactions in legislative or trading situations where all members of a council or cartel must be seen to be present before business can validly be conducted.

SUMMARY AND CONCLUSIONS

There are at least ten requirements for the authentication of data communication that are fulfilled by the foregoing procedures:

1. Means should exist to authenticate hosts, users, messages and channels.

2. Procedures should exist to take advantage for protection of either asymmetric or symmetric ciphers.

3. Host/user authentication should be mutual and simultaneous.

4. The presence of hosts and users, and the integrity of channels should be tested continuously.

5. Authentication procedures should be automatic and transparent to human users.

6. Procedures should present minimal loading to the communications system.

7. Protocols should exist for all communication modes: Two-party or one-to-one (ping-pong), broadcast or one-to-many (round-robin), and conference or many-to-many (race-track).

8. Authentication should not rely on message encryption.

9. Authentication procedures themselves must be resistant to cryptanalysis and traffic analysis.

10. Procedures should not depend upon defined message boundaries (eg. they should accommodate telemetry, videograhics, voice etc.).

Actually these procedures, like all others, fall short of the mark. I still can't be sure I am communicating with a desired person or persons and none other; nor that what we receive is what was intended. Not without 200-digit keys, previously delivered tokens, mediation of "big brother", or all of the above. Perhaps I never will.

ACKNOWLEDGMENT

This work was supported in part by the Natural Science and Engineering Research Coucil, Canada under grant #A-3172. Their support is gratefully acknowledged.

References

[1] Agnew, G.B., *Secrecy and Privacy in a Global Area Network Environment*, Proc. Eurocrypt'84, pp. 349-363.

[2] Anderson, D.P., et al., *A Protocol for Secure Communication and its Performance*, Proc. 7th Int. Conf. on Distributed Computing, 1988, pp. 473-480.

[3] Blum, M. and S. Micali, *How to Generate Cryptographically Secure Sequences of Pseudo Random Bits*, Proc. 23rd Sym. on Foundations of Computer Science, 1982, p. 112-117.

[4] Chaum, D. and T.H. Guenta, *A Secure and Privacy Protecting Protocol for Transmitting Personal Information Between Organisations*, Proc. Eurocrypt'85, pp. 118-167.

[5] Davida, G.I., et al., *Security and Privacy*, Proc. Compsac'78, pp. 194-203.

[6] Davies. D.W. and J.K. Hirst, *Encipherment and Signature in Teletex*, Proc. 6th ICCC, 1982.

[7] Davio, M., J.M. Geothals and J.J. Quisquater, *Authentication Procedures*, Workshop on Cryptography, Bad Feurstein, DBR, May 29 — April 2, 1982.

[8] Even, S., et al., *On the Security of Ping-Pong Protocols when Implemented using the RSA*, Proc. Eurocrypt'84, pp. 58-92.

[9] Ramesh, K., *A Security Imbedded Authentication Protocol*, Proc. IEEE Infocom'88, pp. 105-109.

[10] Kent, S.T., et al., A Personal Authentication System for Access Control to the Defense Data Network, Proc. Eascon'82, pp. 89-93.

[11] Letham, L., et al., *Software Security is Provided by an EPROM that Performs an Authentication Handshake*, 4th Int. Conf. on Computers and Communications, 1985, pp. 122-126.

[12] Mason, A., *A Pay-Per-View Conditional Access System for DBS by Means of Secure Over-Air Credit Transmissions*, Proc. Int. Conf. on Secure Communications Systems, 1984, pp. 66-70.

[13] Muftic, S., *Secure Mechanisms for Computer Networks: Results of the CEC COST-11 Tex Project*, Computer Networks and ISDN Systems, 15, 1988, pp. 67-72.

[14] Perugia, O. et al., *On Encryption and Authentication Procedures for Tele-Surveillance Systems*, Ottawa, 1981.

[15] Price, W., *Encryption in Computer Networks and Message Systems*, Int. Sym. on Computer Systems, Ottawa, 1981.

[16] Purdy, G.B., G.J. Simmonds and James A. Studier, *A Software Protection Scheme*, Proc. Sym. on Security and Privacy, 1982, pp. 99-103.

[17] Rivest, R.L., A. Shamir and L.M. Adleman, *A Method for Obtaining Digital Signatures and Public Key Cryptosystems*, AAAS Annual Meeting, 1980.

[18] Shaumüller-Bichl, I. and E. Piller, *A Method of Software Protection Based on the Use of Smart Cards and Cryptographic Technique*, Proc. Eurocrypt'84, pp. 446-454.

[19] Simmonds, G.J., *Message Authentication Without Privacy*, AAAS Annual Meeting, 1980.

[20] Simmonds, G.J., *The Practice of Authentication*, Proc. Eurocrypt'86, pp. 81-84.

[21] Vazinani, U. and V.V Vazinani, *Efficient and Secure Pseudo-Random Number Generation*, Proc. Crypto'84, pp. 193-201.

SECURE CRYPTOGRAPHIC INITIALISATION
of Remote Terminals in an
Electronic Funds Transfer/Point of Sale System

Mark Ames

Group Information Systems
National Australia Bank
East Melbourne, Victoria 3002, Australia

*Notice: The views presented here are those of the author's based entirely on public
sources reflecting the generalised experience of Australian banks in the operation of
EFT/POS systems. This paper is not an official publication of the National Australia
Bank Limited, nor should it be construed as specifically representing any policies,
practices or products of the National Australia Bank.*

Abstract

As Electronic Funds Transfer at Point of Sale (EFT/POS) systems expand and the need
for more secure key management is recognised, it becomes desirable to develop secure
remote initialisation strategies for terminals across public data networks.

The Rivest, Shamir, Adleman (RSA) and Data Encryption standard (DES) algorithms
are considered with attention to key management, logical security, and implementation
requirements of each.

1. Introduction

A shopper making a purchase at one of Australia's 13,000 EFT/POS terminals hands a
plastic debit card to the cashier and enters a Personal Identification Number (PIN) into
a hand held device called a PIN pad. The PIN identifies the shopper as the person
authorised to access the account indicated on the card.

The PIN is protected by circuitry inside the PIN pad and by encryption as it crosses one
or more networks on its way to the bank or other financial institution that issued the card.
Before the PIN pad can start accepting financial transactions, it has to be initialised with
a cryptographic key and other information. The cryptographic key has to be kept as secret
as possible since its purpose is to protect the customer's PIN.

Managing and controlling the procedures for putting these secret keys into a new PIN
pad is one of the more critical and difficult tasks in operating an EFT/POS network.

Because the keys are secret, it requires two trusted bank officers (each with part of the
key) to enter the key into the PIN pad. The bank officers often have to travel to a PIN
pad supplier's depot or to a retail outlet where the PIN pad is installed, so it can easily
take two people two hours to put the initial key into one PIN pad. And if one of the sixteen
random characters of the key is entered incorrectly, they have to go back and do it again.

Australian banks have been looking for ways to load the keys into a PIN pad remotely (i.e., without having to send bank officers to each PIN pad or each PIN pad to a bank's computer centre). Since each PIN pad is connected to a bank computer by a data network, the banks would like to distribute the keys electronically using the network. The problem is keeping the keys secret during this process of remote initialisation.

Remote cryptographic initialisation for EFT/POS has been discussed within the banks for some years now and is presently being considered by Standards Australia.

Australian banks use the IBM key management system [EhMa78] on their EFT/POS networks and are adopting the new Australian Standard 2805 [Kemp88]. Both systems are based on the Data Encryption Algorithm [ASA85c], [ANSI81] also referred to as the Data Encryption Standard (DES) [NBS77].

The Australian Standard offers a significant improvement in security and key management. The fixed terminal master key of the IBM system is replaced with a 16 digit value that is used along with data entered from magnetic stripe cards to derive terminal keys. This 16 digit value is referred to as Card Value 6 for master-session key initialisation [ASA88a] or as the terminal key for transaction key management [ASA88c]. The value is used as a 64 bit DES key in the key derivation process (which uses less secure non-cryptographic data) for master-session key management and as the initial high level key in transaction key management. In both cases it is the initial cryptographic value or key.

The Standard mandates regular, automatic changing of all keys. Keys are stored in a physically secure, tamper resistant module within the PIN pad. AS2805 refers to the secure module within the PIN pad as the Terminal Cryptographic Unit (TCU).

However, PIN pads will still receive their initial key by manual entry either at the terminal key pad or PIN pad key pad. As mentioned above, manual entry of key values presents a number of problems: device failure due to keying errors, potential for exposure of key values, and high labour costs. Manual entry is often done by contractors, which introduces additional administrative and key management problems.

Published key management standards do not specify methods for electronic distribution of initial cryptographic keys to remote terminals nor do any de facto standards exist.

The banks have had the better part of a decade in the EFT/POS arena. The system has matured with experience and so has public and official awareness. Australia's EFT/POS network now faces a period of rapid expansion [Taka88] [AFRV], and requirements for improved efficiency and security are converging [BANK89]. The banks are actively seeking workable, efficient, and secure alternatives to manual key entry.

Automation of the key loading process is an obvious part of the solution and has been described in principle [Webe89] though no specific scheme has been detailed.

The models presented here deal only with establishing a secure cryptographic channel between the PIN pad and a single network host. Full initialisation procedures include installation and communications data and other procedures that vary amongst financial transaction acquirers. (Acquirers are usually banks, but can be other financial institutions or computer bureaux).

2. Secure Loading Facility

Suppliers of PIN pads to Australian EFT/POS networks load user specific software and data including a unique PIN pad identification value (PPID) into the devices before they are delivered for installation. This is usually done at the supplier's depot or manufacturing centre using a microcomputer connected to the PIN pad via a serial port. Various levels of security and access controls apply to these facilities.

Remote initialisation using DES requires secret user specific data in the form of cryptographic keys. Pilot programs within the banks have loaded secret keys using a microcomputer fitted with a security processing board. These boards are tamper resistant and offer security of stored secret values comparable to that of security modules attached to host systems. (These modules are variously referred to as hardware security modules [MeMa82], tamper resistant modules [DaPr82], cryptographic facilities [EhMa78], etc. The term 'tamper resistant module' (TRM) will be used generically to refer to host, microcomputer, or workstation devices offering similar features.) Many of the features of a TRM are also included in the PIN pad Terminal Cryptographic Unit (TCU). Tamper resistant modules fitted to microcomputers offer secure key generation and storage facilities and may provide a hardware-based access control system.

The keys are loaded through a (cryptographically) insecure line connected to the PIN pad. For this reason, the loading facility must be housed in a physically secure environment that minimises the spread of electromagnetic emissions.

Combining the facilities for loading secret and non-secret user specific data may reduce handling time and costs, and can be integrated into a control structure to prevent unauthorised modification of PIN pad software.

A secure facility for loading cryptographic values is essential: secret keys are part of the initialisation process whether a public key or secret key cryptosystem is used.

3. Key Management

Automated key loading does not solve the problem of secure distribution of the key between the PIN pad and the transaction acquirer's host computer. The distribution of values entered from magnetic stripe cards, especially for master-session key initialisation, also needs to be dealt with: these are not secret values, but need to be handled efficiently and discretely since they are a major input to the cryptographic initialisation process. It is also desirable that devices loaded with working keys not pass outside the control of the transaction acquirer, which is not always the case where contractors or vendors are involved with installation.

The objective of remote initialisation as described here is secure distribution of an initial DES key (KI) to the PIN pad and the acquiring bank's host computer. This is a 16 digit hexadecimal value which replaces the manual key entry of Card Value 6 in AS2805.6.4 (master-session key management) or the terminal key in AS2805.6.2 (transaction key management). Alternatively, the KI could be used in a network specific scheme external to the standard.

Some of the first discussions of remote initialisation looked at options to provide intermediary key domains to isolate the equipment supplier from the operating network. It was realised that suppliers and the users (the banks) have to trust each other and consult closely on security. Moreover, risks are reduced by limiting the number of key domains and concentrating on security and key management in procedures applied to 'open' the secure channel between PIN pad and acquirer host.

The advantages of remote initialisation include reduced key management overheads and installation times plus greater security and control of cryptographic keys. The principal design goals of remote initialisation schemes must be those of the network as a whole: to safeguard secret information and prevent successful attacks on the system.

The assumption is made here that secret keys and any other values that, of themselves or in conjunction with non-secret information, would permit access to secure communications must be kept secret, and that disclosure of a public key will not in itself permit compromise or attack.

A number of other specific management and control objectives are considered in the models set out below. These take the practical aspects of implementation into consideration along with requirements for security:

1. Generation, storage, and computation of secret values to take place in a secure environment -- preferably in a TRM.

2. Key distribution and entry procedures to avoid exposure of clear key parts or components to any person.

3. Minimise requirements for the use of couriers or trusted agents in key distribution and key entry.

4. Exploit and utilise existing or tested and available technologies.

5. Minimise likely development, implementation, maintenance, and other recurrent costs.

6. Allow for introduction of new technologies (e.g IC cards) as they become appropriate to the application.

4. Authentication

The logical security of remote initialisation relies on two principal factors: authentication and cryptographic security. Assuming that both algorithms under discussion provide adequate cryptographic security in a sound key management environment, authentication needs careful consideration.

The acquirer wants to be certain that a PIN pad asking to access the network is an authorised device in an authorised location, and not some attacker or thrill seeker attempting to access the network illegally. An acquirer also wants to make sure that a PIN pad won't start passing transactions to some other host computer masquerading as the acquirer since this could be an attempt to defraud merchants in the EFT/POS system as well as card holders.

The PIN pad initiates access requests and has to identify itself to the acquirer host and 'present its credentials', i.e., prove that it is who it says it is. Likewise the acquirer's host computer has to establish its bona fides before the PIN pad begins to send transactions to it. This is entity authentication [DaPr82].

Cryptographically generated message authentication codes (MACs) [ASA85b] (based on ANSI X9.9) will not be used in the models presented below because they are only mandated for financial messages, and the key management Standard [ASA85a] requires key separation, i.e., a key used to generate a MAC cannot be used for any other purpose. There is ample scope within the models presented to include an initialisation MAC key.

However, other techniques exploiting cryptography and the structure of key management and message systems can be used to verify both messages and entities. This discussion will concentrate on entity authentication and message authentication based on structural elements of the system.

Seberry and Peipzryk point out that entity verification based on shared secret values implies message authentication. In the DES scheme, the shared secret key implies authentication. In RSA, authentication is achieved using the digital signature facility of the algorithm. Of further and critical importance in an RSA implementation is authentication of the public key because of the possibility of an attack based on key substitution in which an impostor masquerades as an authorised entity [MeMa82].

Both DES and RSA authentication methods are based on key management and security of secret keys. These cryptographic methods can make use of collateral information accessible to the acquirer at the time of initialisation. This is typically database information linking a valid PPID to installation schedules, key sets, merchant sites, communications terminals, etc. Collateral information is not as secure as cryptographic information, nor need it be: sound authentication procedures must ultimately rely on the security provided by the key management system.

5. Which Algorithm: RSA or DES?

The Rivest, Shamir, Adleman (RSA) public key cryptography algorithm [RSA78] has been promoted as a more secure and less management intensive means of distributing initial cryptographic keys. Secure key distribution schemes are available under the DES cryptographic system in current use, though they are traditionally demanding of human resources.

It is beyond the scope of this paper to detail comparative advantages of the two algorithms. RSA is attractive because it appears to overcome key distribution problems. However, the description of the implementation will show that this may not be the case for the type of networks currently in use in the Australian EFT/POS system [MeMa82]. Both algorithms can be used in practice to implement remote initialisation procedures.

The long term viability of both algorithms is challenged by predictable improvements in computational machinery, and current implementations have an estimated life-time of only a few years [Beke89]. There is, however, the possibility of replacing DES [SePi89] with a system-compatible (i.e., 64 bit block cipher) algorithm of equal or better strength [GuDa90].

The security of RSA is additionally subject to progress in the mathematics of factoring large integers, which is less predictable but continues to advance [LLMP]. There is no candidate for system-compatible replacement of RSA.

6. Notation

The following abbreviations and conventions of notation are used in the models presented below:

SKsup	Supplier's Secret Key	**PKpp**	PIN Pad Public Key
SKpp	PIN Pad Secret Key	**PPID**	PIN Pad Identification Value
PKhost	Host(Bank) Public Key	**RN**	Random Number
KI	Initial DES Key	**DTS**	Date Time Stamp
e	DES encryption	**c**	RSA encipherment
s	RSA signature		

Values that are encrypted are represented in parentheses ( ) or square brackets [ ] preceded by the key and the operator. For example, a random number encrypted by the DES key KI is represented as **eKI(RN)**. Multiple values included in a cryptographic process are separated by commas: **cPPpp(PPID,DTS)** indicates that the PIN pad identity number and a date-time stamp are encrypted by the PIN pad public key. **sSKsup(cPPpp[PPID,DTS])** means that the PIN pad identity number and a date-time stamp are encrypted by the PIN pad public key, and the result of that encryption has been signed by the secret key of the supplier.

7. DES Remote Initialisation Scheme: Encrypted Key Tables

A DES initialisation key can be loaded into a PIN pad using a secure loading facility. The facility can be located at a secure site operated by the supplier, but logical access to the tamper resistant module remains under control of the acquiring bank.

The tamper resistant module accesses an encrypted file of initialisation keys and decrypts a KI before passing it to the PIN pad via a communications port. The KI must be securely stored in the PIN pad TCU.

The file of Initialisation Keys is generated by the acquiring bank in a cryptographic workstation equipped with a tamper resistant module. A KEK is generated to encrypt the Initialisation Key file, and a single record index generation key (IGK) is generated at the same time. The encrypted file of Initialisation Keys on removable media can be transported to the loading site by normal channels. A similar system of encrypted key distribution has been in use among Australian banks since 1985.

The bank's cryptographic workstation also encrypts the initialisation key file and the IGK under the appropriate variants of the bank's master key for loading onto the host system.

Bank officers will visit the secure loading site to enter the KEK and IGK into the system. The keys can be carried in parts or components on magnetic media and machine read directly into the secure loading facility; this protects the Bank officers from exposure to the key parts or components. The introduction of IC cards could make it possible to carry out this task without a trusted agent.

The KI is selected from the key file by executing a one way function (OWF) using the Unit Serial Number (IBM system) or PIN pad identity value (AS2805) and the Index Generation Key (IGK). The product of the OWF is a pointer to a record in the key file. The selection process remains under the bank's control, and there is no direct correlation between the USN/PPID and the key file record. (See figure 1).

Initialisation takes place after the PIN pad is installed at the retail site: the values that would otherwise be entered manually into the PIN pad are already present and initialisation can proceed as described in the Australian Standard. That is to say, no additional key management network service messages are required under this scheme.

Note that while the secure loading facility may be outside bank premises and operated by the supplier, cryptographic control is maintained by the bank. Using the encrypted key table and the index generation key, bank officers or trusted agents are only required to visit the site when the system is first set up. There is no ongoing requirement to communicate initialisation keys to the bank's host computer.

Figure 1: Linking an Initialisation Key to a PIN pad.

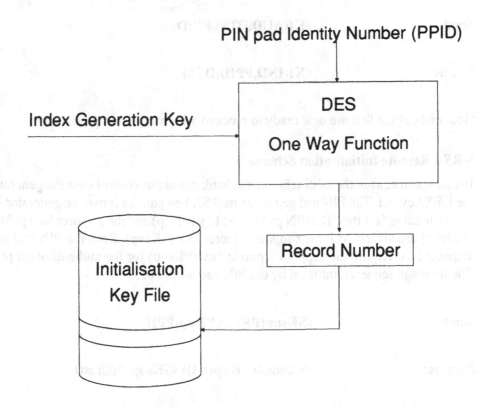

8. An extension of the DES scheme

The scheme can be extended to replace the manual card swipe entry of the initialisation card values or the initialisation transaction values specified in the Australian Standard. In that case, the KI is a data encryption key and the initial exchange between the PIN pad and host should feature entity authentication and a means for verifying the cryptographic link as well as data security.

The clear text PPID is transmitted to the host to identify the PIN pad along with authentication data (AUD) in a sign-on message. The authentication data can include the collateral information noted above (4), some of which is included on the magnetic stripe card used by the retailer in the daily sign-on procedure. An authentication data block can be assembled from the collateral data and a date-time stamp and encrypted under KI. The date-time stamp (DTS) is included for a further message authentication check and to prevent replay and plain text attacks.

If the decrypted authentication data along with other collateral data at the host evidences the PPID, the identity of the PIN pad is verified and the selected key is confirmed as correct. The host responds with the encrypted initialisation data (IND) and authentication data. Since the initialisation data is non-redundant, a recognisable value has to be returned to the PIN pad. This should also be time-variant and link specific. In this example, the PPID and a date time stamp are used.

The basic security components of the initialisation sign on message are shown below:

Send: **eKI(AUD,DTS),PPID**

Receive: **eKI(IND,PPID,DTS)**

Both ends of the link are now ready to proceed with transaction processing.

9.RSA Remote Initialisation Scheme

In this scheme, as in the DES scheme, the bank maintains control over the generation of the DES key, KI. The PIN pad generates its RSA key pair (or it may be generated by the secure loading facility). The PIN pad public key (PKpp) and the acquirer host public key (PKhost) are signed with the supplier's secret key (SKsup). Both the PIN pad and the acquirer host require the supplier's public key (PKsup) for the authentication process. The message sequence initiated by the PIN pad would be:

Send: **sSKsup(PKpp),PKpp,PPID**

Receive: **sSKhost[cPKpp(KI)],sSKsup(PKhost)**

The PIN pad is authenticated by comparing the clear value of the PIN pad public key (PKpp) with the signed value. The host is authenticated by its signed public key (sSKsup[PKhost]).

The message is authenticated by the signed encrypted DES key. This is important because an attacker could substitute another KI if cPKpp(KI) were not signed.

The PIN pad secret key must be securely stored in the PIN pad TCU, and, if loaded by the secure loading facility, should be deleted from the loading facility as soon as it is in place in the PIN pad TCU. A secure loading facility is required to protect the supplier's secret key during the signing process.

To set up the scheme, the acquiring bank and the supplier must arrange the authentic exchange of their public keys, and the signed value (sSKsup[PKhost]) must be returned to the acquiring bank. The signed value does not have to be treated as secret, nor is it vulnerable to substitution.

This scheme differs from the others presented here in one significant respect; there is no acquirer-specific key management information in the PIN pad.

10. Truncated RSA Scheme

In this abbreviated RSA scheme the PIN pad does not require an RSA key pair. It does require the bank's host public key (PKhost) and an authentication value made up of the PPID and a random number (RN). The random number is a padding element to complete a message block that is signed by the supplier's secret key. The signing operation takes place within the secure loading facility.

An initialisation request from the PIN pad includes the clear value of the PPID and RN, and the signed value under the supplier's secret key (sSKsup[PPID,RN]). The message may also include the KI (generated by the PIN pad), and a date-time stamp (DTS) encrypted by the host public key (cPKhost[KI,DTS]).

An attacker who obtains the authentication value (which is not a secure value) and its cryptographically signed value can construct a sign-on request incorporating the attacker's own KI. To prevent disclosure of the signed authentication value in transit to the acquirer, the host modulus is chosen sufficiently larger than the supplier modulus [DaPr82] or a suitable compression method is chosen [SePi89] so the signed authentication value may be contained in the message block encrypted by the host public key. The initialisation request then takes the form:

$$\textbf{cPKhost[KI,DTS,sSKsup(PPID,RN)],PPID,RN}$$

With the authenticated host public key pre-loaded into the PIN pad, only the legitimate host can decrypt the KI and the other encrypted authenticating data. Only the authentic host can proceed with transaction messages.

Note that if this single message is used, the PIN pad is only required to perform one encryption on one RSA block. This is significant because a decryption or signing operation is much more computationally intensive than encryption.

The signed value sSKsup(PPID,RN) must be store securely in the PIN pad TCU, since it permits access to the network. The acquirer must also be satisfied that the PIN pad TCU is capable of generating keys with the same degree of randomness available from a TRM.

The only cryptographic exchange required between the supplier and the bank is the authentic exchange of their respective public keys.

Both of the RSA schemes can be extended as per the DES scheme by the host sending the initialisation data encrypted by KI.

11. Conclusion

Models for remote initialisation of PIN pads in an EFT/POS network have been presented, using symmetric (DES) and asymmetric (RSA) algorithms currently available.

While the DES model for remote initialisation requires more key management work during set up, the RSA models are likely to require greater system software and hardware changes to implement. The ongoing key management effort for each of the schemes is approximate, as is the security offered by the two algorithms.

The models vary more significantly in terms of PIN pad and host processing requirements, communications overheads, and implementation effort.

The decision to implement a DES or an RSA based remote initialisation scheme will ultimately depend on commercial considerations and a strategic assessment of the strength and viability of the respective algorithms.

REFERENCES

[AFRV]   A. McCathie, 'Integrated EFTPOS system up and running this year', *Australian Financial Review*, January 17, 1990, p.31.

[ANSI81]   ANSI X3.92(1981), American National Standard Data Encryption Algorithm, American National Standards Institute.

[ASA85a]   AUSTRALIAN STANDARD AS2805 Electronic Funds Transfer - Requirements for Interfaces Part 3(1985), PIN Management and Security, Standards Association of Australia.

[ASA85b]   AUSTRALIAN STANDARD AS2805 Electronic Funds Transfer - Requirements for Interfaces Part 4(1985), Message Authentication, Standards Association of Australia.

[ASA85c]   AUSTRALIAN STANDARD AS2805 Electronic Funds Transfer - Requirements for Interfaces Part5(1985), Data Encryption Algorithm, Standards Association of Australia.

[ASA88a]   AUSTRALIAN STANDARD AS2805 Electronic Funds Transfer - Requirements for Interfaces Part 6.1(1988), Key Management - Principles, Standards Association of Australia.

[ASA88b]   AUSTRALIAN STANDARD AS2805 Electronic Funds Transfer - Requirements for Interfaces Part 6.2(1988), Key Management - Transaction Keys, Standards Association of Australia.

[ASA88c]   AUSTRALIAN STANDARD AS2805 Electronic Funds Transfer - Requirements for Interfaces Part 6.4(1988), Key Management - Session Keys - Terminal to Acquirer, Standards Association of Australia.

[BANK89]   *Australian Banks EFT Security Code & Practice (1989)*,
            Australia & New Zealand Banking Group Ltd,
            Commonwealth Bank of Australia Ltd,
            National Australia Bank,
            Rural & Industries Bank of Western Australia,
            State Bank of Victoria,
            State Bank of New South Wales,
            State Bank of South Australia,
            Westpac Banking Corporation.

[Beke89]   Henry Beker, 'Management and Control of Systems', *Information Security Guide*, IBC Technical Services, London, 1989.

[Chor88]   D.N. Chorafas, 'EFT/POS as a secure system solution', *Electronic Funds Transfer*, Butterworths, London, 1988, pp. 327-332.

[DaPr82]   D.W. Davies, W.L. Price, *Security for Computer Networks*, Wiley, Chichester, 1982.

[EhMa78]   W. F. Ehrsam, S. M. Matyas, C. H. Meyer and W. L. Tuchman,
           'A cryptographic key management scheme for implementing the data
           encryption standard', *IBM Systems J.*, Vol.17, No.2, 1978, pp.106-125.

[GuDa90]   H. Gustafson, E. Dawson, B. Caelli, 'Comparison of Block Ciphers',
           Abstracts of Auscrypt90, Sydney, 8-11 January 1990, pp. 163-165.

[Kemp88]   E. A. Kemp, 'Encryption in Electronic Funds Transfer Applications',
           Aust. Computer J., Vol.20,No.2, 1988, pp.170-177.

[LLMP]   A.K. Lenstra, H.W. Lenstra, Jr., M.S. Manasse, J.M. Pollard, 'The number
           field sieve', unpublished.

[MeMa82]   C.H. Meyer, S.M. Matyas, *Cryptography: A New Dimension in Computer
           Data Security*, Wiley, New York, 1982.

[NBS77]   National Bureau of Standards, *Data Encryption Standard*, Federal
           Information Processing Standards Publication 46, Jan. 1977.

[RSA78]   R.L. Rivest, A. Shamir, L. Adleman,'A Method for Obtaining Digital
           Signatures and Public Key Cryptosystems', *Communications of the
           ACM*, Vol.21, No.2, 1978, pp. 120-126.

[SePi89]   J. Seberry, J. Piepzryk, *Cryptography: an introduction to computer security*,
           Prentice Hall, Sydney, 1989.

[Taka88]   P. Takac, *Eftpos in Australia: Developments, Trends, and Market Size*,
           Royal Melbourne Institute of Technology Centre for Technology Policy
           & Management, Melbourne, 1988.

[Webe89]   R. Weber, 'Controls in Electronic funds Transfer Systems: a Survey and
           Synthesis', *Computers & Security*, Vol.8, No. 2,1989, pp.123-137.

Vol. 408: M. Leeser, G. Brown (Eds.),Hardware Specification, Verification and Synthesis: Mathematical Aspects. Proceedings, 1989. VI, 402 pages. 1990.

Vol. 409: A. Buchmann, O. Günther, T. R. Smith, Y.-F. Wang (Eds.), Design and Implementation of Large Spatial Databases. Proceedings, 1989. IX, 364 pages. 1990.

Vol. 410: F. Pichler, R. Moreno-Diaz (Eds.), Computer Aided Systems Theory – EUROCAST '89. Proceedings, 1989. VII, 427 pages. 1990.

Vol. 411: M. Nagl (Ed.), Graph-Theoretic Concepts in Computer Science. Proceedings, 1989. VII, 374 pages. 1990.

Vol. 412: L. B. Almeida, C. J. Wellekens (Eds.), Neural Networks. Proceedings, 1990. IX, 276 pages. 1990.

Vol. 413: R. Lenz, Group Theoretical Methods in Image Processing. VIII, 139 pages. 1990.

Vol. 414: A.Kreczmar, A. Salwicki, M. Warpechowski, LOGLAN '88 – Report on the Programming Language. X, 133 pages. 1990.

Vol. 415: C. Choffrut, T. Lengauer (Eds.), STACS 90. Proceedings, 1990. VI, 312 pages. 1990.

Vol. 416: F. Bancilhon, C. Thanos, D. Tsichritzis (Eds.), Advances in Database Technology – EDBT '90. Proceedings, 1990. IX, 452 pages. 1990.

Vol. 417: P. Martin-Löf, G. Mints (Eds.), COLOG-88. International Conference on Computer Logic. Proceedings, 1988. VI, 338 pages. 1990.

Vol. 418: K. H. Bläsius, U. Hedtstück, C.-R. Rollinger (Eds.), Sorts and Types in Artificial Intelligence. Proceedings, 1989. VIII, 307 pages. 1990. (Subseries LNAI).

Vol. 419: K. Weichselberger, S. Pöhlmann, A Methodology for Uncertainty in Knowledge-Based Systems. VIII, 136 pages. 1990 (Subseries LNAI).

Vol. 420: Z. Michalewicz (Ed.), Statistical and Scientific Database Management, V SSDBM. Proceedings, 1990. V, 256 pages. 1990.

Vol. 421: T. Onodera, S. Kawai, A Formal Model of Visualization in Computer Graphics Systems. X, 100 pages. 1990.

Vol. 422: B. Nebel, Reasoning and Revision in Hybrid Representation Systems. XII, 270 pages. 1990 (Subseries LNAI).

Vol. 423: L. E. Deimel (Ed.), Software Engineering Education. Proceedings, 1990. VI, 164 pages. 1990.

Vol. 424: G. Rozenberg (Ed.), Advances in Petri Nets 1989. VI, 524 pages. 1990.

Vol. 425: C. H. Bergman, R. D. Maddux, D. L. Pigozzi (Eds.), Algebraic Logic and Universal Algebra in Computer Science. Proceedings, 1988. XI, 292 pages. 1990.

Vol. 426: N. Houbak, SIL – a Simulation Language. VII, 192 pages. 1990.

Vol. 427: O. Faugeras (Ed.), Computer Vision – ECCV 90. Proceedings, 1990. XII, 619 pages. 1990.

Vol. 428: D. Bjørner, C. A. R. Hoare, H. Langmaack (Eds.), VDM '90. VDM and Z – Formal Methods in Software Development. Proceedings, 1990. XVII, 580 pages. 1990.

Vol. 429: A. Miola (Ed.), Design and Implementation of Symbolic Computation Systems. Proceedings, 1990. XII, 284 pages. 1990.

Vol. 430: J. W. de Bakker, W.-P. de Roever, G. Rozenberg (Eds.), Stepwise Refinement of Distributed Systems. Models, Formalisms, Correctness. Proceedings, 1989. X, 808 pages. 1990.

Vol. 431: A. Arnold (Ed.), CAAP '90. Proceedings, 1990. VI, 285 pages. 1990.

Vol. 432: N. Jones (Ed.), ESOP '90. Proceedings, 1990. IX, 436 pages. 1990.

Vol. 433: W. Schröder-Preikschat, W. Zimmer (Eds.), Progress in Distributed Operating Systems and Distributed Systems Management. Proceedings, 1989. V, 206 pages. 1990.

Vol. 435: G. Brassard (Ed.), Advances in Cryptology – CRYPTO '89. Proceedings, 1989. XIII, 634 pages. 1990.

Vol. 436: B. Steinholtz, A. Sølvberg, L. Bergman (Eds.), Advanced Information Systems Engineering. Proceedings, 1990. X, 392 pages. 1990.

Vol. 437: D. Kumar (Ed.), Current Trends in SNePS – Semantic Network Processing System. Proceedings, 1989. VII, 162 pages. 1990. (Subseries LNAI).

Vol. 438: D. H. Norrie, H.-W. Six (Eds.), Computer Assisted Learning – ICCAL '90. Proceedings, 1990. VII, 467 pages. 1990.

Vol. 439: P. Gorny, M. Tauber (Eds.), Visualization in Human-Computer Interaction. Proceedings, 1988. VI, 274 pages. 1990.

Vol. 440: E.Börger, H. Kleine Büning, M.M. Richter (Eds.), CSL '89. Proceedings, 1989. VI, 437 pages. 1990.

Vol. 441: T. Ito, R. H. Halstead, Jr. (Eds.), Parallel Lisp: Languages and Systems. Proceedings, 1989. XII, 364 pages. 1990.

Vol. 442: M. Main, A. Melton, M. Mislove, D. Schmidt (Eds.), Mathematical Foundations of Programming Semantics. Proceedings, 1989. VI, 439 pages. 1990.

Vol. 443: M. S. Paterson (Ed.), Automata, Languages and Programming. Proceedings, 1990. IX, 781 pages. 1990.

Vol. 444: S. Ramani, R. Chandrasekar, K. S. R. Anjaneyulu (Eds.), Knowledge Based Computer Systems. Proceedings, 1989. X, 546 pages. 1990. (Subseries LNAI).

Vol. 445: A. J. M. van Gasteren, On the Shape of Mathematical Arguments. VIII, 181 pages. 1990.

Vol. 446: L. Plümer, Termination Proofs for Logic Programs. VIII, 142 pages. 1990. (Subseries LNAI).

Vol. 447: J. R. Gilbert, R. Karlsson (Eds.), SWAT 90. 2nd Scandinavian Workshop on Algorithm Theory. Proceedings, 1990. VI, 417 pages. 1990.

Vol. 449: M. E. Stickel (Ed.), 10th International Conference on Automated Deduction. Proceedings, 1990. XVI, 688 pages. 1990. (Subseries LNAI).

Vol. 450: T. Asano, T. Ibaraki, H. Imai, T. Nishizeki (Eds.), Algorithms. Proceedings, 1990. VIII, 479 pages. 1990.

Vol. 451: V. Mařík, O. Štěpánková, Z. Zdráhal (Eds.), Artificial Intelligence in Higher Education. Proceedings, 1989. IX, 247 pages. 1990. (Subseries LNAI).

Vol. 452: B. Rovan (Ed.), Mathematical Foundations of Computer Science 1990. Proceedings, 1990. VIII, 544 pages. 1990.

Vol. 453: J. Seberry, J. Pieprzyk (Eds.), Advances in Cryptology – AUSCRYPT '90. Proceedings, 1990. IX, 462 pages. 1990.